AF361745

Measurement of the Environmental Impact of Materials

Measurement of the Environmental Impact of Materials

Editors

Franz-Georg Simon
Ute Kalbe

MDPI • Basel • Beijing • Wuhan • Barcelona • Belgrade • Manchester • Tokyo • Cluj • Tianjin

Editors
Franz-Georg Simon
Contaminant Transfer and
Environmental Technologies
Federal Institute for Materials
Research and Testing (BAM)
Berlin
Germany

Ute Kalbe
Contaminant Transfer and
Environmental Technologies
Federal Institute for Materials
Research and Testing (BAM)
Berlin
Germany

Editorial Office
MDPI
St. Alban-Anlage 66
4052 Basel, Switzerland

This is a reprint of articles from the Special Issue published online in the open access journal *Materials* (ISSN 1996-1944) (available at: www.mdpi.com/journal/materials/special_issues/Measurement_Environmental).

For citation purposes, cite each article independently as indicated on the article page online and as indicated below:

LastName, A.A.; LastName, B.B.; LastName, C.C. Article Title. *Journal Name* **Year**, *Volume Number*, Page Range.

ISBN 978-3-0365-5984-1 (Hbk)
ISBN 978-3-0365-5983-4 (PDF)

Contents

About the Editors

Franz-Georg Simon

Franz-Georg Simon is the head of BAM's division Contaminant Transfer and Environmental Technologies. Before joining BAM in 1998, he worked at the ABB Corporate Research Center in Baden (Switzerland). He received a degree as Diplom-Chemiker from University of Frankfurt/Main and Dr. rer. nat. from University of Mainz. His main research areas are the recovery of resources from waste and the assessment of the environmental compatibility of secondary raw materials.

Ute Kalbe

Ute Kalbe earned her Master's degree in 1983 and PhD in 1995 in crystallography, both obtained from Humboldt University, Berlin. For seven years of her career, she worked for the Department of Soil Science within the Technical University of Berlin. Since 1997, she has been working in the division Contaminant Transfer and Environmental Technologies at the Federal Institute for Materials Research and Testing (BAM) in Berlin, Germany. Her research interests include soil and site assessment, recycling of waste and resource recovery, particularly from incineration bottom ash, long-term behavior of landfill liners, soil pore structure characterization, and environmental compatibility of synthetic sports surfaces. In this context, she has actively been involved especially in the development, standardization and validation of leaching procedures at the national, European and international levels. She is currently the head of the Expert Advisory Board on Soil Testing Procedures of the German Federal Ministry for the Environment and Consumer Protection.

Preface to "Measurement of the Environmental Impact of Materials"

Throughout their life cycles—from production, usage, through to disposal—materials and products interact with the environment (water, soil, and air). At the same time, they are exposed to environmental influences and, through their emissions, have an impact on the environment, people, and health. Accelerated experimental testing processes can be used to predict the long-term environmental consequences of innovative products before these actually enter the environment.

We are living in a material world. Building materials, geosynthetics, wooden toys, soil, nanomaterials, composites, wastes and more are research subjects examined by the authors of this book. The interactions of materials with the environment are manifold. Therefore, it is important to assess the environmental impact of these interactions. Some answers to how this task can be achieved are given in this Special Issue.

Franz-Georg Simon and Ute Kalbe
Editors

Editorial

Measurement of the Environmental Impact of Materials

Franz-Georg Simon * and **Ute Kalbe**

BAM Bundesanstalt für Materialforschung und-prüfung, 12200 Berlin, Germany; ute.kalbe@bam.de
* Correspondence: franz-georg.simon@bam.de

Citation: Simon, F.-G.; Kalbe, U. Measurement of the Environmental Impact of Materials. *Materials* **2022**, *15*, 2208. https://doi.org/10.3390/ma15062208

Received: 10 March 2022
Accepted: 14 March 2022
Published: 17 March 2022

Publisher's Note: MDPI stays neutral with regard to jurisdictional claims in published maps and institutional affiliations.

Global material use has increased by a factor of eight in the 20th century, and has reached more than 10 tons per capita per year [1]. Negative impacts on the environment can occur in all phases of a material's life cycle, i.e., from its production from natural resources, to use, and end-of-life. Quantification of the environmental impact of a material is not easy to assess. One approach to identify this driving force is the so-called IPAT formula, defined by Ehrlich and Holdren in 1971 [2]. The environmental impact (I) is the product of population (P), affluence (A) and a technology term (T) where T is understood as the reciprocal of the efficiency or material productivity ($T = 1/e$). Population is still increasing, and economic development is usually accompanied by an increase in material productivity.

Materials can be man-made from natural resources or recycled from waste. The performance of materials, e.g., mechanical properties or maximum service lifetime, can be improved. This can be attained, for example, by using coatings to protect against corrosion, by additives to enhance stability and processability, or by the establishment of composites.

This Special Issue, 'Measurement of the Environmental Impact of Materials', is focused on the impact that materials have on the environmental compartments of soil, water and air. Leaching and emission processes, including underlying mechanisms, are a recurring topic in most published articles in the present Special Issue. Contributions have come from three continents and numerous countries (USA, Germany, France, Latvia, Poland, Russia, Japan, Korea and Thailand) indicating the global dimension of the subject.

The release of substances from materials due to contact with water has been simulated and quantified by the analysis of eluates for various materials, such as polymer-based materials and primary or secondary construction materials and reported in this Special Issue. Emissions into the environment caused by leaching can affect surface waters, groundwater and soils. Some papers show that, notably, it is not sufficient to refer to the solid matter content of harmful substances, and that it is necessary to consider the processes that lead to their release. Thereby, not only chemical processes such as simple dissolution take place, but also physical mechanisms such as sorption/desorption and diffusion. The mobility of very fine particles, e.g., colloids or nanoparticles, must also be taken into account.

The environmental impact in the use phase of materials and products can be simulated using laboratory experiments that consider relevant use scenarios under accelerated conditions. A better understanding of the material's underlying processes can be obtained when these tests are combined with exposure experiments using artificial weathering corresponding to field conditions. It is very important to study the complex and dynamic leaching processes of substances from various materials to enable regulators to set limit values and allow producers to modify their products in order to minimize the release of hazardous substances. Different exposure tests and mass transfer processes for polymer materials are reviewed in the article from Bandow et al. [3]. Weiler et al. studied the leaching behavior of irrigated building structures made of carbon-reinforced concrete with accelerated simulations at a laboratory scale [4]. Evaluation concepts are discussed in the second part of the work [5] to align simulation results to field conditions and enable the establishment of threshold values.

Construction products in contact with water and soil are subject to the possible influence of leachates on organisms. Ecotoxicological tests are becoming increasingly established as tools for understanding the impact of released hazardous substances from materials to the environment. An example of a test needing only moderate time and effort was given by von Wolff and Stephan [6], using the observation of the reproduction of worms under the influence of leachates from construction products.

Another aspect of environmental simulations regarding materials used in construction products is the need for the acquisition of reliable data on pollutant emissions, particularly regarding volatile organic compounds (VOC) to indoor air. Appropriate testing procedures are required for this purpose, in order to enable a comparable basis for the declaration of the environmental performance of products (e.g., for CE marking) by the producers. Emissions to indoor air from natural building materials using large emission chambers were studied by Richter et al. [7], and were found to be uncritical. However, the case of formaldehyde emissions from wooden toys investigated by Even et al. was different [8]. Here, the use of miniaturized test chambers was effective for the exposure assessment.

Materials may have a certain function in the environment. For example, the application of zeolitic materials as soil conditioners and as slow-release fertilizers was investigated by Szatanik-Kloc et al. [9]. It was found that low doses had a negligible effect only and the functionality has obviously been overestimated so far. Products for coastal protection can be produced from geosynthetic materials. This offers advantages such as low weight and a long service lifetime, but also presents the risk to the environment of introducing an additional source of microplastic particles. This was studied in the article by Scholz et al. [10]. The main factor influencing the ageing of geotextiles was identified, and a half-life of 330 years for the loss of 50% of the strain was calculated from their simulations. Additionally, no harmful ecotoxicological effects by leachates were observed.

A technically important function of magnesium-based alloys can be the storage of hydrogen in sustainable energy systems. The environmental burden of CaO addition to enhance performance was evaluated through a life cycle assessment (LCA) by Shin et al. [11]. More research is necessary in order to identify materials that increase performance, and the consideration of their environmental impacts should be in the foreground.

Key aspects of a sustainable circular economy have been addressed in some of the papers. The environmental impact of raw material extraction from natural deposits can be lowered when waste materials are used instead of new materials being constantly required. An example of this is the recycling of seafood waste to mono- and tricalcium phosphates [12].

One of the largest waste streams in industrialized countries originates from construction and demolition waste (CDW). The wet processing of CDW allows the utilization of the main fractions of the treatment as a substitute for natural sand and gravel [13] because pollutants are enriched in smaller sized fractions. Environmental compatibility was tested with column leaching tests, which were found to be useful for characterization as well as for quality control.

Leaching tests represent a common thread in the submissions to this Special Issue [3–6,13–16]. The underlying mass transfer principles of column leaching tests are discussed by Liu et al. [14], and Sakanakura et al. [15] showed that adsorption parameters can be determined in batch leaching tests with varying liquid–solid ratios. Ash from sewage sludge incineration (SSA) could partly replace phosphate from natural resources such as fertilizer components. However, sewage sludge contains pollutants which are transferred to the ash. Meisterjahn et al. [16] investigated the leaching of engineered nanomaterials from SSA which could be a component of sewage sludge as a result of residues from consumer products in waste water.

The direct environmental impact of materials can be evaluated by determining the release of hazardous substances during production, use, or in the end-of-life phase. Suitable measurement and testing methods for the assessment of this impact are available, as shown in many of the papers. The indirect impact of materials to the environment is caused by associated energy consumption, all emissions during manufacturing, and the depletion of natural resources. The quantification of this impact is complex but can be assessed by LCA. The sink function of nature for all negative impacts, and thus the ecosystem service [17], is limited. Hence, the use of recycled waste materials may be advantageous, in this respect, when these materials are properly characterized regarding their environmental compatibility.

It is a materials world. The rise in global material consumption will only continue in the future. Therefore, the identification of environmentally benign materials is indispensable for regulators, manufacturers, and users.

Author Contributions: Conceptualization, writing and editing, F.-G.S.; writing and editing, U.K. All authors have read and agreed to the published version of the manuscript.

Funding: This research received no external funding.

Conflicts of Interest: The authors declare no conflict of interest.

References

1. Krausmann, F.; Gingrich, S.; Eisenmenger, N.; Erb, K.H.; Haberl, H.; Fischer-Kowalski, M. Growth in global materials use, GDP and population during the 20th century. *Ecol. Econ.* **2009**, *68*, 2696–2705. [CrossRef]
2. Ehrlich, P.A.; Holdren, J.P. Impacts of population growth. *Science* **1971**, *171*, 1212–1217. [CrossRef] [PubMed]
3. Bandow, N.; Aitken, M.D.; Geburtig, A.; Kalbe, U.; Piechotta, C.; Schoknecht, U.; Simon, F.-G.; Stephan, I. Using environmental simulations to test the release of hazardous substances from polymer-based products: Are realism and pragmatism mutually exclusive objectives? *Materials* **2020**, *13*, 2709. [CrossRef] [PubMed]
4. Weiler, L.; Vollpracht, A. Leaching of carbon reinforced concrete—Part 1: Experimental investigations. *Materials* **2020**, *13*, 4405. [CrossRef] [PubMed]
5. Weiler, L.; Vollpracht, A.; Matschei, T. Leaching of carbon reinforced concrete—Part 2: Discussion of evaluation concepts and modelling. *Materials* **2020**, *13*, 4937. [CrossRef] [PubMed]
6. von Wolff, M.; Stephan, D. Testing of eluates from waterproof building materials for potential environmental effects due to the behavior of Enchytraeus albidus. *Materials* **2021**, *14*, 294. [CrossRef] [PubMed]
7. Richter, M.; Horn, W.; Juritsch, E.; Klinge, A.; Radeljic, L.; Jann, O. Natural building materials for interior fitting and refurbishment—What about indoor emissions? *Materials* **2021**, *14*, 234. [CrossRef] [PubMed]
8. Even, M.; Wilke, O.; Kalus, S.; Schultes, P.; Hutzler, C.; Luch, A. Formaldehyde emissions from wooden toys: Comparison of different measurement methods and assessment of exposure. *Materials* **2021**, *14*, 262. [CrossRef] [PubMed]
9. Szatanik-Kloc, A.; Szerement, J.; Adamczuk, A.; Józefaciuk, G. Effect of low zeolite doses on plants and soil physicochemical properties. *Materials* **2021**, *14*, 2617. [CrossRef] [PubMed]
10. Scholz, P.; Putna-Nimane, I.; Barda, I.; Liepina-Leimane, I.; Strode, E.; Kileso, A.; Esiukova, E.; Chubarenko, B.; Purina, I.; Simon, F.-G. Environmental impact of geosynthetics in coastal protection. *Materials* **2021**, *14*, 634. [CrossRef] [PubMed]
11. Shin, H.-W.; Hwang, J.-H.; Kim, E.-A.; Hong, T.-W. Evaluation of hydrogenation kinetics and life cycle assessment on Mg2NiHx– CaO composites. *Materials* **2021**, *14*, 2848. [CrossRef] [PubMed]
12. Seesanong, S.; Boonchom, B.; Chaiseeda, K.; Boonmee, W.; Laohavisuti, N. Conversion of bivalve shells to monocalcium and tricalcium phosphates: An approach to recycle seafood wastes. *Materials* **2021**, *14*, 4395. [CrossRef] [PubMed]
13. Prieto-Espinoza, M.; Susset, B.; Grathwohl, P. Long-term leaching behavior of organic and inorganic pollutants after wet processing of solid waste materials. *Materials* **2022**, *15*, 858. [CrossRef] [PubMed]
14. Liu, B.; Finkel, M.; Grathwohl, P. Mass transfer principles in column percolation tests: Initial conditions and tailing in heterogeneous materials. *Materials* **2021**, *14*, 4708. [CrossRef] [PubMed]
15. Sakanakura, H.; Ito, K.; Tang, J.; Nakagawa, M.; Ishimori, H. Determining adsorption parameters of potentially contaminant-releasing materials using batch tests with differing liquid-solid ratios. *Materials* **2021**, *14*, 2534. [CrossRef] [PubMed]
16. Meisterjahn, B.; Schröder, N.; Oischinger, J.; Hennecke, D.; Weinfurtner, K.; Hund Rinke, K. Leaching of titanium dioxide nanomaterials from agricultural soil amended with sewage sludge incineration ash: Comparison of a pilot scale simulation with standard laboratory column elution experiments. *Materials* **2022**, *15*, 1853. [CrossRef] [PubMed]
17. Millennium Ecosystem Assessment. *Ecosystems and Human Well-Being: Synthesis*; Island Press: Washington, DC, USA, 2005.

 *materials*

Review

Using Environmental Simulations to Test the Release of Hazardous Substances from Polymer-Based Products: Are Realism and Pragmatism Mutually Exclusive Objectives?

Nicole Bandow [1], Michael D. Aitken [2], Anja Geburtig [3], Ute Kalbe [3,*], Christian Piechotta [3], Ute Schoknecht [3], Franz-Georg Simon [3] and Ina Stephan [3]

[1]   German Environment Agency, Corrensplatz 1, 14195 Berlin, Germany; nicole.bandow@uba.de
[2]   Department of Environmental Sciences and Engineering, University of North Carolina, Chapel Hill, NC 27599-7431, USA; mike_aitken@unc.edu
[3]   Bundesanstalt für Materialforschung und-prüfung (BAM), 12200 Berlin, Germany; anja.geburtig@bam.de (A.G.); Christian.piechotta@bam.de (C.P.); ute.schoknecht@bam.de (U.S.); franz-georg.simon@bam.de (F.-G.S.); ina.stephan@bam.de (I.S.)
*   Correspondence: ute.kalbe@bam.de

Received: 15 May 2020; Accepted: 9 June 2020; Published: 15 June 2020

Abstract: The potential release of hazardous substances from polymer-based products is currently in the focus of environmental policy. Environmental simulations are applied to expose such products to selected aging conditions and to investigate release processes. Commonly applied aging exposure types such as solar and UV radiation in combination with water contact, corrosive gases, and soil contact as well as expected general effects on polymers and additional ingredients of polymer-based products are described. The release of substances is based on mass-transfer processes to the material surfaces. Experimental approaches to investigate transport processes that are caused by water contact are presented. For tailoring the tests, relevant aging exposure types and release quantification methods must be combined appropriately. Several studies on the release of hazardous substances such as metals, polyaromatic hydrocarbons, flame retardants, antioxidants, and carbon nanotubes from polymers are summarized exemplarily. Differences between natural and artificial exposure tests are discussed and demonstrated for the release of flame retardants from several polymers and for biocides from paints. Requirements and limitations to apply results from short-term artificial environmental exposure tests to predict long-term environmental behavior of polymers are presented.

Keywords: environmental simulations; polymer-based products; artificial weathering; degradation; leaching; soil contact

---

1. Introduction

Many materials are exposed to the ambient environment, so that there is a need to understand how such exposure might affect the environment and vice versa. To come up to growing demands, plastics must be optimized using several additives. As long as such additives stay within the materials over the whole life cycle, they do not pose a problem to the environment, even if they are classified as hazardous. Whether, and if so in what quantity, such substances are released into the environment during a product's service life, depends on the conditions of use. For example, building materials can be exposed to the ambient atmosphere, sunlight, and precipitation, can be in direct contact with soil, or can be submerged in groundwater or marine environments. Government regulatory agencies can be concerned with the release of potentially harmful substances embedded in or on the surface of a material, such as a biocide in paint. Manufacturers, on the other hand, might be interested in the reciprocal impact of environmental exposure on the material itself, which could influence its longevity and/or efficacy. The interface of these two perspectives—how releasing a key component to the environment affects the material and how changes to material properties due to environmental exposure influence the release of a chemical of concern—can also be important. For purposes of this review, we use the term "chemical" to mean a substance that either was added to a material by the manufacturer, is a residual constituent of the manufacturing process, or is a product resulting from transformation of the material upon environmental exposure. From an environmental perspective, such chemicals might be regarded as contaminants, whereas to the manufacturer they may be key ingredients necessary to achieve the desired material properties, unanticipated byproducts of material manufacture, or unavoidable transformation products resulting from normal service in the field.

Regulatory agencies often require testing to evaluate the potential environmental impact of a material exposed to the ambient environment. Standardized tests have been developed for such purposes, either by the agency itself or by standardization bodies (such as the International Organization for Standardization, ISO). In some cases, there might not be a standardized test, but protocols have been developed that are widely adopted. Apart from regulatory constraints, manufacturers concerned about effects of the environment on a particular material or an ingredient in a material can have more freedom to develop de novo tests, though more in the context of research and development than in creating a standard protocol.

The development of a standardized test is typically subject to considerable stakeholder involvement, with the aim to optimize relevance while accounting for user concerns regarding time and cost. Although environmental exposures occur over a period of years, testing is by necessity conducted over much shorter time scales. Short-term tests can also be useful to quickly compare the potential impacts on two or more materials.

A typical question, then, in developing a standard testing method is how to condense the time frame for a certain type of environmental exposure (such as exposure to ambient precipitation) into a time frame that is both practical and cost-effective. Doing this inherently involves tradeoffs between expediency and the extent to which the resulting test fits the purpose for which it was designed. The results of such tests are the basis for regulatory processes. How to assess the results, e.g., development of criteria for assessment, will not be considered here.

Plastics are currently in the focus of environmental policy due to their long-term behavior and therefore to their persistence. Not only do they appear as visible particles in the sea and on the beach, the almost unknown behavior of their additives and the related transformation products are of environmental concern. Therefore, this review focuses on the simulation of the environmental behavior of polymer-based products. Examples of such studies carried out by the authors are given. To our knowledge the combination of environmental exposure of polymer-based products to induce aging processes with the transfer of chemicals to environmental compartments in one experiment is quite rare. We believe that this combination is crucial for a better understanding and risk assessment and is therefore highlighted in this review article.

Photochemical, chemical, and biological processes have the potential to transform a chemical in a material into one or more products whose identities and properties might not be known yet, but might be of environmental concern [1–3] (Figure 1). Such products are often overlooked because material-testing methods are usually designed to test known (target) substances. In some cases, bioassays [4] intended to detect the potential effect(s) of such products might be required by a regulatory agency [5]. Important are not only the chemicals that polymers originally consist of, but also the compounds that are processed via structural transformation in the environment or by organisms [6]. Abiotic processes (e.g., photolysis, hydrolysis, oxidation) can modify or degrade substances to structurally altered species. Biodegradation and biodeterioration are the interactions of microorganisms or other organisms with materials [7]. The biodegradation of organic matter is the result of the action of a multitude of different organisms. In all stages, microorganisms and their metabolism play an important role as they both use organic matter as a source of nutrients and their enzymes can result in other chemical changes not linked directly to their use of a substrate. They can also cause degradation of a material by colonizing its surface and causing physicochemical changes (e.g., in pH, moisture conditions) that leads to abiotic degradation. The huge diversity within biological systems allows them to contribute to the degradation of a wide range of materials, including many synthetic polymers.

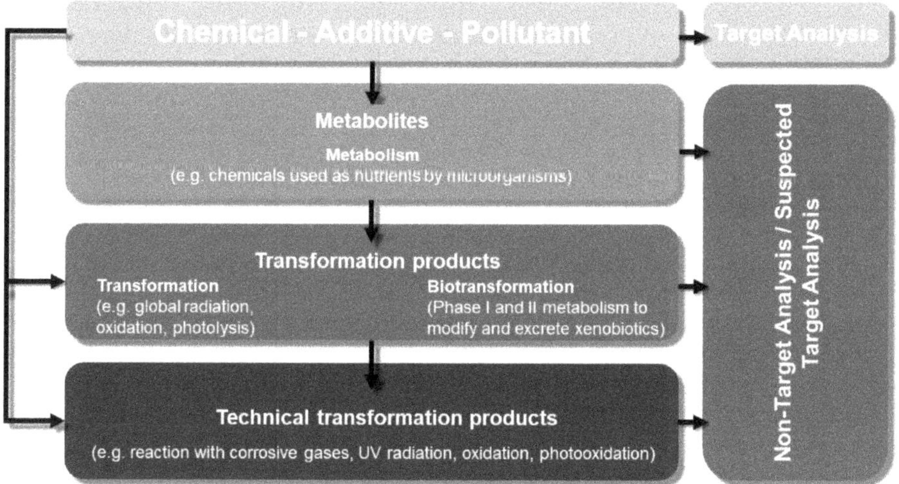

Figure 1. Possible consequence for chemicals under environmental relevant conditions in general.

The degradation of polymers is a known reason for the release of additives and other chemicals from the material into the environment [5,8]. For the quantification of released chemicals that are known, for example, additives such as stabilizers, a target-analytical approach must be established. For the identification, characterization, and subsequent quantification of possible transformation products or metabolites, a non-target- or suspected-target-analysis strategy is essential [9].

2. Aging Exposure Types

Often, the release of hazardous substances changes with the aging status of the plastic material. Aging—which is defined as the entirety of irreversible chemical and physical processes within a material over time [10]—can be caused by several natural environmental conditions. This chapter summarizes the nature-oriented simulations of such processes.

2.1. Weathering

For polymeric systems, weathering exposure causes mainly photochemical aging [11–13]. Solar UV radiation can split organic bonds. In the presence of oxygen, this typically leads to photo-oxidation, including e.g., chain scissions. The kinetics of the photochemical processes greatly depends on the temperature. Besides UV irradiance and temperature, also rain and humidity can have a major influence on the weathering results. Both can lead to hydrolytic degradation of some polymeric materials. Moreover, cyclic moisture conditions can generate mechanical stresses and accelerate migration processes.

There are several concepts for simulating outdoor weathering effects in artificial laboratory tests; each has specific advantages and disadvantages.

In Xenon arc devices [14]—with a spectral irradiance distribution closely resembling that of solar radiation—the high noon scenario is mapped, due to radiation heating mainly from the IR and VIS radiation amounts [15]. On the one hand, this reflects the maximum differences in surface temperatures between different colors [16]; on the other hand, it neglects the moisture parameter, which can be important for e.g., migration, crack formation or abrasion due to mechanical stresses, or PVC discoloration. Here—only in the presence of moisture—the temporarily generated polyenes are oxidized by the photocatalytic action of TiO_2 under wet conditions; under the too-dry Xenon arc conditions, these polyenes remain, noticeable as an unrealistic, strong discoloration [17]. In Xenon arc devices, the temperature gradient over the thickness occurs similarly to high noon natural weathering conditions, but it varies with many parameters. These include not only IR radiation and IR absorption, but also the material's thickness and thermal conductivity, as well as fluid dynamic processes, which greatly depend on the installation position and environment.

In various UV fluorescent lamp devices [18], a temperature gradient is induced by rear-contacting to the cooler laboratory temperature, in order to generate dew during the dark wetting times—a concept that works quite well for metal coating panels, but not for thicker plastic specimens. During the irradiation, the specimen's surface is heated by convection. Unfortunately, this can lead to huge temperature variations, depending on the kind of temperature control and the sample thickness [19].

However, due to the negligible radiation heating, the use of UV fluorescent lamps enables a very accurate temperature control over the whole specimen (surface as well as bulk) by integrating the UV lamps into a climate chamber, excluding the complex fluid dynamic effects. On the one hand, this concept enables accurate control of the surface climate. On the other hand, a large temperature and humidity range can be adjusted at the sample surface [20].

Due to the strong and individually different interactions, standardized artificial weathering tests are quite pragmatic. Although generally aiming at acceleration, no acceleration factor compared with any outdoor exposure can be generalized for different polymers. There are various objectives in the conceptual design of such artificial weathering tests: e.g., the determination of acceleration factors [21], the comparison of the ranking of different materials in regard to aging effects [22,23], or the identification of potential failure locations for components.

2.2. UV Radiation in Terms of Photocatalysis or Photocatalytic Effects

The degradation of emerging pollutants in the environment especially in the context of so-called AOP (advanced oxidation processes) is of great interest and was investigated in many scientific studies. Consequently, chemicals or pollutants released from aged materials can be degraded into distinct transformation products.

For photocatalytic reactions in terms of AOP titanium oxide is the most common compound. Here, titanium (IV) dioxide is the species of interest. It is chemically inert and can only be dissolved in sulfuric of hydrofluoric acid. When illuminated with UV radiation, photocatalytic radical reactions can take place by generating free charge carriers, electrons in the conduction band and holes in the valence band. Normally, these charge carrier pairs recombine very quickly; however, due to the band bending in the area of the surface, charge carriers can be separated. These generally react with adsorbed oxygen and water to form hydroxyl and peroxy radicals to degrade or damage polymer-based materials and their additives [24].

Polypropylene (PP), for example, is used in food packaging. Due to its inertness, it is commonly not biodegradable and resists microbial or enzymatic degradation. Titanium dioxide can be used to accelerate the photodegradation of synthetic polymers, resulting in the release of other components or additives in the polymer with subsequent transformation processes, releasing a various number of transformation products into the environment [25,26].

2.3. Ozone as an Example of a Corrosive Gas

Ozone cracking, or the interaction of ozone with materials such as polymers, will attack double bonds in rubber chains or other polymers [27]. Natural rubber and polyunsaturated polymers like, polybutadiene can be corroded and finally degraded. Here, the concentration of ozone acting as a corrosive gas is in direct correlation to the extend or degree of degradation. This kind of degradation or decomposition of polymers enables the release of imbedded chemicals such as stabilizers, additives, or other pollutants to the environment. This test can be performed according to the standard ASTM D1149-18 [28].

2.4. Soil Bed

Soil bed tests are used when the materials in their use phase will either be directly exposed to the soil or heavy organic soiling has to be expected (e.g., outdoor textiles [29,30], buffers of cars). Any soil bed is basically a black box concerning the soil organisms (e.g., bacteria, fungi, algae, mites, nematodes). Depending on the initial pH of the soil and material, the degree of moisture, and the temperature, these organisms are more or are less active. While some might be in a dormant stage (e.g., as spores), others might find their ecological niche under precisely these circumstances.

The greater the biological diversity in a soil bed, the more likely it is that a material inserted in such an environment will be colonized or even deteriorated. Qualifying and quantifying the different groups of organisms in a soil bed is an immensely time-consuming and expensive undertaking, and the results are short-lived because adding a new material can change the community at the interface of material and soil. Communities also fluctuate when events such as rain or dryness leave pores in the soil structure, whether aerobic or anaerobic. On a locally very small scale, changes in pH can occur, as fungi in particular emit organic acids as secondary metabolites.

To gain some control over such a complex test setup, it is common practice to include either degradable materials (e.g., cotton textile, degradable wood species) and to monitor their deterioration or to measure sum parameters such as O_2 consumption or CO_2 output during an experiment to document metabolism. When using degradable materials, these serve as a reference benchmark to tell of a certain capacity the soil showed over a longer period, as most soil bed tests are performed for from at least 4 weeks up to several years. Withdrawing reference materials after time intervals makes it possible to monitor some aspects of the capacity of a soil bed. Mainly, strength loss, mass loss, or changes of the appearance of the surface are measured to rate or to quantify the impact of a soil bed on a material.

Soil containing a high degree of variability of organisms is required for this test, and the biological activity must be maintained during the test. To achieve an acceleration of effects between material and soil, two basic principles are important: (i) keep the moisture level in the soil so that neither dryness nor waterlogging occurs, (ii) do not allow temperatures below 15 °C or above 35 °C. Fine-tuning temperature and wetness makes it possible to tailor the soil bed to different use conditions. If comparable results are to be obtained, it is necessary to use soil of the same physical and chemical description regarding e.g., particles size, water holding capacity, and pH.

3. Mass-Transfer Processes

Material-testing protocols are generally designed to evaluate the transfer of a chemical from the material to the environmental compartment of interest (air, water, soil). A leaching procedure, for example, might be developed to reasonably approximate the total volume of water to which the material might be exposed over a defined period of interest, but condensed into a short-term test designed to reproduce the total volume per unit surface area or sample mass.

To analyze the release, two different approaches are possible: (i) as a direct proof, the amount of the released substance during the exposure can be measured, or (ii) as an indirect proof, the remaining concentration after several exposures is compared to the substance's original amount within the specimen. Such residual analysis is not always possible, e.g., the release of bisphenol A only develops during polycarbonate aging [31–33]. Either the water samples can be directly analyzed or the analytes can be enriched by passive sampling techniques in the experiments [34] or afterwards by techniques such as solid phase extraction or stir bar sorptive extraction [35].

To investigate the release during the exposure of polymer sheets in a weathering chamber, the accumulated concentration within the circulated rainwater can be measured, using a circulation system to spray water. For the investigation of polar substances, the circulated water can be sampled and analyzed after various periods. A main disadvantage of this method is that sorption of the substance somewhere within the device, such as at seals or pipes, is hardly preventable. To avoid this, as many stainless steel sections as possible should be used, e.g., the water tank. Another possibility is the separation of aging and leaching into two consecutive steps. To investigate nonpolar species, silicone-based passive samplers can be placed in the water drainage to enrich the analytes [36].

Transformation processes of analytes in the material itself or after release in the eluates further complicates the analysis. The method of sample preparation or analysis may be not suitable for transformation products and can be adapted only after the identification of chemicals. This identification procedure is time-consuming and usually does not identify all unknown compounds in a sample [37]. If both the residual content within the specimen and the release during the exposure can be measured with sufficient precision, a mass balance can be established; here, all other possible losses, such as sorption [38] or degradation [39], should be considered.

Various leaching tests, such as tank tests, batch tests, and column percolation tests, have been developed to measure the mass transfer by contact with water. The choice of test depends on the material properties, as well as on the intended application of the material under investigation. Monolithic and sheet-like products are usually tested in tank tests (e.g., [40–42]), while batch tests (agitation by shaking in a bottle, e.g., [43,44]) and up-flow column percolation tests have been successfully established (e.g., [45,46]) for granular materials. However, some tests, in particular batch tests and one-stage tank tests developed for compliance testing, are less suitable for the investigation of long-term mass-transfer processes because the conditions in the test are not comparable to field scenarios (e.g., contact time between sample and leachant, liquid-to-solid ratio). For granular materials, column percolation tests are an appropriate tool for investigating the long-term contaminant transfer as e.g., organic compounds [47] from a variety of materials as reconstruction products [48], waste [49] and fly ash and slag [50] to soil and water.

Tank tests may include permanent immersion (e.g., dynamic surface leaching test, DSLT, CEN/TS 16637-2 [40]) or intermittent immersion at defined ratios between the exposed surface area and the volume of water used for leaching (EN 16105 [41]). Eluate concentrations can be transferred to emissions per surface area if needed to assess results. The test design of the DSLT includes increasing leaching intervals and enables the identification of diffusion-controlled leaching processes. Intermittent immersion tests with a sequence of dry and wet cycles are closer to realistic exposure conditions and reproduce mass-transfer processes in the material also during dry periods [51].

Mass transfer to water includes various mechanisms that determine the concentration's time profile. Often not only a single mechanism but rather a shift from one dominating mechanism to another during longer leaching experiments is observed [52,53]. As an example, in a CEN TC 351 standard [40], different mechanisms are described: for compounds with low water solubility, maximum release is limited, which is often indicated by almost constant eluate concentrations (Figure 2a). At the beginning, a first flush or surface wash-off is observed if compounds with high water solubility are loosely attached to material surfaces (Figure 2b). High eluate concentrations for the first fraction are accompanied by rapidly decreasing concentrations, especially in cases of a rapid depletion due to limited stock in the material. Depletion proceeds from the surface to the inner layers, establishing a concentration gradient responsible for diffusion processes in the material itself. Diffusion-controlled processes are proportional to the square root of contact time and are indicated by the pattern of the eluate concentrations after the defined periods of water contact (Figure 2c). Figure 2d shows a combination of first flush and diffusion. Diffusion in bulk material is often slow. Inhomogeneous materials hamper a mechanistic description of diffusion processes, as only overall diffusion rates for the bulk material are determined [54]. Pores filled with water enhance diffusion within the material, but important for mass transfer is not only the total porosity, but also the structure of the pore network e.g., dead ends and surface area [55], linkage between the pores [56]. A higher influence on such processes by the shape of pores than their sizes was found by Al-Raoush [57]. Molecular size, existing water-filled pores, and water solubility are substance parameters influencing the diffusion rate.

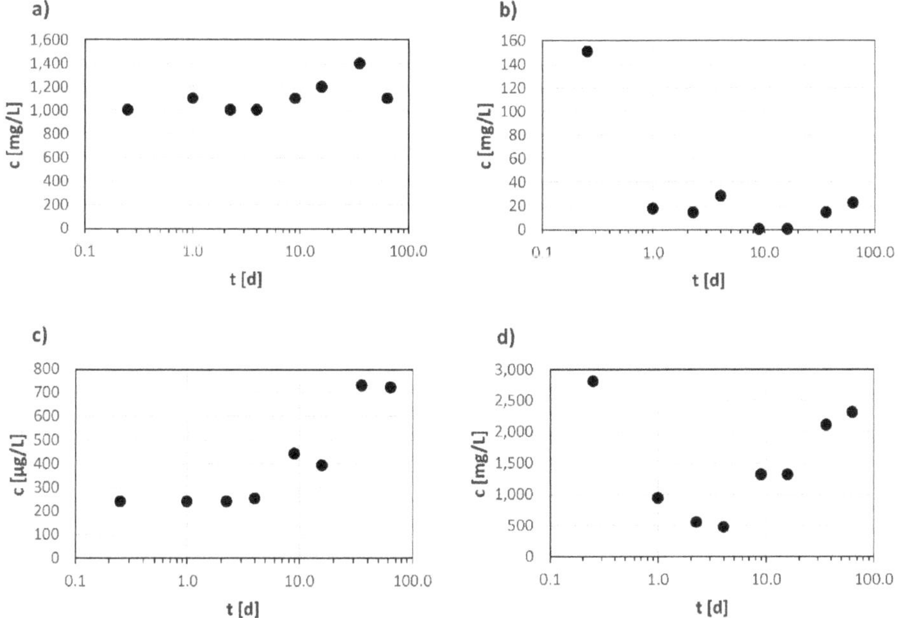

Figure 2. Examples of concentration patterns in elute fractions from the dynamic surface leaching test that indicate different release mechanisms as described in [40]. On the *x*-axis, the period of water exchange in days is shown for (**a**) a process controlled by water solubility, (**b**) first flush (**c**) a diffusion-controlled process, and (**d**) first flush combined with a diffusion-controlled process.

The velocity of mass transfer at the solid/water interface is driven mainly by the concentration gradient between the two phases and will slow when equilibrium is reached [58–60]. The thickness of the water boundary layer between the two phases depends on the turbulence in the system. In situations where the liquid phase is not agitated, diffusion through this layer may become the rate-limiting step [59]. For column percolation tests, equilibrium is rapidly established at the beginning of a test for many materials in a broad range of particle size distributions and flow velocities [61,62], but mostly cannot be maintained throughout the test [63]. Achievement of equilibrium is important if the results from laboratory leaching tests are to be comparable to field studies [61,64].

4. Tailoring Exposure Tests—Examples

To accelerate the release of a substance in an artificial environmental test, some knowledge is needed of the release mechanism, the degradation behavior of the material, and the limits of the relevant environmental factors. Based on this knowledge, several exposure types can be combined in one exposure test, e.g., in weathering chambers.

For the release of additives or aging products during weathering exposure of polymeric systems [65], two effects must be considered. On the one hand, the polymeric matrix degrades due to chemical aging, probably uncovering embedded ingredients. On the other hand, cyclic stress can greatly increase the migration effects of various components.

To accelerate the release of additives or aging products, an artificial weathering test should respect both effects, but still keep the exposure parameters within natural limits.

To increase mechanical stresses, long wetting periods can be included, which increase the moisture penetration depth. Additionally, thermal stress can be applied in artificial weathering tests by varying the temperature levels. To accelerate polymer matrix degradation, high temperatures should be chosen. Doing so, both maximum outdoor temperatures and possible transition temperatures of the polymeric system should be considered, to avoid unrealistic natural effects.

An example of an exposure scheme that includes cyclic thermal mechanical stress is demonstrated in (Figure 3). It has been applied to different materials [36,66–69]. This 24-h weathering cycle was adapted from a German standard for traffic signs [70]. Long wetting and drying periods of four hours each guarantees deep penetration of rainwater as well as sufficient drying, resulting in the transport of water-soluble extracts out of the polymeric system. At the same time, mechanical stresses are generated on the surface, which can lead to surface crack formation and thus increased diffusion. Similar effects can be achieved with strong temperature cycles. Therefore, high temperature levels are alternated with frost periods. During the whole 24 h, continuous UV irradiance is applied, using UVA-340 nm fluorescent lamps, in accordance with ISO 4892-3 type 1A.

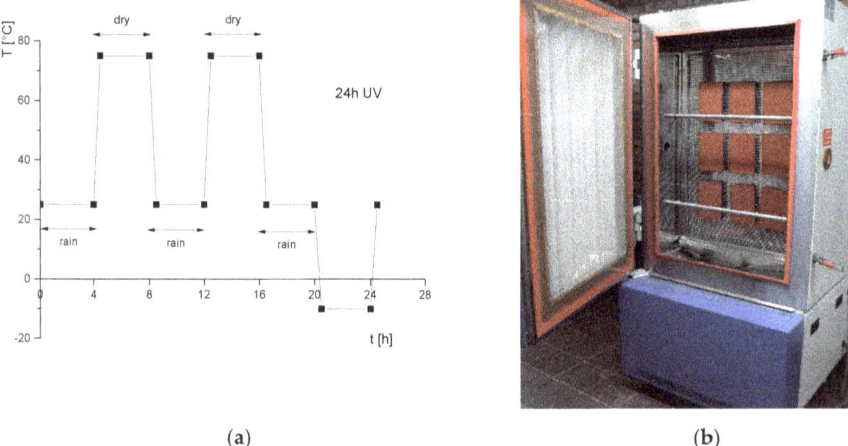

(a)

(b)

Figure 3. (**a**) 24-h weathering cycle with continuous irradiation [70] and (**b**) weathering chamber prepared for exposing samples of synthetic athletic tracks.

4.1. Metals and PAH from Synthetic Sports Surfaces

Artificial turf and other synthetic surfaces have widely been established as replacements for natural grass or cindered pitches and tracks on sports facilities. This is mainly due to substantial improvements in sports performance over the years of development in engineering of such surfaces and less requirements to maintain the desired conditions to enable a more intense usage. Polyethylene, polypropylene, polyamide, polyurethane (PUR), Nylon, styrene butadiene rubber (granules from discarded tires), ethylene propylene diene monomer rubber are frequently used materials for the production of the synthetic components of artificial sports surfaces. Numerous additives (e.g., ZnO as vulcanization agent, PAH from carbon black as UV stabilizer and phenols as antioxidants) are used to tailor the desired properties for this specific application. Due to degradation of the materials throughout its life cycle, there are, however, concerns about posing a risk to the environment by transferring hazardous substances from outdoor sports facilities to percolating rainwater and soil.

The leaching behavior with water of individual polymeric components of artificial turf systems and sports surfaces was investigated using batch tests and column tests with and without previous weathering [67,68]. For rubber granules, particularly granules from discarded car tires (i.e., SBR, styrene butadiene rubber), the release of Zn [71,72] and other organic substances [73] was reported. It was found that increasing water exposure decreases the release of inorganic and organic substances, such as Zn and PAH [74]. Due to degradation of the polymeric matrix and the uncovering of leachable components during weathering, Zn release partly increases again [68,75,76]. The topmost layer used for athletic tracks usually consists of SBR and/or EPDM (ethylene propylene diene monomer rubber) granules and PUR as a binding agent. Combining weathering exposure, using a cycle and a weathering chamber as displayed in Figure 3, and leaching tests revealed the long-term contaminant release [68,74] and points to the possible increase of some substances when the polymer matrix begins to degrade (see Figure 4).

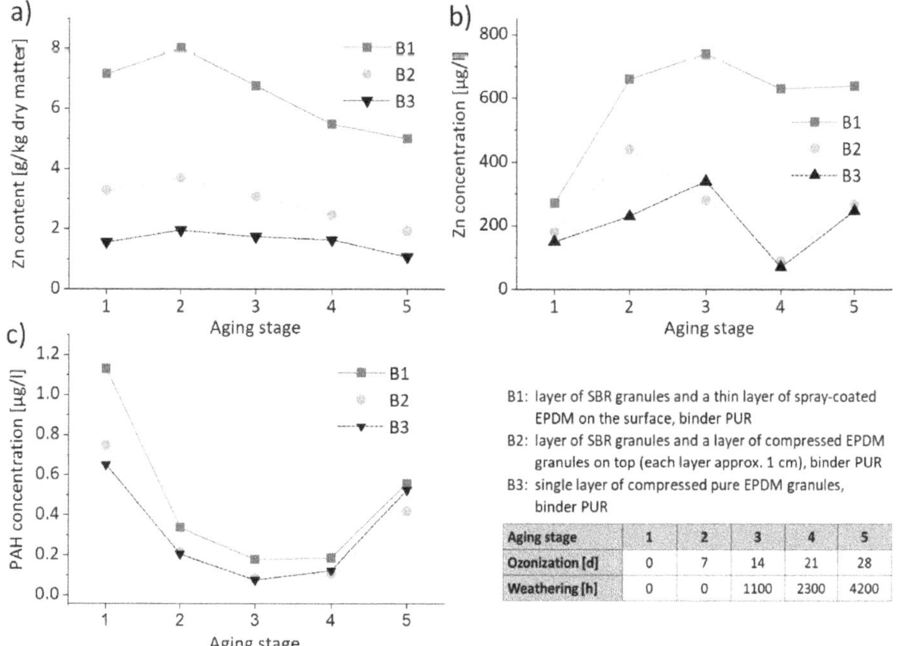

Aging stage	1	2	3	4	5
Ozonization [d]	0	7	14	21	28
Weathering [h]	0	0	1100	2300	4200

Figure 4. Trend of (**a**) Zn content in solid matter in the course of the artificial weathering of three types of sports floorings (B1 to B3), as well as release of (**b**) Zn and (**c**) PAH measured in the eluates of intermittent batch leaching tests at L/S (liquid-to-solid ratio) of 2 L/kg during artificial weathering.

With granules coated with PUR to imitate the color of grass, the release of Zn was found to be retarded [77]. In addition to the investigation of the leaching behavior of Zn and organic substances, Li et al. [78] studied the emission of volatile organic substances in the gaseous phase under laboratory conditions as well as natural weathering conditions and found a significant reduction in the latter case.

In contrast to concerns about the environmental compatibility of discarded tire granules, Edil et al. [79] also point to the applicability of this material for the remediation of contaminated water such as landfill leachate, which can be attributed to the high sorption capacity of hazardous volatile organic compounds and was found by performing large scale laboratory tests and field tests.

The increase in leached substances from rubber granules after simulation of mechanical stress using several approaches was reported, e.g., by [80,81].

The complexity of synthetic surfaces on outdoor installations of sports facilities poses a challenge to realism when studying the environmental compatibility, e.g., of polymer-based turf systems or athletic tracks. The mutual influence among the individual components and therefore the appropriate test setup to simulate actual conditions in the field must be taken into account. A concept has been developed to investigate the leaching behavior of such complete synthetic sports surface installations using column tests [67]. All components were assembled in glass columns in accordance with the known or expected field conditions and percolated using demineralized water up to a certain liquid-to-solid ratio. The contaminant release measured in the resulting eluate fractions gives a more realistic picture than do investigations of the individual components. A set of typical outdoor sports surface installations was studied using this tailored test procedure. The contaminant transfer to groundwater was modeled and assessed in terms of the potential environmental impact of selected regulated contaminants [73,82]. Nevertheless, for production control purposes, simpler test methods and separate limit values are needed for the sake of practicability; these are currently being developed. The release of further substances of environmental concern from polymeric sports surfaces and upcoming alternative materials remains to be investigated [83].

4.2. Antioxidants from PE

Polymer materials such as polyethylene contain a wide variety of organic additives to improve their stability and their typical properties regarding manufacturing processes and the interaction with environmentally relevant impact factors. Additives used for these applications and requirements include stabilizers, such as antioxidants and other organic compounds. Here, investigations of the residual content of substances within a weathered polymer such as polyethylene for use in water pipelines were performed to evaluate the stabilizer's migration behavior, where the remaining antioxidant concentration was measured by means of HPLC-DAD and GC-MS, after various aging exposure scenarios [84,85].

4.3. Flame Retardants from Thermoplastics

Observations of the effects on the environment of critical substances released by weathering, such as fire retardants, were expressed as a blooming out at the surface, checked by FTIR spectroscopy, and as a loss of surface-related flame-retarding functionality [66].

4.4. Carbon Nanotubes from Polymer Nanocomposites

The release of carbon nanotubes (CNT) due to various aging exposures was quantified by radiolabeling the CNTs [86,87]. It was found that irradiation increases the tendency of the studied polymer nanocomposites to release CNTs more than do other environmental impacts, such as shaking in water or rapid temperature changes. The reason for this is suggested to be the photochemical degradation of the polymeric matrix, uncovering CNT.

4.5. Additives from Polymer Recyclates

Polyvinyl chloride, polystyrene, and polyethylene samples from a plastics recycling plant were exposed to 1000 and 2000 h of artificial aging by photo-oxidation (continuous irradiation without wetting periods) and thermo-oxidation (80 °C) to simulate the fate of potential pollutants in the environment. In general, the investigated plastic particles were smaller than 5 mm and showed increased leaching of organic and inorganic substances after a combined thermal and UV irradiation exposure, see Figure 5 [88]. The release of substances was measured with column percolation tests with water as leachant. Thermal exposure at 80 °C alone was much less effective in accelerating the aging of the polymers, as shown by the release from the samples.

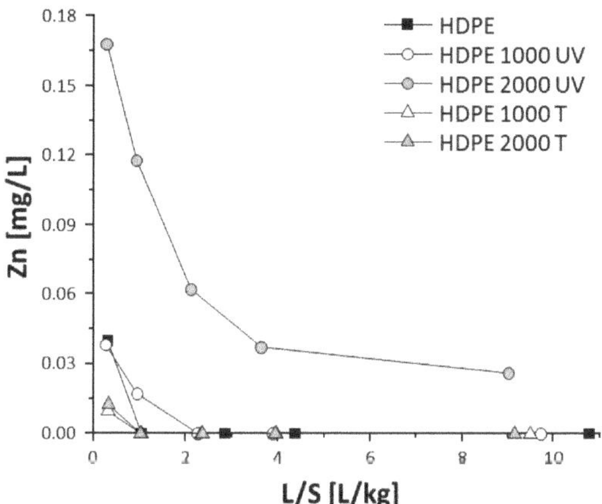

Figure 5. Release of Zn as a function of the liquid-to-solid ratio from high-density polyethylene (HDPE) granules in a column leaching experiment. The greatest release was observed in HDPE particles after 2000 h combined thermal and UV irradiation exposure. Converted to cumulative release, the displayed concentrations correspond to approx. 1% of the total Zn content [88].

5. Comparison of Natural and Artificial Exposure Tests

Natural environmental exposure tests show a broad local and seasonal or annual variability. In contrast, artificial environmental exposure tests enable reproducible results. To simulate the release of critical substances from a material, one can adjust the parameters, aiming at potentially critical environmental exposure conditions, to accelerate or maximize the mass transfer. Thus, to evaluate the ecological menace of a material, one would focus on such a worst-case scenario.

Compared to long-lasting natural environmental exposures, there is a tendency in short-term test approaches to accelerate the exposure in question into a time frame as short as possible. For example, water contact is intensified in leaching tests. However, intra-material mass transfer will affect the release of a chemical into the water. The time frame over which this occurs depends on water uptake and the thickness of the material. With reactive processes, such as biological, chemical, or photochemical processes, it is more difficult to condense the phenomena in question into shorter time frames. Photochemical exposures can be made continuous rather than diurnal, but it would be difficult to increase irradiance without also influencing reactivity, and the time scale of interest is not shortened by very much. Similarly, rates and extents of chemical reactions depend on the concentrations of the reactants, so that increasing the concentration of the reactant could affect the nature of the reaction. Microbial processes, in particular, especially if they involve microbial communities, are the most difficult to artificially accelerate without altering the outcome. Along with microbial processes, increasing the temperature at which a test is conducted can accelerate the phenomena in question without significantly influencing the outcome qualitatively.

Because the mass-transfer processes described above have such an important role in the ultimate release of chemicals into the environment, the effect of exposure cycles on these mass-transfer processes is important to understand and, therefore, might be necessary to reproduce in material-testing protocols. This can include wet/dry cycles, sunlight, freeze/thaw cycles, and ambient atmospheric chemistry that can vary diurnally over scales of days to weeks and seasonally in the same location. For example, leaching tests that involve long periods of contact between the leaching liquid and the material will decrease interfacial mass transfer as the substance in question accumulates in the liquid phase, in turn decreasing intra-material diffusion. In this case, cycling between short wetting periods and longer dry periods might actually increase the release of substances, thus inherently accelerating the process of primary concern while introducing more realism into the testing procedure. In general, any protocol that maximizes intra-material diffusion and/or interfacial mass transfer will accelerate the release of a substance of concern into the relevant environmental compartment. Conversely, any protocol that allows the substance of concern to accumulate in the environmental compartment near the interface with the material will slow its release from the material.

5.1. Real Weathering Versus the Weathering Chamber and Soil Bed Test in the Example of the Release of Brominated Flame Retardants

An accelerated aging concept was developed to investigate the release of brominated flame retardants (BFR), which are known for their persistent bio-accumulative and toxic properties, from polymer products to the water and soil compartments [89]. Polystyrene (PS) containing hexabromocyclododecane (HBCD) and polypropylene (PP) containing bromodiphenyl ether BDE-209 were used. These additives are known as substances of very high concern [5,90,91].

The studied PS and PP samples (pieces sized 10 cm × 10 cm × 1 cm) were exposed to a defined weathering schedule in a climate chamber in accordance with a quality requirement of RAL (German National Committee for Delivery and Quality Assurance) [70] (as in Figure 3). BFR were analyzed in the rainwater collected in the climate chamber. An extraction with hexane was performed and the obtained aliquots of the extracts were concentrated and analyzed by GC-MS (BDE 209) or LC-MS/MS (HBCD). Additionally, the total bromine contents were monitored for the aged and untreated samples using laser ablation inductively coupled plasma mass spectrometry (LA-ICP-MS) and X-ray fluorescence analysis (XRF) as a non-destructive and rapid method.

Soil bed tests were conducted in an irrigated concrete basin operated indoors at a defined temperature and humidity using a reference soil with high microbial activity. The amount of rainwater was adjusted to promote the growth of a wide range of aerobic microorganisms and varied between 4% and 9% of the soil's dry mass. The water content was additionally monitored by the weight of the vessel holding the specimens capturing water from raining periods. Correct humidity is a fundamental parameter for biological activity. Five test specimens of each polymer were inserted to a depth of up to half of their length in the soil in the vessel. The microbial activity of the soil was monitored by a reference polyurethane bar, whose haze and tensile strength were analyzed. The release of HBCD and BDE-209 was captured by passive samplers (silicon bars) placed in a distinct distance to the test pieces. After toluene was used to extract the corresponding soil samples and the passive samplers, the extracts were analyzed by GC-MS (BDE 209) or LC-MS/MS (HBCD) [36]. Due to the poor water solubility of the selected polybrominated flame retardants only trace amounts were found by extracting the soil and water samples and the passive samplers. The amount of bromine in the test pieces (aged and stored references) was analyzed by LA-ICP-MS and XRF. Here, an up to 20% decrease of BDE-209 in polypropylene and of HBCD in polystyrene was observed (see Figure 6).

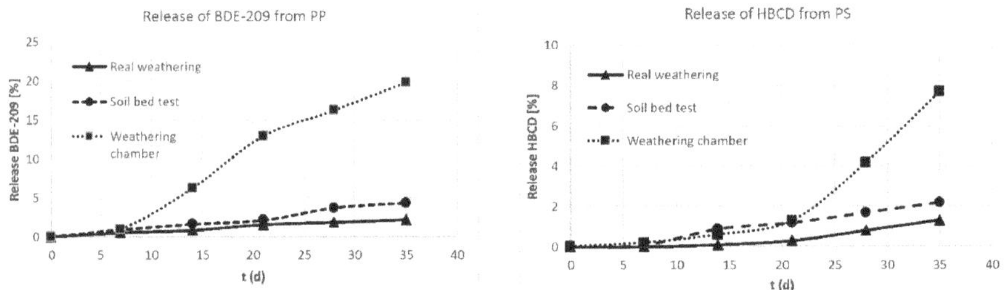

Figure 6. Release of selected polybrominated flame retardants from PP (**left**) and PS (**right**) samples under different environmentally relevant conditions.

For comparison, real time weathering experiments were performed by placing the samples on a weathering rack starting in July 2017 (in a location southeast of Berlin); the rack was aligned in a 45°

angle to the south ASTM D1435-13 [92]. The weathering data were regularly recorded by the German Weather Service (DWD). An accelerated release was evaluated by comparing real weathering to climate chamber weathering and soil bed experiments as displayed in Figure 7 [36].

(a)   (b)

Figure 7. Application of different weathering scenarios: (**a**) soil bed test with flame-protected PP and PS samples and (**b**) real weathering experiments to evaluate the release of polybrominated flame retardants.

5.2. Release of Biocides from Coatings

Coating materials can contain biocides (film preservatives) that are intended to protect the coating film against biodeterioration caused by fungi and algae. These biocides usually act in a water film that is formed on the coatings surface by precipitation or condensed water. Once on the surface, the biocides can be washed off by runoff water.

Reports of the mass transfer of biocides from façade coatings into urban surface waters [93] spurred detailed investigation of biocide leaching from paints and renders. Studies of coatings were performed under laboratory conditions in either simple immersion tests [94,95] or weathering chambers [93,96], in field experiments [39,97–100], and in settlement areas [93,101,102]. Most of the studies were performed on coatings containing polymeric organic binders, but some studies also included mineral products [103].

The European standard EN 16105 [41] was developed to determine the release of substances from coatings in intermittent contact with water under laboratory conditions [104]. The transport of biocides within the wet coatings is influenced by diffusion [96,104,105], desorption from the coating material [106], and partition between polymeric binder and water [107].

It can be assumed that desorption and diffusion processes follow the same physicochemical principles under laboratory and field conditions. However, experiments under natural weathering conditions include additional, highly variable influencing factors that affect leaching. Therefore, the progression of emission curves from field studies is less consistent and less repeatable than the progression of emission curves from laboratory experiments [39,99,100,108]. UV radiation can cause the photodegradation of biocides [39,99]. In addition, actual exposure of vertical facades to driving rain, but also to sunlight and microclimate, must be considered when results from field experiments are applied to real buildings in an urban environment [109].

The evaluation of risks that can be caused by release of biocides into the environment is a fundamental request of the European Biocidal Products Regulation [110]. For environmental risk assessments, reliable estimation of the expected release of biocides under service conditions is required. This presupposes concepts to use information on mass transfer that is gathered under laboratory or field test conditions to predict chemical release under service conditions. "Transfer functions" are required, which not only need to consider varying exposure to water, but also additional processes, e.g., degradation.

Emissions of biocides from paints were compared for laboratory tests and field experiments [100,108]. Emissions of the investigated biocides from the paints were much greater in the laboratory experiment performed in accordance with EN 16105 [41] than emissions observed during about two years of outdoor exposure in Berlin (see Figure 8).

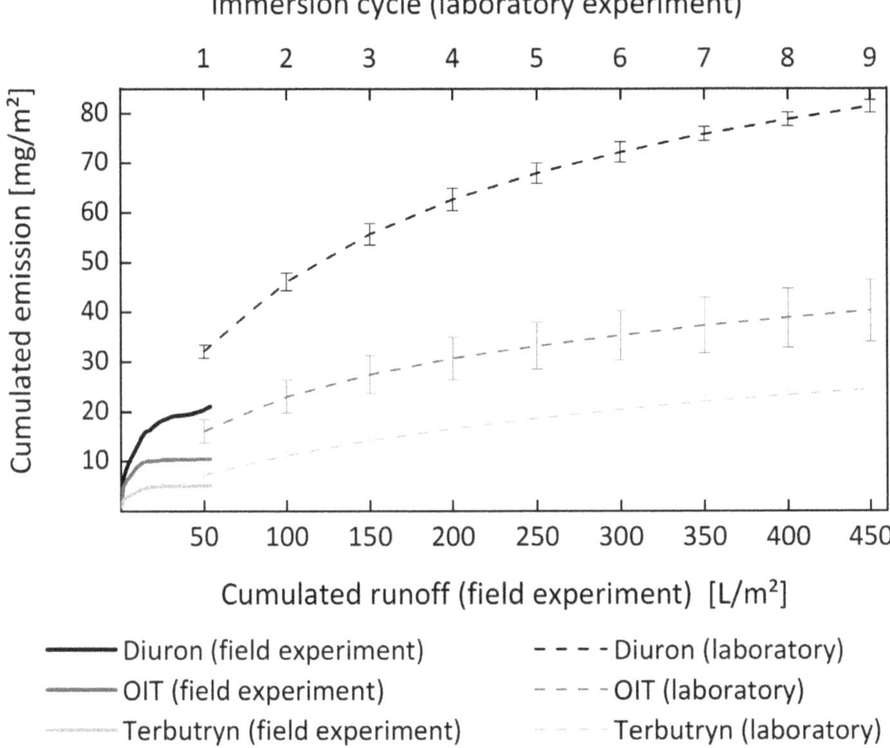

Figure 8. Cumulative emission curves for the biocides diuron, OIT (2-octyl-2H-isothiazol-3-one), and terbutryn from a vinyl acetate-based paint on wood. The amount of contact water is used to compare results from laboratory (EN 16105) and field experiments. During the laboratory experiment, each immersion cycle that consists of two immersion periods of 60 min represents 50 L/m² contact water (upper x-axis). Emissions during the field experiment are related to the collected runoff. The laboratory data represent mean values from four experiments, and the error bars indicate standard deviation. The laboratory data represent mean values from four experiments, and the error bars indicate standard deviation.

Emission curves from six independent replications of a field experiment demonstrate the influence of variable exposure conditions. However, it was possible to reproduce the different curves using a preliminary regression model that includes the different actual weather data, i.e., exposure to wind-driven rain, temperature, relative humidity, and global radiation (see Figure 9) [108]. Although this can help us understand the effects of weather on leaching processes, it remains a challenging task to apply information on the material properties determined in laboratory tests to service conditions and to include this knowledge in appropriate models and emission scenarios [96,111].

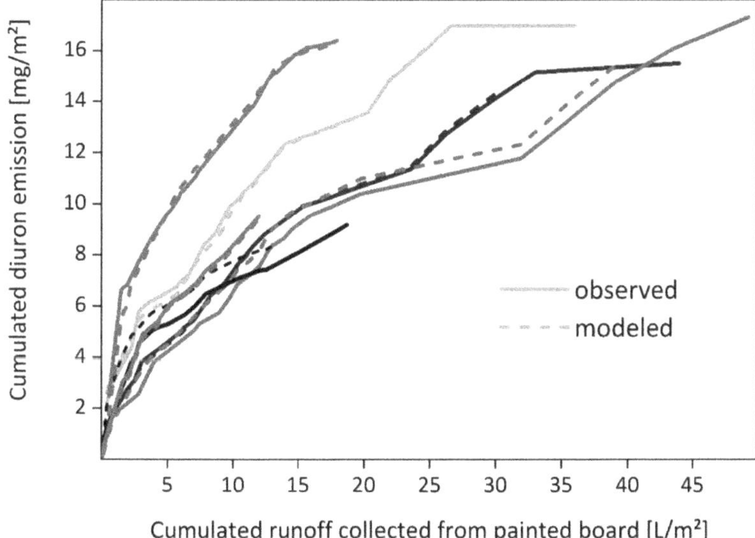

Figure 9. Comparison of experimental emission curves for the biocide diuron and emission curves obtained from a regression model that considers the various actual weather conditions. Boards painted with a vinyl acetate-based paint were exposed to natural weathering at two different sites in Berlin and 60 km north of Berlin and during different periods—resulting in six independent experiments. Runoff was collected after all rain events and analyzed for diuron. Area-related diuron emissions were calculated and demonstrated as cumulative emission curves.

5.3. The Time Scale Conundrum

Despite efforts to accelerate the release of a chemical from a material during material testing, there will always be a difference between time scales relevant to actual environmental exposures and the time scale necessary for testing. This might be less important if the test captures the release of most of the chemical over the shorter time scale; in this case, it provides a reasonable estimate of impact, particularly acute impact. If intra-material diffusion governs the rate at which a substance of concern is released; however, the time scale of a test procedure might be too short to estimate environmental impact, depending on the thickness of the material.

Short-term tests can provide useful information even if only a small fraction of a substance of interest is released over the duration of the test.

5.4. Long-Term Validation

It is not common for testing methods to be subject to validation procedures to verify the efficacy of the test relative to actual, longer-term environmental exposures. Leaching tests for soil and waste materials, for example, have been compared with results from field lysimeters operated over longer periods [112–114], and studies have been conducted to collect rainwater in contact with building surfaces after a longer period in service than would be captured in a standard leaching test [97]. In general, it is worth considering standardizing the field testing of materials after a period in service as part of a quality-improvement strategy to refine test methods. However, such time-consuming and costly studies shall be designed to allow general conclusions, rather than to investigate individual products. It is also important to consider that no straightforward validation is possible, as exposure scenarios are highly dependent on location. Even in a single location, two outdoor exposures at different times may vary greatly. Despite the lack of a general conclusion, the investigation of long-term behavior is important—also from a regulatory point of view—to evaluate if release is falling below a critical level after which no further significant release is to be expected. This decrease of release can be caused either by depletion of the material or by a change in availability.

6. Conclusions

Several laboratory tests have been developed in recent decades to investigate the effects of environmental exposure on the release of possible hazardous substances from polymer-based products. As they can focus only on a single or a limited number of test parameters, artificial environmental exposure tests can show high acceleration factors, but always risk decreasing relevance to reality. Furthermore, acceleration factors are sometimes reached under unrealistic conditions, which may allow different chemical reactions or degradation mechanisms than those observed under natural conditions. In contrast to this, natural environmental tests are close to reality per se, but are very specific regarding the exposure site and time. Especially because they are hardly reproducible even in the same location, much more effort should be made to document all possibly necessary exposure parameters to facilitate the interpretation of results.

Both approaches are necessary to assess the release of hazardous substances and are best combined to obtain complementary information, see Figure 10. Almost more important than the laboratory test itself is reliable planning before starting, and some basic knowledge about the following is necessary: starting points are the considered substances, the polymer itself, and the expected natural environmental exposure. The release mechanism is substance-specific, and knowledge about it facilitates the choice of a proper analytical method suitable for the examined matrix and sufficiently sensitive and selective. Depending on the polymeric material, one or more degradation mechanism can be relevant for the aging behavior. This is strongly connected to usage scenarios and the parameter limits of the resulting natural environmental exposure.

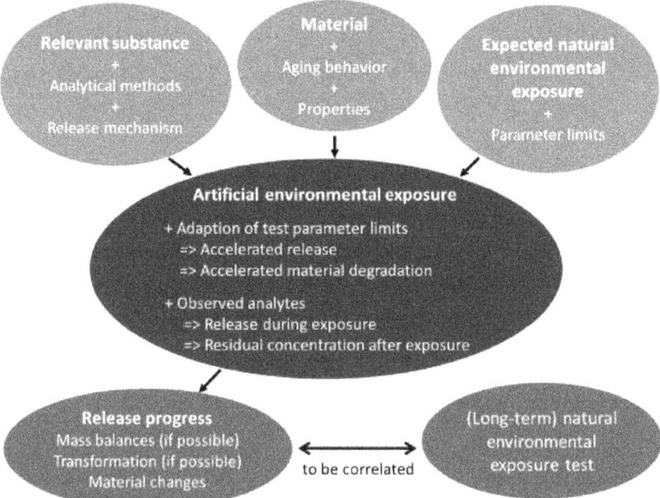

Figure 10. Flow chart for the design of artificial environmental exposure tests. Top: required previous knowledge; center: options of exposure tests; bottom: expected results.

The artificial environmental exposure tests should be defined based on this knowledge. The test parameter limits should be adapted (within reasonable limits), aiming at accelerated material degradation and maximum release. If the test setup makes it possible to measure both the release during the exposure and the residual concentration within the material after various exposures, mass balances can be established. With useful test durations, the progress of the release, which need not be monotonic, can be followed.

Such release progress from artificial environmental exposure tests must be correlated with (long-term) natural environmental tests to get an impression of time scales in natural environment.

Author Contributions: Conceptualization, Writing—Original Draft, M.D.A. and N.B., Visualization, Data Interpretation, Writing—Review & Editing, M.D.A., N.B., A.G., U.K., C.P., U.S., F.-G.S., and I.S. All authors have read and agreed to the published version of the manuscript.

Funding: This research received no external funding.

Conflicts of Interest: The authors declare no conflict of interest.

References

1. Koch, C.; Sures, B. Degradation of brominated polymeric flame retardants and effects of generated decomposition products. *Chemosphere* **2019**, *227*, 329–333. [CrossRef] [PubMed]
2. Rosato, A.; Barone, M.; Negroni, A.; Brigidi, P.; Fava, F.; Xu, P.; Candela, M.; Zanaroli, G. Microbial colonization of different microplastic types and biotransformation of sorbed PCBs by a marine anaerobic bacterial community. *Sci. Total Environ.* **2020**, *705*, 135790. [CrossRef] [PubMed]
3. Urbanczyk, M.M.; Bester, K.; Borho, N.; Schoknecht, U.; Bollmann, U.E. Influence of pigments on phototransformation of biocides in paints. *J. Hazard. Mater.* **2019**, *364*, 125–133. [CrossRef] [PubMed]
4. Drewes, J.E.; Hemming, J.; Ladenburger, S.J.; Schauer, J.; Sonzogni, W. An assessment of endocrine disrupting activity changes during wastewater treatment through the use of bioassays and chemical measurements. *Water Environ. Res.* **2005**, *77*, 12–23. [CrossRef]
5. Koch, C.; Schmidt-Kotters, T.; Rupp, R.; Sures, B. Review of hexabromocyclododecane (HBCD) with a focus on legislation and recent publications concerning toxicokinetics and—Dynamics. *Environ. Pollut.* **2015**, *199*, 26–34. [CrossRef]
6. Carstens, L.; Cowan, A.R.; Seiwert, B.; Schlosser, D. Biotransformation of Phthalate Plasticizers and Bisphenol A by Marine-Derived, Freshwater, and Terrestrial Fungi. *Front. Microbiol.* **2020**, *11*, 317. [CrossRef]
7. Stephan, I.; Askew, P.; Gorbushina, A.; Grinda, M.; Hertel, H.; Krumbein, W.E.; Müller, R.J.; Pantke, M.; Plarre, R.; Schmitt, G.; et al. Biogenic impact on materials. In *Springer Handbook of Metrology and Testing*; Czichos, H., Saito, T., Smith, L., Eds.; Springer: Berlin/Heidelberg, Germany, 2011; pp. 769–844.
8. Koch, C.; Sures, B. Environmental concentrations and toxicology of 2,4,6-tribromophenol (TBP). *Environ. Pollut.* **2018**, *233*, 706–713. [CrossRef]
9. Bletsou, A.A.; Jeon, J.; Hollender, J.; Archontaki, E.; Thomaidis, N.S. Targeted and non-targeted liquid chromatography-mass spectrometric workflows for identification of transformation products of emerging pollutants in the aquatic environment. *Trac.-Trend. Anal. Chem.* **2015**, *66*, 32–44.
10. German Institute for Standardization. *DIN 50035: 2012-09 Begriffe auf dem Gebiet der Alterung von Materialien—Polymere Werkstoffe (Terms and Definitions Used on Ageing of Materials—Polymeric Materials)*; German Institute for Standardization: Berlin, Germany, 2012.
11. Simon, F.-G.; Geburtig, A.; Wachtendorf, V.; Trubiroha, P. Materials and the Environment. In *Springer Handbook of Metrology and Testing*; Czichos, H., Saito, T., Smith, L., Eds.; Springer: Berlin/Heidelberg, Germany, 2011; pp. 845–886.
12. Wypych, G. *Handbook of Material Weathering*, 3rd ed., ChemTec Publishing: New York, NY, USA, 2003
13. Ranby, B.; Rabek, J.F. *Photodegradation, Photo-Oxidation and Photostabilization of Polymers*; Wiley-Interscience: New York, NY, USA, 1975.
14. International Standardization Organization. *ISO 4892-2: 2013-03 Plastics—Methods of Exposure to Laboratory Light Sources—Part 2: Xenon-Arc Lamps*; International Standardization Organization: Geneva, Switzerland, 2013.
15. Boxhammer, J.; Rudzki, T. The Importance of Spectral Distribution and Intensity of Artificial-Light Sources in UV and IR Region of Radiation for Accelerated Aging of Polymers. *Angew. Makromol. Chem.* **1985**, *137*, 15–27. [CrossRef]
16. Schönlein, A.; Haillant, O.; Senff, S. Surface Temperatures of Colour Painted Specimen in Natural and Artificial Weathering with Different Laboratory Light Sources for Optimized Testing and Investigations. In *4th European Weathering Symposium—Natural and Artificial Ageing of Polymers*; Reichert, T., Ed.; Gesellschaft für Umweltsimulation e.V.: Karlsruhe, Germany, 2009; pp. 47–59.

17. Trubiroha, P. The Influence of the Atmospheric Humidity to the Decoloration of PVC during Weathering and during the Following Inexposed Phase. *Angew. Makromol. Chem.* **1988**, *158*, 141–150. [CrossRef]

18. International Standardization Organization. *ISO 4892-3: 2016-02 Plastics—Methods of Exposure to Laboratory Light Sources—Part 3: Fluorescent UV Lamps*; International Standardization Organization: Geneva, Switzerland, 2016.

19. Francis, A.; Fowler, S.; Tobin, B. Temperature Control during Fluorescent UV Weathering Testing of Plastic Materials. In *8th European Weathering Symposium—Natural and Artificial Ageing of Polymers*; Reichert, T., Ed.; Gesellschaft für Umweltsimulation e.V.: Karlsruhe, Germany, 2017; pp. 165–180.

20. Trubiroha, P.; Geburtig, A.; Wachtendorf, V. 40 years of Global-UV-Test weathering device with fluorescent UV lamps and a precise microclimatical control into the future. In *9th European Weathering Symposium, Natural and Artificial Ageing of Polymers*; Reichert, T., Ed.; Gesellschaft für Umweltsimulation GUS: Basel, Switzerland, 2019; pp. 37–47.

21. Boxhammer, J. Shorter test times for thermal- and radiation-induced ageing of polymer materials 1: Acceleration by increased irradiance and temperature in artificial weathering tests. *Polym. Test.* **2001**, *20*, 719–724. [CrossRef]

22. Schulz, U.; Trubiroha, P.; Schernau, U.; Baumgart, H. The effects of acid rain on the appearance of automotive paint systems studied outdoors and in a new artificial weathering test. *Prog. Org. Coat.* **2000**, *40*, 151–165. [CrossRef]

23. Wachtendorf, V.; Schulz, U.; Trubiroha, P. Adaption of the Acid Dew and Fog (ADF) Test for Service Life Predictions of Aircraft Coatings. In *1st European Weathering Symposium—Natural and Artificial Ageing of Polymers*; Reichert, T., Ed.; Gesellschaft für Umweltsimulation e.V.: Karlsruhe, Germany, 2003; pp. 301–318.

24. Goedecke, C.; Sojref, R.; Nguyen, T.Y.; Piechotta, C. Immobilization of photocatalytically active TiO_2 nanopowder by high shear granulation. *Powder Technol.* **2017**, *318*, 465–470. [CrossRef]

25. Horikoshi, S.; Serpone, N.; Hisamatsu, Y.; Hidaka, H. Photocatalyzed degradation of polymers in aqueous semiconductor suspensions. 3. Photooxidation of a solid polymer: TiO_2-blended poly (vinyl chloride) film. *Environ. Sci. Technol.* **1998**, *32*, 4010–4016. [CrossRef]

26. Kamrannejad, M.M.; Hasanzadeh, A.; Nosoudi, N.; Mai, L.; Babaluo, A.A. Photocatalytic Degradation of Polypropylene/TiO_2 Nano-composites. *Mater. Res.* **2014**, *17*, 1039–1046. [CrossRef]

27. Khanin, S.E.; Angert, L.G.; Kuleznev, V.N.; Maloshchuk, Y.S. Ozone Resistance of Rubbers Based on Polymer Mixtures. *Colloid J. USSR* **1975**, *37*, 79–83.

28. ASTM International. *ASTM D1149-18 Standard Test Methods for Rubber Deterioration—Cracking in an Ozone Controlled Environment (2018)*; ASTM International: West Conshohocken, PA, USA, 2018.

29. European Committee for Standardization. *prEN 12225: 2019-09 Geotextiles and Geotextile-Related Products—Method for Determining the Microbiological Resistance by a Soil Burial Test*; European Committee for Standardization: Brussels, Belgium, 2019.

30. International Standardization Organization. *ISO 846: 2019-03 Plastics—Evaluation of the Action of Microorganisms*; International Standardization Organization: Geneva, Switzerland, 2019.

31. Akbay, I.K.; Ozdemir, T. Monomer migration and degradation of polycarbonate via UV-C irradiation within aquatic and atmospheric environments. *J. Macromol. Sci. A* **2016**, *53*, 340–345. [CrossRef]

32. Collin, S.; Bussiere, P.O.; Therias, S.; Lambert, J.M.; Perdereau, J.; Gardette, J.L. Physicochemical and mechanical impacts of photo-ageing on bisphenol a polycarbonate. *Polym. Degrad. Stab.* **2012**, *97*, 2284–2293. [CrossRef]

33. Diepens, M.; Gijsman, P. Photodegradation of bisphenol a polycarbonate. *Polym. Degrad. Stab.* **2007**, *92*, 397–406. [CrossRef]

34. Enell, A.; Lundstedt, S.; Arp, H.P.H.; Josefsson, S.; Cornelissen, G.; Wik, O.; Berggren Kleja, D. Combining Leaching and Passive Sampling to Measure the Mobility and Distribution between Porewater, DOC, and Colloids of Native Oxy-PAHs, N-PACs, and PAHs in Historically Contaminated Soil. *Environ. Sci. Technol.* **2016**, *50*, 11797–11805. [CrossRef]

35. Krüger, O.; Christoph, G.; Kalbe, U.; Berger, W. Comparison of stir bar sorptive extraction (SBSE) and liquid-liquid extraction (LLE) for the analysis of polycyclic aromatic hydrocarbons (PAH) in complex aqueous matrices. *Talanta* **2011**, *85*, 1428–1434. [CrossRef] [PubMed]

36. Piechotta, C.; Becker, R.; Köppen, R.; Traub, H.; Ostermann, M. Analytical elucidation of the release of the polybrominated flame retardants hexabromocyclododecane (HBCD) and decabromo-diphenyl ether (decaBDE-209) out of polymers under environmental conditions. *Materials* **2020**. in preparation.

37. Hollender, J.; Schymanski, E.L.; Singer, H.P.; Ferguson, P.L. Nontarget Screening with High Resolution Mass Spectrometry in the Environment: Ready to Go? *Environ. Sci. Technol.* **2017**, *51*, 11505–11512. [CrossRef] [PubMed]

38. Ackerman, A.H.; Hurtubise, R.J. The effects of adsorption of solutes on glassware and teflon in the calculation of partition coefficients for solid-phase microextraction with 1PS paper. *Talanta* **2000**, *52*, 853–861. [CrossRef]

39. Bollmann, U.E.; Minelgaite, G.; Schlüsener, M.; Ternes, T.A.; Vollertsen, J.; Bester, K. Photodegradation of octylisothiazolinone and semi-field emissions from facade coatings. *Sci. Rep.* **2017**, *7*, 41501. [CrossRef] [PubMed]

40. European Committee for Standardization. *CEN/TS 16637-2: 2014-11 Construction Products—Assessment of Release of Dangerous Substances—Part 2: Horizontal Dynamic Surface Leaching Test*; European Committee for Standardization: Brussels, Belgium, 2014.

41. European Committee for Standardization. *EN 16105: 2011-12 Paints and Varnishes—Laboratory Method for Determination of Release of Substances from Coating in Intermittent Contact with Water*; European Committee for Standardization: Brussels, Belgium, 2011.

42. Märkl, V.; Pflugmacher, S.; Reichert, A.; Stephan, D.A. Leaching of Polyurethane Systems for Waterproofing Purposes Whilest Curing. *Water Air Soil Pollut.* **2017**, *228*, 280. [CrossRef]

43. German Institute for Standardization. *DIN 19529:2009-01 Elution von Feststoffen—Schüttelverfahren mit einem Wasser-/Feststoffverhältnis von 2 L/kg zur Untersuchung der Elution von Anorganischen Stoffen für Materialien mit einer Korngröße bis 32 mm—Übereinstimmungsuntersuchung (Leaching of Solid Materials—Batch Test at a Liquid to Solid Ratio of 2 l/kg for the Examination of the Leaching Behaviour of Inorganic Substances for Materials with a Particle Size Upto 32 mm—Compliance Test)*; German Institute for Standardization: Berlin, Germany, 2009.

44. European Committee for Standardization. *EN 12457-2: 2002-09 Characterization of Waste—Leaching; Compliance Test for Leaching of Granular and Sludges—Part 2: One Stage Batch Test at a Liquid to Solid Ratio of 10 L/kg with Particle Size below 4 mm (without or with Size Reduction)*; European Committee for Standardization: Brussels, Belgium, 2002.

45. German Institute for Standardization. *DIN 19528: 2009-01 Elution von Feststoffen—Perkolationsverfahren zur Gemeinsamen Untersuchung des Elutionsverhaltens von Organischen und Anorganischen Stoffen für Materialien mit einer Korngröße bis 32 mm—Grundlegende Charakterisierung mit einem Ausführlichen Säulenversuch und Übereinstimmungsuntersuchung mit einem Säulenschnelltest*; German Institute for Standardization: Berlin, Germany, 2009.

46. European Committee for Standardization. *CEN/TS 16637-3: 2016-06 Construction Products—Assessment of Release of Dangerous Substances—Part 3: Horizontal up-Flow Percolation Test*; European Committee for Standardization: Brussels, Belgium, 2016.

47. Banzhaf, S.; Hebig, K.H. Use of column experiments to investigate the fate of organic micropollutants—A review. *Hydrol. Earth Syst. Sci.* **2016**, *20*, 3719–3737. [CrossRef]

48. Hjelmar, O.; Hyks, J.; Wahlström, M.; Laine-Ylijoki, J.; van Zomeren, A.; Comans, R.; Kalbe, U.; Schoknecht, U.; Krüger, O.; Grathwohl, P.; et al. *Robustness Validation of TS-2 and TS-3 Developed by CEN/TC351/WG1 to Assess Release from Products to Soil, Surface Water and Groundwater*; Final Report; NEN: Delft, The Netherlands, 2013.

49. Quina, M.J.; Bordado, J.C.M.; Quinta-Ferreira, R.M. Percolation and batch leaching tests to assess release of inorganic pollutants from municipal solid waste incinerator residues. *Waste Manag.* **2011**, *31*, 236–245. [CrossRef]

50. Tiwari, M.K.; Bajpai, S.; Dewangan, U.K.; Tamrakar, R.K. Suitability of leaching test methods for fly ash and slag. A review. *J. Radiat. Res. Appl. Sci.* **2015**, *8*, 523–537. [CrossRef]

51. Heisterkamp, I.; Gartiser, S.; Kalbe, U.; Bandow, N.; Gloßmann, A. Assessment of leachates from reactive fire-retardant coatings by chemical analysis and ecotoxicity testing. *Chemosphere* **2019**, *226*, 85–93. [CrossRef]

52. van der Sloot, H.A.; Heasman, L.; Quevauviller, P. Harmonization of Leaching/Extraction Tests. In *Environmental Science*; Elsevier: Amsterdam, The Netherlands, 1997; Volume 70.

53. Voglar, G.E.; Leštan, D. Equilibrium leaching of toxic elements from cement stabilized soil. *J. Hazard. Mater.* **2013**, *246*, 18–25. [CrossRef]

54. Finkel, M.; Grathwohl, P. Impact of pre-equilibration and diffusion limited release kinetics on effluent concentration in column leaching tests: Insights from numerical simulations. *Waste Manag.* **2017**, *63*, 58–73. [CrossRef]

55. Xiong, Q.; Baychev, T.G.; Jivkov, A.P. Review of pore network modelling of porous media: Experimental characterisations, network constructions and applications to reactive transport. *J. Contam. Hydrol.* **2016**, *192*, 101–117. [CrossRef] [PubMed]

56. Al-Raoush, R.I. Experimental investigation of the influence of grain geometry on residual NAPL using synchrotron microtomography. *J. Contam. Hydrol.* **2014**, *159*, 1–10. [CrossRef] [PubMed]

57. Prodanović, M.; Lindquist, W.B.; Seright, R.S. Porous structure and fluid partitioning in polyethylene cores from 3D X-ray microtomographic imaging. *J. Colloid Interface Sci.* **2006**, *298*, 282–297. [CrossRef] [PubMed]

58. Rowe, R.K.; Saheli, P.T.; Rutter, A. Partitioning and diffusion of PBDEs through an HDPE geomembrane. *Waste Manag.* **2016**, *55*, 191–203. [CrossRef]

59. Schwarzenbach, R.P.; Gschwend, P.M.; Imboden, D.M. *Environmental Organic Chemistry*; John Wiley: New York, NY, USA, 2003.

60. Leal, A.M.M.; Blunt, M.J.; LaForce, T.C. Efficient chemical equilibrium calculations for geochemical speciation and reactive transport modelling. *Geochimica et Cosmochimica Acta* **2014**, *131*, 301–322. [CrossRef]

61. Grathwohl, P. On equilibration of pore water in column leaching tests. *Waste Manag.* **2014**, *34*, 908–918. [CrossRef]

62. Hyks, J.; Astrup, T.; Christensen, T.H. Leaching from MSWI bottom ash: Evaluation of non-equilibrium in column percolation experiments. *Waste Manag.* **2009**, *29*, 522–529. [CrossRef] [PubMed]

63. Wehrer, M.; Totsche, K.U. Detection of non-equilibrium contaminant release in soil columns: Delineation of experimental conditions by numerical simulations. *J. Plant Nutr. Soil Sci.* **2003**, *166*, 475–483. [CrossRef]

64. Kosson, D.S.; Van der Sloot, H.A.; Garrabrants, A.; Seignette, P. *EPA/600/R-14/061 Leaching Test Relationships, Laboratory-to-Field Comparisons and Recommendations for Leaching Evaluation Using the Leaching Environmental Assessment Framework (LEAF)*; US Environmental Protection Agency: Cincinnati, OH, USA, 2014.

65. Luo, H.; Li, Y.; Zhao, Y.; Xiang, Y.; He, D.; Pan, X. Effects of accelerated aging on characteristics, leaching, and toxicity of commercial lead chromate pigmented microplastics. *Environ. Pollut.* **2020**, *257*, 113475. [CrossRef] [PubMed]

66. Braun, U.; Wachtendorf, V.; Geburtig, A.; Bahr, H.; Schartel, B. Weathering resistance of halogen-free flame retardance in thermoplastics. *Polym. Degrad. Stab.* **2010**, *95*, 2421–2429. [CrossRef]

67. Kalbe, U.; Krüger, O.; Wachtendorf, V.; Berger, W.; Hally, S. Development of Leaching Procedures for Synthetic Turf Systems Containing Scrap Tyre Granules. *Waste Biomass Valoriz.* **2013**, *4*, 745–757. [CrossRef]

68. Wachtendorf, V.; Kalbe, U.; Krüger, O.; Bandow, N. Influence of weathering on the leaching behaviour of zinc and PAH from synthetic sports surfaces. *Polym. Test.* **2017**, *63*, 621–631. [CrossRef]

69. Wachtendorf, V.; Kalbe, U.; Krüger, O.; Bandow, N.; Geburtig, A. Laboratory Weathering Test for Simulating the Leaching from Artificial Sporting Grounds. In *Natural and Artificial Ageing of Polymers*; Reichert, T., Ed.; Gesellschaft für Umweltsimulation e.V.: Karlsruhe, Germany, 2015; pp. 195–208.

70. RAL Güteschutzgemeinschaft Verkehrszeichen und Verkehrseinrichtungen e.V. *Güteanforderungen an Standard-Verkehrszeichen für Ortsfeste Beschilderung*; Verfahren B (Bewitterung) (Quality Auditing Commission for Traffic Signs and Transport Facilities); RAL: Hagen, Germany, 1989.

71. Bocca, B.; Forte, G.; Petrucci, F.; Costantini, S.; Izzo, P. Metals contained and leached from rubber granulates used in synthetic turf areas. *Sci. Total Environ.* **2009**, *407*, 2183–2190. [CrossRef]

72. Hofstra, U. *Leaching of Zinc from Rubber Crumb from Shredded Tyres in Artificial Turf Fields. Comprehensive Review of Dutch Studies*; SGS INTRON Report A866570/R20120345; SGS INTRON: Sittard, The Netherlands, 2012.

73. Menichini, E.; Abate, V.; Attias, L.; De Luca, S.; Di Domenico, A.; Fochi, I.; Fiorte, G.; Iacovella, N.; Iamiceli, A.L.; Izzo, P.; et al. Artificial-turf playing fields: Contents of metals, PAHs, PCBs, PCDDs and PCDFs, inhalation exposure to PAHs and related preliminary risk assessment. *Sci. Total Environ.* **2011**, *409*, 4950–4957. [CrossRef]

74. Kalbe, U.; Krüger, O.; Wachtendorf, V.; Berger, W. *Umweltverträglichkeit von Kunststoff—Und Kunststoffrasenbelägen auf Sportfreianlagen*; Schriftenreihe des Bundesinstitutes für Sportwissenschaft; Sportverlag Strauß: Hellenthal, Germany, 2012; ISBN 978-3-86884-522-8.

75. Verschoor, A.J. *Leaching of Zinc from Rubber Infill on Artificial Turf (Football Pitches)*; Laboratory for Ecological Risk Assessment: Bilthoven, The Netherlands, 2007; RIVM Report 601774001.

76. Verschoor, A.J.; Cleven, R.F.M.J. *Risk Assessment of Leaching of Substances from Synthetic Polymeric Matrices*; RIVM Report 711701096; National Institue for Public Health and the Environment: Bilthoven, The Netherlands, 2009.

77. Gomes, J.; Mota, H.; Bordado, J.; Cadete, M.; Sarmento, G.; Ribeiro, A.; Baiao, M.; Fernandes, J.; Pampulin, V.; Custodio, M.; et al. Toxicological Assessment of Coated versus Uncoated Rubber Granulates Obtained from Used Tires for Use in Sport Facilities. *J. Air Waste Manag. Assoc.* **2010**, *60*, 741–746. [CrossRef]

78. Li, X.; Berger, W.; Musante, C.; Mattina, M.I. Characterization of substances released from crumb rubber material used on artificial turf fields. *Chemosphere* **2010**, *80*, 279–285. [CrossRef]

79. Edil, T.B. A review of environmental impacts and environmental applications of shredded scrap tires. In *Scrap Tire Derived Geomaterials—Opportunities and Challenges*; Hazarika, H., Yasuhara, K., Eds.; Taylor & Francis Group: London, UK, 2008; pp. 3–18.

80. Söver, A.; Frormann, L.; Kipscholl, R. High impact-testing machine for elastomers investigation under impact loads. *Polym. Test.* **2009**, *28*, 871–874. [CrossRef]

81. Unice, K.M.; Bare, J.L.; Kreider, M.L.; Panko, J.M. Experimental methodology for assessing the environmental fate of organic chemicals in polymer matrices using column leaching studies and OECD 308 water/sediment systems: Application to tire and road wear particles. *Sci. Total Environ.* **2015**, *533*, 476–487. [CrossRef]

82. Kalbe, U.; Susset, B.; Bandow, N. *Umweltverträglichkeit von Kunststoffbelägen auf Sportfreianlagen—Modellierung der Stofffreisetzung aus Sportböden auf Kunststoffbasis zur Bewertung der Boden—Und Grundwasserverträglichkeit*; Schriftenreihe des Bundesinstituts für Sportwissenschaft; Sportverlag Strauß: Hellenthal, Germany, 2016; Volume 2016/05, ISBN 978 3-86884-536-5

83. Cheng, H.; Hu, Y.; Reinhard, M. Environmental and Health Impacts of Artificial Turf: A Review. *Environ. Sci. Technol.* **2014**, *48*, 2114–2129. [CrossRef] [PubMed]

84. Lundback, M.; Hedenqvist, M.S.; Mattozzi, A.; Gedde, U.W. Migration of phenolic antioxidants from linear and branched polyethylene. *Polym. Degrad. Stab.* **2006**, *91*, 1571–1580. [CrossRef]

85. Brocca, D.; Arvin, E.; Mosbaek, H. Identification of organic compounds migrating from polyethylene pipelines into drinking water. *Water Res.* **2002**, *36*, 3675–3680. [CrossRef]

86. Rhiem, S.; Barthel, A.K.; Meyer-Plath, A.; Hennig, M.P.; Wachtendorf, V.; Sturm, H.; Schaffer, A.; Maes, H.M. Release of (14)C-labelled carbon nanotubes from polycarbonate composites. *Environ. Pollut.* **2016**, *215*, 356–365. [CrossRef]

87. Wachtendorf, V.; Anik, Y.; Geburtig, A.; Sturm, H.; Meyer-Plath, A. CarboLifeCycle—Weathering-Induced Degradation of Nanoparticle-Functionalised Composites. In *Natural and Artificial Ageing of Polymers*; Reichert, T., Ed.; Gesellschaft für Umweltsimulation e.V.: Karlsruhe, Germany, 2011; pp. 125–135.

88. Bandow, N.; Will, V.; Wachtendorf, V.; Simon, F.-G. Contaminant release from aged microplastic. *Environ. Chem.* **2017**, *14*, 394–405. [CrossRef]

89. Piechotta, C.; Iznaguen, H.; Traub, H.; Feldmann, I.; Köppen, R.; Witt, A.; Jung, C.; Becker, R.; Oleszak, K.; Bücker, M.; et al. Bewitterungsszenarien im Vergleich—Veränderungen in der Oberflächenmorphologie von Polypropylen (PP) und Polystyrol (PS) unter dem Aspekt des Austrags von polybromierten Flammschutzmitteln. In *47 Jahrestagung der Gesellschaft für Umweltsimulation, Umwelteinflüsse Erfassen, Simulieren, Bewerten*; Ziegahn, K.-F., Ed.; GUS e.V.: Blankenloch-Stutensee, Germany, 2018; pp. 115–128.

90. Robertson, L.W.; Ludewig, G. Toxicity of polychlorinated biphenyls and polybrominated diphenyl ethers and their metabolites. *Drug Metab. Rev.* **2006**, *38*, 21.

91. Vonderheide, A.P.; Mueller, K.E.; Meija, J.; Welsh, G.L. Polybrominated diphenyl ethers: Causes for concern and knowledge gaps regarding environmental distribution, fate and toxicity. *Sci. Total Environ.* **2008**, *400*, 425–436. [CrossRef]

92. ASTM International. *ASTM D1435-13: 2013 Standard Practice for Outdoor Weathering of Plastics*; ASTM International: West Conshohocken, PA, USA, 2013.

93. Burkhardt, M.; Zuleeg, S.; Schmid, P.; Hean, S.; Lamani, X.; Bester, K.; Boller, M. Leaching of additives from construction materials to urban storm water runoff. *Water Sci. Technol.* **2011**, *63*, 1974–1982. [CrossRef]

94. Jungnickel, C.; Stock, F.; Brandsch, T.; Ranke, J. Risk assessment of biocides in roof paint. *Environ. Sci. Pollut. Res.* **2008**, *15*, 258–265. [CrossRef]

95. Schoknecht, U.; Gruycheva, J.; Mathies, H.; Bergmann, H.; Burkhardt, M. Leaching of biocides used in facade coatings under laboratory test conditions. *Environ. Sci. Technol.* **2009**, *43*, 9321–9328. [CrossRef]

96. Wangler, T.P.; Zuleeg, S.; Vonbank, R.; Bester, K.; Boller, M.; Carmeliet, J.; Burkhardt, M. Laboratory scale studies of biocide leaching from facade coatings. *Build. Environ.* **2012**, *54*, 168–173. [CrossRef]

97. Burkhardt, M.; Zuleeg, S.; Vonbank, R.; Bester, K.; Carmeliet, J.; Boller, M.; Wangler, T. Leaching of Biocides from Facades under Natural Weather Conditions. *Environ. Sci. Technol.* **2012**, *46*, 5497–5503. [CrossRef] [PubMed]

98. Breuer, K.; Mayer, F.; Scherer, C.; Schwerd, R.; Sedlbauer, K. Wirkstoffauswaschung aus hydrophoben Fassadenbeschichtungen: Verkapselte versus unverkapselte Biozidsysteme. *Bauphysik* **2012**, *34*, 19–23. [CrossRef]

99. Bollmann, U.E.; Minelgaite, G.; Schlüsener, M.; Ternes, T.; Vollertsen, J.; Bester, K. Leaching of terbutryn and its photodegradation products from artificial walls under natural weather conditions. *Environ. Sci. Technol.* **2016**, *50*, 4289–4295. [CrossRef] [PubMed]

100. Schoknecht, U.; Mathies, H.; Wegner, R. Biocide leaching during field experiments on treated articles. *Environ. Sci. Eur.* **2016**, *28*, 1–10. [CrossRef] [PubMed]

101. Bollmann, U.E.; Tang, C.; Eriksson, E.; Jönsson, K.; Vollertsen, J.; Bester, K. Biocides in urban wastewater treatment plant influent at dry and wet weather: Concentrations, mass flows and possible sources. *Water Res.* **2014**, *60*, 64–74. [CrossRef]

102. Bollmann, U.E.; Vollertsen, J.; Carmeliet, J.; Bester, K. Dynamics of biocide emissions from buildings in a suburban stormwater catchment—Concentrations, mass loads and emission processes. *Water Res.* **2014**, *56*, 66–76. [CrossRef]

103. Breuer, K.; Hofbauer, W.; Krueger, N.; Mayer, F.; Scherer, C.; Schwerd, R.; Sedlbauer, K. Effectiveness and durability of biocidal ingredients in facade coatings. *Bauphysik* **2012**, *34*, 170–182. [CrossRef]

104. Schoknecht, U.; Sommerfeld, T.; Borho, N.; Bagda, E. Interlaboratory comparison for a laboratory leaching test procedure with facade coatings. *Prog. Org. Coat.* **2013**, *76*, 351–359. [CrossRef]

105. Wangler, T.P. Modelling biocide release from architectural coatings. *Chem. Today* **2011**, *29*, 14–16.

106. Styszko, K.; Bollmann, U.E.; Wangler, T.P.; Bester, K. Desorption of biocides from renders modified with acrylate and silicone. *Chemosphere* **2014**, *95*, 187–191. [CrossRef]

107. Bollmann, U.E.; Ou, Y.; Mayer, P.; Trapp, S.; Bester, K. Polyacrylate—Water partitioning of biocidal compounds: Enhancing the understanding of biocide partitioning between render and water. *Chemosphere* **2015**, *119*, 1021–1026. [CrossRef] [PubMed]

108. Schoknecht, U.; Mathies, H.; Wegner, R.; Uhlig, S.; Baldauf, H.; Colson, B. *Emissions of Material Preservatives into the Environment—Realistic Estimation of Environmental Risks through the Improved Characterization of the Leaching of Biocides from Treated Materials Used Outdoors*; Federal Environmental Agency: Dessau-Roßlau, Germany, 2016.

109. Coutu, S.; Wyrsch, V.; Rossi, L.; Emery, P.; Golay, F.; Carneiro, C. Modelling wind-driven rain on buildings in urbanized area using 3-D GIS and LiDAR datasets. *Build. Environ.* **2013**, *59*, 528–535. [CrossRef]

110. EU. Regulation No 528/2012 of the European Parliament and of the Council Concerning the Making Available on the Market and Use of Biocidal Products (2012). Available online: https://echa.europa.eu/regulations/biocidal-products-regulation/legislation (accessed on 11 June 2020).

111. Schoknecht, U.; Bagda, E. On the face of it—Leaching of actives from facades—Current knowledge and actions. *Eur. Coat. J.* **2014**, *3*, 18–23.

112. Butera, S.; Hyks, J.; Christensen, T.H.; Astrup, T.F. Construction and demolition waste: Comparison of standard up-flow column and down-flow lysimeter leaching tests. *Waste Manag.* **2015**, *43*, 386–397. [CrossRef] [PubMed]

113. Di Gianfilippo, M.; Hyks, J.; Verginelli, I.; Costa, G.; Hjelmar, O.; Lombardi, F. Leaching behaviour of incineration bottom ash in a reuse scenario: 12 years-field data vs. lab test results. *Waste Manag.* **2018**, *73*, 367–380. [CrossRef] [PubMed]

114. Kluge, B.; Werkenthin, M.; Wessolek, G. Metal leaching in a highway embankment on field and laboratory scale. *Sci. Total Environ.* **2014**, *493*, 495–504. [CrossRef]

Article

Leaching of Carbon Reinforced Concrete—Part 1: Experimental Investigations

Lia Weiler * and Anya Vollpracht

Institute of Building Materials Research, RWTH (Rheinisch-Westfälische Technische Hochschule) Aachen University, 52062 Aachen, Germany; vollpracht@ibac.rwth-aachen.de
* Correspondence: weiler@ibac.rwth-aachen.de; Tel.: +49-241-8095134

Received: 7 September 2020; Accepted: 30 September 2020; Published: 2 October 2020

Abstract: The composite material 'carbon concrete composite (C^3)' is currently capturing the building sector as an 'innovative' and 'sustainable' alternative to steel reinforced concrete. In this work, its environmental compatibility was investigated. The focus of this research was the leaching behavior of C^3, especially for the application as irrigated façade elements. Laboratory and outdoor exposure tests were run to determine and assess the heavy metal and trace element emissions. In the wake of this work, the validity of laboratory experiments and the transferability to outdoor behavior were investigated. The experimental results show very low releases of environmental harmful substances from carbon concrete composite. Most heavy metal concentrations were in the range of <0.1–8 µg/L, and higher concentrations (up to 32 µg/L) were found for barium, chromium, and copper. Vanadium and zinc concentrations were in the range of 0.1–60 µg/L, boron and nickel concentrations were clearly exceeding 100 µg/L. Most of the high concentrations were found to be a result of the rainfall background concentrations. The material C^3 is therefore considered to be environmentally friendly. There is no general correlation between laboratory leaching data and outdoor emissions. The results depend on the examined substance and used method. The prediction and evaluation of the leaching of building elements submitted to rain is therefore challenging. This topic is debated in the second part of this publication.

Keywords: leaching; carbon concrete composite; irrigated construction elements; environmental compatibility; irrigated building materials

---

1. Introduction

The use of composite building materials offers a wide range of advantages for the building industry as for example new functions, savings in weight and costs, or new design options. This leads to an increasing use of new materials or material combinations in construction with uncertain recycling methods and unknown emission behavior: Physical and chemical bonds between the components, for example, might impede the materials separation for recovery of recyclables after their service life and therefore lower the resource efficiency. Mutual influences or reactions of newly combined substances can change emissions despite an unchanged content level [1,2]. In the context of an existing and rising environmental awareness, the environmental compatibility of building materials is also rising in importance. To reconcile both, it is necessary to consider and avoid a possibly harmful release of substances from the outset.

An important environmental aspect is the leaching of potentially harmful substances from construction elements that are in permanent or temporary contact with water, followed by an entry into the environmental compartments soil, ground- and/or surface waters. Depending on the type of material used, these substances can be inorganic (heavy metals and trace elements) or organic (e.g., unreacted monomers, additives, impurities, degradation products or biocides). The use of construction products without prior testing can therefore lead to environment and health risks. So far, these effects are often not considered in an appropriate way. A detailed study including 100 organic and inorganic substances in the rainwater discharge of Berlin has shown that the discharge of rainwater into rivers can raise concentrations of some contaminants by a factor of ten [3]. The heavy metals chromium, nickel and vanadium are named as construction material, mainly concrete, born emissions [3,4]. This also applies to organic emissions such as nonylphenolic compounds [5]. Zinc and copper from metal roofs, façade coatings and renders were also found to be relevant. It has to be mentioned that a precise allocation of substance leaching to its sources is certainly difficult as concentrations are measured in the runoffs and possible sources are identified by comparison of catchments.

The annual mean concentrations of zinc, copper, lead and sometimes also cadmium in German, French and Austrian rainwater discharge are exceeding the European EQS values (environmental quality standards) [6], and, in case of zinc and copper, the recommendations of the German Federal Environment Agency for the EQS. Irrigated façades and roofs were identified as one source [3,7,8].

Throughout their whole life irrigated construction elements are exposed to an intermittent wet–dry stress. Compared to construction elements that are in permanent contact with water, this can cause a deviating leaching behavior with increased or lower release, depending on the building material and the leached substance. Chromium, for example, seems to be leached in higher amounts in outdoor and irrigation experiments, and barium and selenium are leached to a higher extend in permanent water contact [9]. At the moment there is no broad database available concerning the correlations between standardized laboratory and laboratory irrigation tests; even less is known about correlations with outdoor leaching behavior.

Table 1 shows a summary of previous research projects with laboratory irrigation with a focus on simulated rain events. Studies describing the leaching behavior of biocides from paints and varnishes after EN 16105:2011 [10] or using permanent water contact to replace irrigation are not outlined as they do not meet the desired 'rain' criteria.

Table 1. Overview of previous research projects with laboratory irrigation.

Leached Material	Investigated Emissions	Irrigation Unit	Intensity and Irrigation Procedure	Reference
concrete façades/exposed concrete	chromium, zinc	plastic tray with perforated bottom	25 mm/h,1.5 h rain, 2 h break, 1.5 h rain, 37 h drying time, total: 650 mm	2005 [11]
mineral construction materials	sulfate, chloride, fluoride, cyanide, 16 trace elements	spray mechanism (compressed air + nozzle)	0.7–5 mm/h, finely dispersed droplets (mist)	2008 [9]
synthetic resin render with biocides	biocides	weathering chamber	85 mm/h, 20 irrigation intervals of 1 h in 5 d, T = 50 °C–60 °C, total: 6800 mm	2009 [12]
renders/render-paint-systems with biocides	biocides	"water pressure and flow rate were controlled"	75 mm/h, 2 min/d	2009 [13]
metal building materials (stainless steel, copper, zinc)	heavy metals, trace elements	spray mechanism (compressed air + nozzle)	0.7 mm/h–3.5 mm/h, finely dispersed droplets (mist)	2011 [14]
renders, mortars	sodium, potassium, sulfate, chloride, 10 trace elements	spray mechanism (compressed air + nozzle)	0.7–5 mm/h, total: 60 mm, finely dispersed droplets (mist)	2012 [15]

It is obvious that a broad variety of testing conditions and examined substances was chosen by the researchers. This leads to difficulties in the assessment and aggregation of the available data. It is furthermore revealed (and was also stated by Schoknecht [16] and Nebel [17]) that there are difficulties in adjusting realistic rain intensities, respectively, rain drop sizes at a laboratory scale.

Corresponding conclusions from different studies concerning influencing parameters on leaching of intermittent moistened construction materials pertain to:

- Wet–dry stress: In porous building materials, the drying phases cause a transport of water and dissolved substances to the surface by capillary transport during drying dissolved substances precipitate at the surface. This leads to an increased availability for leaching or wash-off in the following rain [11,13]. The microstructure plays a major role for the transport processes, especially under wet–dry stress. This might lead to changing release patterns over time, as moisture transport additionally fosters transformations of the matrix structure by causing pore changes through, e.g., micro-cracking or shrinkage [11,13,15,18].

- Material composition: Different substances contained in a building material can interact, which can cause bonding but also an increase in the release. The content itself is not decisive for the release of substances [2,11,13,19,20].

- Substance-dependent leaching: Leaching rates depend on physical and chemical properties of the substances and the building material, especially the solubility of the particular substances, which can be pH dependent [9,17,19,21,22]. For intermittent moistened construction elements, the pH of the material, not of the leachate, is pivotal [13,19]. Wet–dry stress, temperature changes and air contact can alter the matrices over time; for example, by carbonation, which leads to decreasing pH values. This influences the long-term leaching [11,19,21,22].

Further parameters and processes such as rain intensities, orientation of test specimen, temperature or leachate composition are relevant too, but were not investigated in detail in laboratory irrigation, whereas several studies, e.g., [13,23,24] are discussing these factors for outdoor exposure. Standardized laboratory leaching tests are not appropriate methods to consider these factors. The disregarding may lead to an over- or underestimation of the real leachate concentrations, depending on the composition of the building material [9,11,13,15,21].

In this paper, investigations concerning the leaching behavior of carbon textile reinforced concrete (C^3) are described. The use of carbon textiles for concrete reinforcement introduces not only a new material to the system but also allows a fundamental change in the use of the matrix raw materials. In [25], the opportunities of new binder systems and concrete compositions are described. Unlike steel reinforcement, carbon reinforcement does not need an alkaline environment and concrete covering to avoid corrosion [25]. New materials like the organic polymer coating of the reinforcement on the one hand and a potentially lower pH or thin concrete covering on the other hand may cause increasing emissions by leaching.

Due to the novelty of the composite material and the complexity of collecting and assessing long time data concerning, e.g., durability, the development of new matrix compositions is still in progress [25]. This work focuses on C^3 with already technically approved components. C^3 as a new composite material combines inorganic and organic components. Heavy metals and trace elements leached from the concrete might interact with organic substances, especially monomers, from the coating of the reinforcement. Especially dissolved organic carbon (DOC) can influence the emission process [2,13]. Biocides for example were found to be retained by the organic reinforcement of fiber cement sheets in Lupsea et al. [1].

To investigate the emission behavior, different test specimens are irrigated using a new laboratory irrigation stand; the released substances are determined and compared to the results of a standardized laboratory test [26], namely the dynamic surface leaching test (DSLT) and outdoor weathering data.

Within the framework of a joint research project, the DSLT data were collected by our partner Verein Deutscher Zementwerke (VDZ) gGmbH and the test specimens were centrally manufactured by the partner Hentschke Bau GmbH.

2. Materials and Methods

2.1. Materials

Preliminary tests were conducted to identify a matrix-reinforcement combination with comparably high leaching. Based on the results, a fine-grained, ready-mixed concrete for the matrix and a carbon fiber grid for the reinforcement were chosen.

2.1.1. Concrete Composition

The ready-mixed dry concrete includes fine grained aggregate with a nominal maximum size of 1 mm. Figure 1 shows the sieving analysis results.

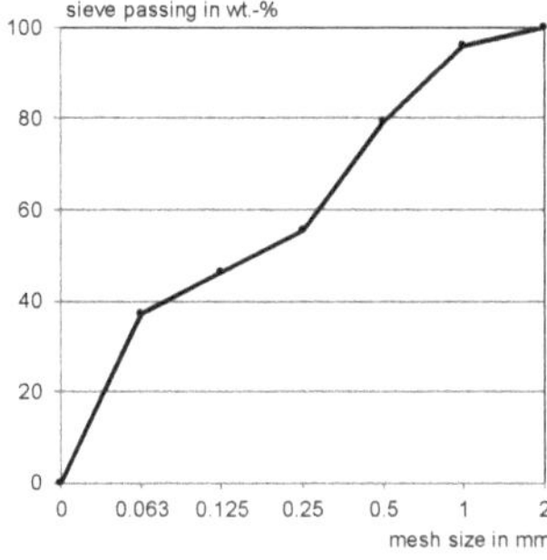

Figure 1. Grading curve of the ready-mixed concrete.

The dry mix consists of approximately 57 wt.% sand and 43% binder. The compositions of each fraction determined by X-ray diffraction are given in Table 2.

Table 2. Mineralogical analysis of the sieve fractions of the ready-mixed dry concrete.

Component	Content of the Sieving Fractions in wt.%					
	0–0.063 mm	0.063–0.125 mm	0.125–0.25 mm	0.25–0.5 mm	0.5–1 mm	1–2 mm
C_3S	30	31.4	–	–	–	–
C_2S	13.8	20	–	–	–	–
C_3A	5.8	6.7	–	–	–	–
Brownmillerite	1.3	1.2	–	–	–	–
Quartz	1.8	25.6	85.3	90.9	91.9	93.9
Ca-langbeinite	0.5	–	–	–	–	–
Anhydrite	4.3	2.5	–	–	–	–
Hemihydrate	0.6	0.3	–	–	–	–
Gypsum	0.8	0.9	–	–	–	–
Calcite	1	1.4	–	–	–	–
Portlandite	0.2	0.4	–	–	–	–
Mullite	3	6.3	–	–	–	–
Hematite	0.4	0.9	–	–	–	–
Microcline	–	2.3	9.7	7.2	5.8	4.1
Fluorphlogopite	–	0.3	0.3	0.5	0.3	–
Anorthite	–	–	3.8	1.4	2.1	2
Clinochlore	–	–	0.8	–	–	–
Pseudobrookite	–	–	0.1	–	–	–
Amorphous	36.5	–	–	–	–	

The binder consists of Portland cement clinker and amorphous components. The presence of Mullite and Hematite indicates that the dry mix contains fly ash, which was confirmed by scan electron microscopy (SEM) in combination with energy dispersive X-ray spectroscopy (EDX). Figure 2 shows the typical spherical shape of fly ash particles with the corresponding EDX spectrum for silicon, aluminum and iron. Moreover, silica fume was found to be part of the sample. Figure 3 shows the also spherical-shaped particles in typical agglomerates and with a significant size difference to fly ash. The EDX spectrum shows no iron and less aluminum content, which might also be a result of the excited environment.

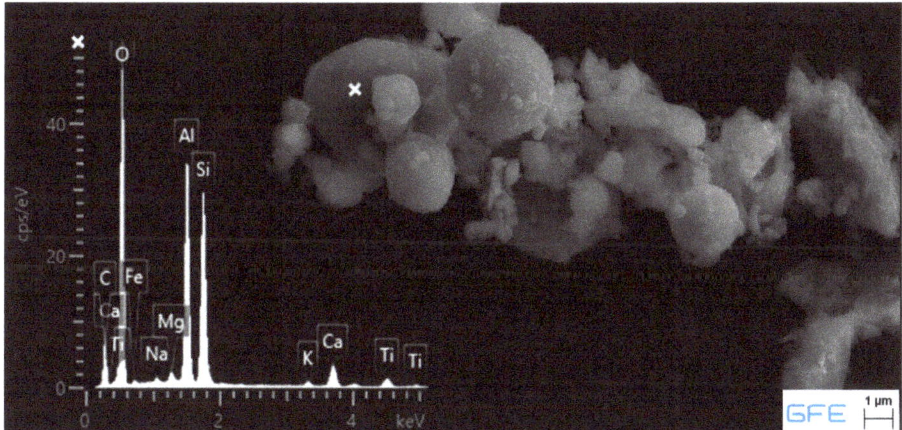

Figure 2. Scan electron microscopy (SEM) picture and corresponding energy dispersive X-ray (EDX) spectrum of ready dry mix <125 µm, fly ash.

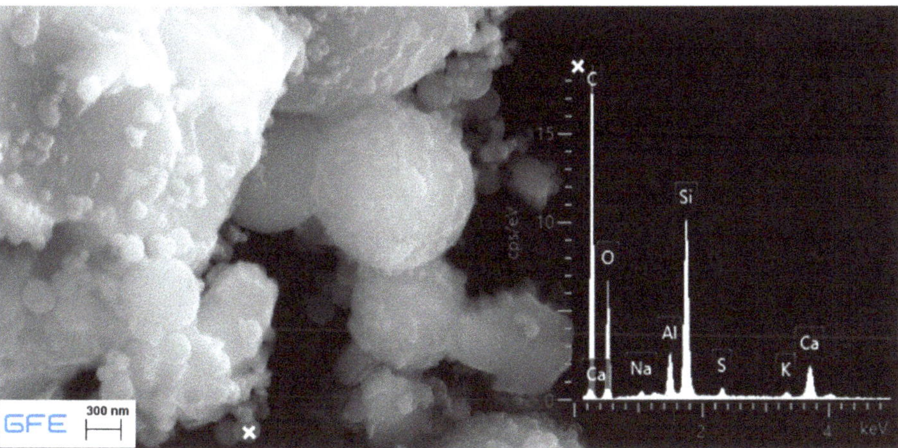

Figure 3. SEM picture and corresponding EDX spectrum of ready dry mix <125 µm, silica fume.

Since no granulated blast-furnace slag was found by selective dissolving with HNO_3/EDTA and only traces were assumed under light microscopy, the whole amorphous share determined by X-ray diffraction is assigned to fly ash and silica fume.

The concrete mixture was obtained with 14 L water per 100 kg dry mix, which corresponds to a water binder ratio of approximately 0.32. During mixing, liquefaction could be observed which most likely traces back to the effect of a superplasticizer in the dry mix.

The fraction <0.125 mm consists mainly of the cementitious binder. This fraction was analyzed for its chemical composition by X-ray fluorescence spectroscopy (XRF), carbon/sulfur analyzer (CSA), and silver nitrate titration (for chloride content). Its trace element and heavy metal contents were analyzed by inductively coupled plasma mass spectrometry (ICP-MS) after aqua regia digestion. The acquired data are presented in Tables 3 and 4.

Table 3. Main components of the <125 μm fraction of the concrete ready mix.

Parameter	Content in wt.%
loss on ignition	2.47
insoluble in HCl	26.24
SO_3	2.08
Na_2O	0.39
K_2O	1.13
chloride	0.036
MgO	0.96
Al_2O_3	9.59
SiO_2	36.57
P_2O_5	0.29
CaO	44.37
TiO_2	0.48
MnO	0.05
Fe_2O_3	2.63

Table 4. Contents of heavy metals and trace elements of the raw materials.

Parameter	Concrete Mixture [27]	Fine Fraction (<125 μm)	Reinforcement Textile Incl. Coating [27]	Average of German Cements [28]
			mg/kg	
antimony	1.1	1.50	0.9	2.9
arsenic	7.2	13.0	2.1	7
barium	252	295	11.0	-
lead	26.5	19.9	0.9	17
cadmium	0.3	0.226	<0.1	0.4
chromium	64.4	27.2	206.0	41
cobalt	6.9	6.84	2.4	8.7
copper	32.1	15.5	45.7	31
molybdenum	3.7	4.39	4.8	-
nickel	38.9	20.5	57.9	23
mercury	0.1	0.068	<0.02	0.06
thallium	1.5	0.275	<0.05	0.4
vanadium	52.8	60.7	1.25	50
zinc	63.1	44.4	93.8	192

Table 4 shows the content of heavy metals and trace elements of both materials determined by ICP-MS after aqua regia digestion. Compared to the average of German cements, the cement used for the ready mix shows no significant deviations in trace elements and heavy metal content. The reinforcement textile contains considerable amounts of copper, nickel and especially chromium in relation to the cement.

2.1.2. Reinforcement

The textile is a carbon fiber grid with a mesh size of 10.7 mm in warp and 14.3 mm in the weft direction. The textile is coated with 13 to 18 wt.% of a coating agent on a styrene-butadiene rubber base. The corresponding IR spectrum is shown in Figure 4.

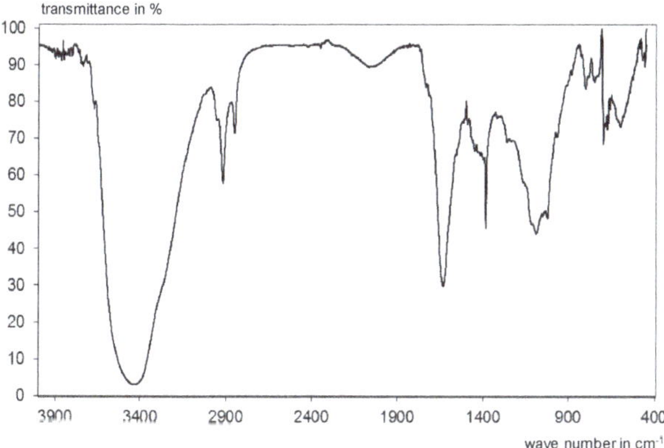

Figure 4. Infrared spectrum of the styrene-butadiene coating of the textile [29].

2.2. Test Specimens

The test specimens were manufactured by our project partner Hentschke Bau GmbH. For all tests, outdoor, laboratory irrigation, and DSLT tests, the test specimens were laminated into a stainless steel and plastic formwork using pre-cut textile pieces for the reinforcement and polytetrafluorethylene (PTFE) spray as a release agent. After demolding, the samples were wrapped airtight in polyethylene (PE) foil and sealed with adhesive tape to prevent carbonation or moisture loss. The wrapped specimens were stored at 20 °C until the beginning of the testing. Figure 5 summarizes the casting process. Table 5 shows the dimensions and production conditions of the investigated test specimens.

(a)

(b)

(c)

(d)

Figure 5. *Cont.*

(e)

Figure 5. Photo documentation of the test specimens molding process [30]; (**a**) empty formwork and cut textiles; (**b**) first concrete layer; (**c**) textile layer; (**d**) filled formwork; (**e**) sealed (left) and unpacked specimens in formwork.

Table 5. Dimensions and production conditions of the investigated test specimens [30].

Testing Method	Label *	Width/ Length	Thickness	Layers of Reinforce-Ment	Concrete Cover	Room Temp	Concrete Temp	Flow Spread
		mm	mm	n	mm	°C	°C	mm
DSLT	D1-1-2 **	150/150	5.50	1	2	20.2	16.5	270
	D1-4-2 **	150/150	16.0	4	2	21.3	17.3	260
	D1-1-20 **	150/150	40.0	1	20	22.0	17.8	260
lab irrigation	L 1-2 A	300/400	5.50	1	2	20.2	16.8	275
	L 1-2 B	300/400	5.50	1	2	19.8	16.6	250
	L 4-2 A	300/400	16.0	4	2	21.6	17.0	255
	L 4-2 B	300/400	16.0	4	2	21.3	17.3	260
	L 1-20 A	300/400	40.0	1	20	22.3	17.5	270
	L 1-20 B	300/400	40.0	1	20	22.8	17.1	265
	L 1-20 C	300/400	40.0	1	20	22.5	24.2	230
	L 1-20 D	300/400	40.0	1	20	23.1	23.3	237.5
outdoor testing	F 1 A	1000/600	20.0	1	2	20.2	21.9	205
	F 1 B	1000/600	20.0	1	2	21.7	24.2	210
	F 1 C	1000/600	20.0	1	2	n. d.	n. d.	n. d.
	F 1g A	1000/600	20.0	1	2	22.0	23.5	247.5
	F 1g B	1000/600	20.0	1	2	21.9	23.1	242.5
	F 4g A	1000/600	20.0	4	2	21.9	23.2	235
	F 4g B	1000/600	20.0	4	2	21.2	23.4	230

* Label systematics: Exposition—Level of Reinforcement—Concrete Cover (no number $\hat{=}$ 2 mm)—Replicate. ** Three test specimens each.

2.3. Methods

2.3.1. pH Dependence Test (pH$_{stat}$)

The leaching characteristics of the relevant trace elements from the concrete ready-mix mortar have been determined as a function of the pH value in the range from pH 3.5 to the natural pH of 12.4 according to EN 14429 [31].

Six mortar prisms of $40 \times 40 \times 160$ mm^3 were produced from the ready mixed concrete and stored sealed in PE foil for 28 days at 20 °C and 65% RH. The prisms were then ground to <1 mm for the testing.

The leaching was carried out using 30 g of the crashed mortar, deionized water and 5 molar HNO$_3$ for titration. The eluates were filtered through a 0.45 μm syringe filter from PET and then analyzed by ICP-OES, flame photometry and ion chromatography.

2.3.2. Dynamic Surface Leaching Test (DSLT)

The tank leaching tests were carried out by our project partner VDZ gGmbH, based on the European harmonized standard CEN/TS 16637-2 [26]. As pictured in Figure 6, the test specimens were positioned in glass chromatography vessels and then covered with 2.3 L of deionized water and closed with a lid. The ratio of eluate volume divided by the surface of the test specimens (V/A) therefore differed, depending on the thickness of each specimen. The samples of D 1-2 were leached with 47.6 L/m^2, D 4-2 with 42.1 L/m^2 and D 1-20 with 29.5 L/m^2 in each step.

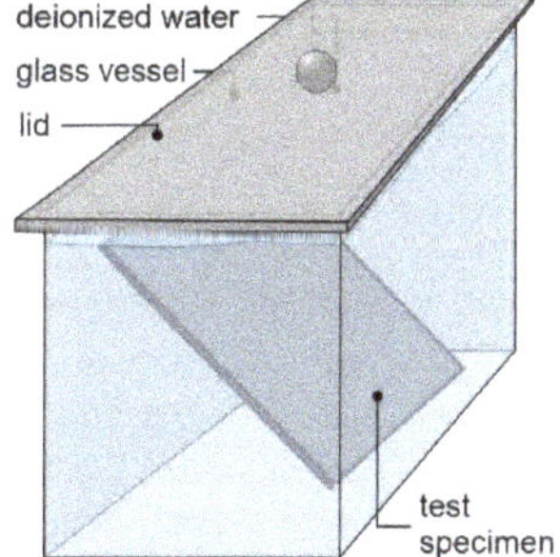

Figure 6. Schematic presentation of the dynamic surface leaching test (DSLT) setup.

The leaching water was renewed in intervals of 0.25, 1, 2.25, 4, 9, 16, 36 and 64 days, as specified in CEN/TS 16637-2 [26]. From each eluate fraction samples were taken and analyzed. The raw data were provided by the VDZ for further comparison calculations in this work.

2.3.3. Laboratory Irrigation

A typical European rain has intensities of 2.5–10 mm/h with drop diameters of 0.5–5 mm, but 1–4 mm for most rain events [32–34]. To reach realistic rain scenarios meeting these conditions, the irrigation stand, pictured in Figure 7, has been developed [35].

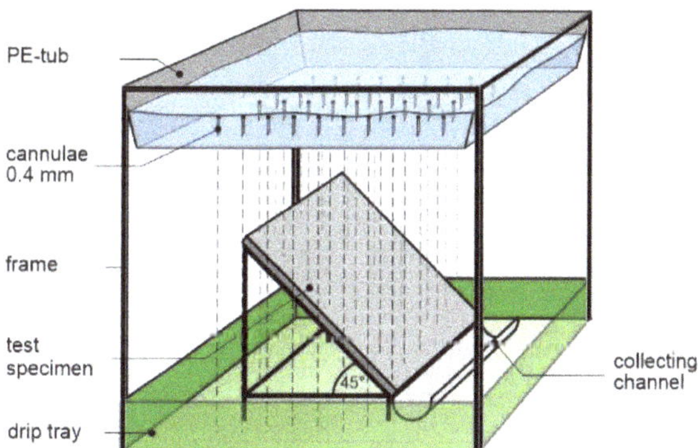

Figure 7. Schematic presentation of the laboratory irrigation stand.

It consists of an irrigation unit made of a PE-tub, equipped with 70, respectively, 100 cannulae evenly spaced out over a base area of 283–300 mm^2 and a frame to place the sample and collect the runoff. The tub is filled with deionized water to a predefined level. The filling level and the number of cannulae determine the rain intensities. An electric motor moves the irrigation unit every 30 min over

20 mm in a horizontal direction to ensure an equal wetting across the whole test specimen (see Figure 8). The test specimens with dimensions of 300–400 mm^2 are located in a 45° angle below the irrigation unit.

Figure 8. Drop distribution over the test specimen of the laboratory irrigation.

Pre-tests showed that cannulae of 0.4 mm diameter and 20 mm length are producing constant drops of 2.2 mm diameter and are applicable to create intensities between 1 and 5 mm/h.

The runoff is saved in PE gutters that lead the water to 15 L collection containers made of glass.

The target irrigation cycle of the investigations of this work is shown in Figure 9. The cycle has been developed using meteorological data from all over Germany in a previous study [35]. The chosen intensities and amounts of rain are based on weather dates of Holzkirchen for the month of July so the rainiest time and area of Germany is reflected in a testing period of one month.

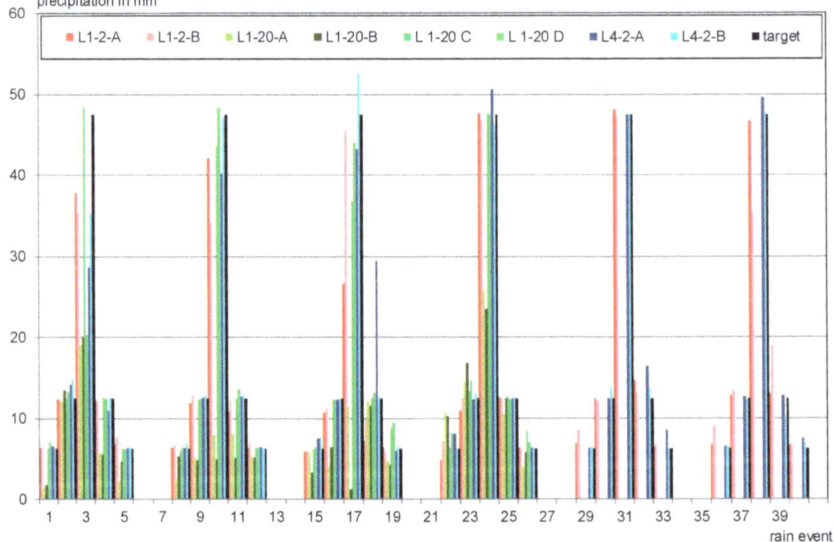

Figure 9. Wetting cycle for the laboratory irrigation.

The desired rain intensities of 1, 2 and 5 mm/h were achieved using filling levels of 4.5 (4.5 mbar pressure), 7 (7 mbar) and 12 cm (12 mbar). Some deviations cannot be avoided due to decreasing pressure over the rain events (especially for high intensities) and due to blocked needles primarily at low pressures. Figure 9 therefore also shows the actual rain cycles that occurred during testing.

The developed method for laboratory irrigation showed well reproducible rain amounts and can be used for further investigations. It turned out to be a relatively simple and cost-efficient test for simulating rain in intensions and drop sizes close to reality. The aimed rain amounts were met with an average deviation of 9.0% (median 3.5%), whereby the high intensities give more weight with 14% as they are not always reached (see Figure 9).

The eluates were homogenized and samples of 3 × 50 mL were taken in PE tubes. One subsample was used to determine pH value and conductivity as well as the concentrations of sodium, potassium and calcium (by flame photometry) and chloride and sulfate (by ion chromatography). The other two subsamples were acidified with 2.5 vol.% supra pure nitric acid. One was used for the analysis of antimony, arsenic, barium, boron, cadmium, chromium, cobalt, copper, lead, molybdenum, mercury, nickel, selenium, thallium, vanadium and zinc (by ICP-MS) the other one was kept as a retain sample. All samples were stored at 4 ± 2 °C until analysis.

2.3.4. Outdoor Testing

For the outdoor testing, the specimens were exposed on a roof in Aachen, Germany (position: 50°46′52.2″ N 6°02′56.8″ E), facing west with a 45° angle to the ground.

Each specimen was attached to the sample holder pictured in Figure 10. For the cracked specimens (see also Figure 11), a special clamping was used to keep the cracks open, what should create a worst case scenario with direct contact of rainwater to the coated reinforcement.

Figure 10. Outdoor testing stands on the roof of the RWTH Aachen University.

Figure 11. Cracked test specimen F1B.

The sample holders are made of stainless steel. The runoff from the test specimens is collected in a stainless steel gutter covered with an adhesive PTFE film leading to 15 L, respectively, 25 L glass bottles that are protected from sunlight by a plastic cover.

The leachates of the outdoor weathering were collected in fractions of one week from which the laboratory samples were taken. In case of long-lasting heavy rainfalls, the glass bottles were changed earlier and kept closed and cool until the scheduled change date. Both fractions were then combined for one weekly eluate and the laboratory samples were taken from this mixture.

The analysis preparation and analytics were conducted analogous to the laboratory irrigation procedure.

For the evaluation of the outdoor testing, data from the weather station 'Aachen Hörn' (resolution: 10 min) were used [36,37]. The weather data are compared to the amount of leachate collected for every test specimen (wide bar ≙ glass panel).

The wind-driven rain amounts were calculated based on the recorded normal rain amounts after Equation (1) derived from ISO 15927-3 [38].

$$R_D = \frac{2}{9} \cdot v \cdot R_N^{\frac{8}{9}} \cdot \cos(D - \Theta) \tag{1}$$

where R_D = hourly wind-driven rain amount in mm, v = hourly wind speed average in m/s, R_N = hourly normal rain amount in mm, D = hourly average of the wind direction with reference to North in °, Θ = wall orientation with reference to North in °. Figure 12 shows the recorded weekly average temperature and the weekly calculated rain amounts per inclined positioned test specimen over the testing period of one year.

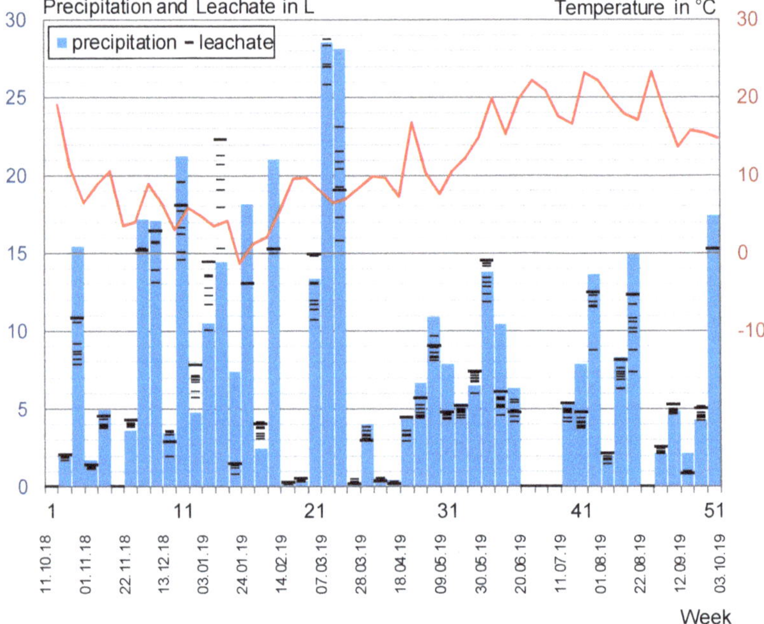

Figure 12. Average temperature, weekly precipitation and collected leachate in L per test specimen.

Due to the prominent position and the proximity to the weather station, no topography factors were applied. Furthermore, as they are not hitting the surface and therefore are not contributing to leaching, negative wind-driven rain amounts were considered as zero.

The test specimens were irrigated 9.6% of the time. The yearly normal rainfall sums up to a total amount of 718 mm, leading to a calculated amount of 398 L/a per test specimen of which 305 L are attributed to normal and 93 L to wind-driven rain.

The collected leachate amounts varied from 75% (F1A) to 90% (glass panel), on average 80%, of the calculated precipitation, showing a decreasing tendency over the testing period. Moreover, 0% was collected for weekly precipitation sums of <0.2 L/m^2; Eluate amounts of >100% of the calculated rain amounts are attributed to mainly snowfall (week 12 to 14) and probably imprecise calculated wind-driven rain. The detailed results for the amounts and analysis of the collected eluates are shown in the electronic supplement.

Apart from one incident of frost-damaged bottles (Series No. 14) and three overflow events (Series No. 16, 20, 49), all eluates were collected steadily over the testing period of one year from 11 October 2018 to 10 October 2019. The conformity of the collected leachate amounts and analyzed concentrations is considered as appropriate for an outdoor experiment.

2.4. Testing Program

To determine the influence of the reinforcement, especially its coating, test specimens with different concrete coverings were examined in double determination for both laboratory tests. Under outdoor conditions also cracked test specimens (F 1g and F 4g) were investigated. The direct contact of rainwater with the reinforcement might have an influence on the leaching behavior. Table 6 gives an overview over the testing programs conditions.

Table 6. Testing program conditions.

Characteristic	DSLT	Laboratory Irrigation	Outdoor Testing
blank	empty leaching vessel	irrigated glass panel	irrigated glass panel
duration	64 d	28 d (+12 d)	365 d
total amount of water	240 to 380 L/m^2	344 L/m^2	663 L/m^2 *
conditions	permanent water contact	scheduled irrigation and drying phases; drop size: about 2.2 mm; intensities: 1, 2, 5 mm/h	outdoor conditions 45° angle to ground facing west

* calculated amount of water, that hit the test specimens.

Due to technical reasons the specimens of the laboratory irrigation were tested at different ages: L1-20 A + B, 15 weeks; L1-2 20 weeks; L4-2 25 weeks. To detect differences caused by age and to investigate the repeatability of the method, a second run with L 1-20 test specimens (samples C and D) at an age of 8 weeks was conducted. In addition, the test specimens L 1-2 and L 4-2 were irrigated for two more weeks after a drying phase of half a year to find out about the long-term behavior and possible influences of carbonation.

3. Results

3.1. Preliminary Note

To give a first statement concerning the environmental compatibility of C^3, the concentrations in the collected eluates and leachates, which could have an immediate impact on the environment, are taken into account. The concentration development is investigated regarding indications for prevalent leaching mechanisms and influencing factors determining the emissions for the respective leaching experiment. The cumulated emissions and further assessment approaches are discussed in part 2 of this publication.

3.2. Overview of Eluate Concentrations in the Different Tests

Four variants of the same material have been leached using four different methods, resulting in different eluate concentrations. Table 7 gives an overview over the range of the respective concentrations.

The detailed test results and further comparisons are given in the following chapters. All measured eluate and leachate concentrations are given in the electronic supplement, Tables S1–S48.

Table 7. Concentration range in the leaching tests.

Substance	Concen-Tration	pH_{stat}		DSLT		Laboratory Irrigation		Outdoor Irrigation		Outdoor Background	
		Min	Max	Min	Max	Min	Max	Min	Max	Min	Max
Na		68	83	1.46	10.6	<0.2	12.7	0.4	334	<0.1	18.4
K		136	174	4.69	41.2	0.8	35.4	1.0	302	<0.1	2.2
Ca	mg/L	4830	12,700	32.0	88.7	1.6	17.6	1.1	17.2	<0.1	9.4
Cl^-		0.2	14.3	<0.1	11.3	<0.1	0.4	<0.1	117	<0.1	39.4
SO_4^{2-}		1.4	742	2.1	5.3	1.2	10.9	<0.1	245	<0.1	19.0
Sb		n.d.	n.d.	<0.1	11.8	<0.1	2.58	<0.1	1.52	<0.1	1.03
As		<2	26.9	<0.02	1.33	<0.1	0.361	<0.1	8.04	<0.1	1.25
Ba		408	2178	12.8	74.1	0.350	14.3	0.520	20.9	0.560	20.8
Pb		<1	172	<0.05	10.3	0.046	0.270	0.079	7.46	<0.1	15.8
B		<5	2,532	n.d.	n.d.	<1	11.4	<1	147	0.62	15.7
Cd		<0.1	6.76	<0.01	0.080	<0.1	<0.1	<0.1	0.210	<0.1	0.660
Cr		3.21	320	0.960	5.02	<0.1	4.71	<0.1	23.3	<0.1	2.26
Co	µg/L	<1	153	<0.01	0.050	<0.1	<0.1	<0.1	3.88	<0.1	0.980
Cu		<1	451	<0.04	2.19	0.046	27.5	<0.1	32.0	0.290	29.1
Mo		5.10	90.4	<0.2	1.50	<0.1	1.44	<0.1	6.20	<0.1	1.55
Ni		<0.5	486	<0.04	2.75	<0.2	8.11	<0.1	419	<0.1	6.96
Hg		n.d.	n.d.	<0.02	<0.02	<0.01	0.023	<0.01	0.028	<0.01	0.020
Se		5.87	30.9	n.d.	n.d.	0.079	0.331	<0.1	3.19	<0.1	1.14
Tl		<1	8.44	<0.01	0.110	<0.1	<0.1	<0.1	<0.1	<0.1	<0.1
V		2.31	240	<0.7	14.5	0.358	5.55	<0.1	64.2	<0.1	2.19
Zn		<10	1423	<0.7	11.9	<1	15.7	0.578	90.0	3.02	154

n.d. = not determined.

As expected, the maximum concentrations of all substances are obtained in the pH_{stat} test. This test method is used to determine the respective solubility. The concentrations are significantly higher than in the other leaching experiments, indicating a dilution in the leaching tests on the monolithic samples. Outdoor and laboratory irrigation as well as the DSLT show different relations among each other, dependent on the observed substance.

High concentrations in comparison to the average, especially for the outdoor irrigation, are often reached only once or twice a year and mainly at the beginning of the test. This is emphasized by Figure 13, showing the average runoff concentrations and their value distribution for the analyzed elements. Some elements are not displayed in Figure 13 and are also excluded from further evaluation due to low prevalence or high concentrations in the blind tests:

- Cadmium and thallium (determination limit: 0.1 µg/L) could not be detected in any eluate in the laboratory irrigation and were detected in maximum 10% of the leachates due to the background concentration outdoor. They were at no time leached from the specimens.

- Cobalt (determination limit: 0.1 µg/L) could not be detected in any eluate in the laboratory irrigation and was found in 92% of the outdoor leachates, always in the range of the background concentrations. A slightly increasing tendency was observed outdoor.

- In the laboratory mercury was detected in concentrations close to the detection limit ($0.01 < c \leq 0.023$ µg/L) in 29% of the eluates. Five of eight test specimens (L 4-2 A + B, L 1-2 A, and L 1-20 C + D) and the outdoor experiments showed no detectable mercury leaching at all.

- Outdoor antimony (detected in 74% of the eluates), lead (93%) and zinc (100%) were released from C^3 in only 6 to 10% of the samples.

- Chloride (determination limit: 0.1 mg/L) was detected in concentrations of $0.1 < c \leq 0.4$ mg/L in only 13% of the laboratory eluates. In 94% of the outdoor leachates, emissions with an average concentration of 2.28 mg/L were measured.

- Copper and zinc showed relatively high releases in the laboratory irrigations' blind testing, which has to be considered in the interpretation of the results. It is assumed that zinc as a ubiquitous element is introduced through several pathways. Copper could be leached from the cannulae that are indicated as stainless steel. This steel might have copper residues or is alloyed with it to enhance corrosion and acid resistance [39].

- Arsenic, lead and selenium were detected in 18% (As, Se), respectively, 7% (Pb) of the laboratory eluates in concentrations of $0.1 < c < 0.3$ µg/L. Outdoors, arsenic is leached and lead is adsorbed. Selenium is mainly leached in concentrations below 0.1 µg/L.
- Sodium, potassium, calcium, barium, boron, chromium, and vanadium were leached in 70 to 100% of the outdoor samples and detected in 100% of the laboratory eluates.

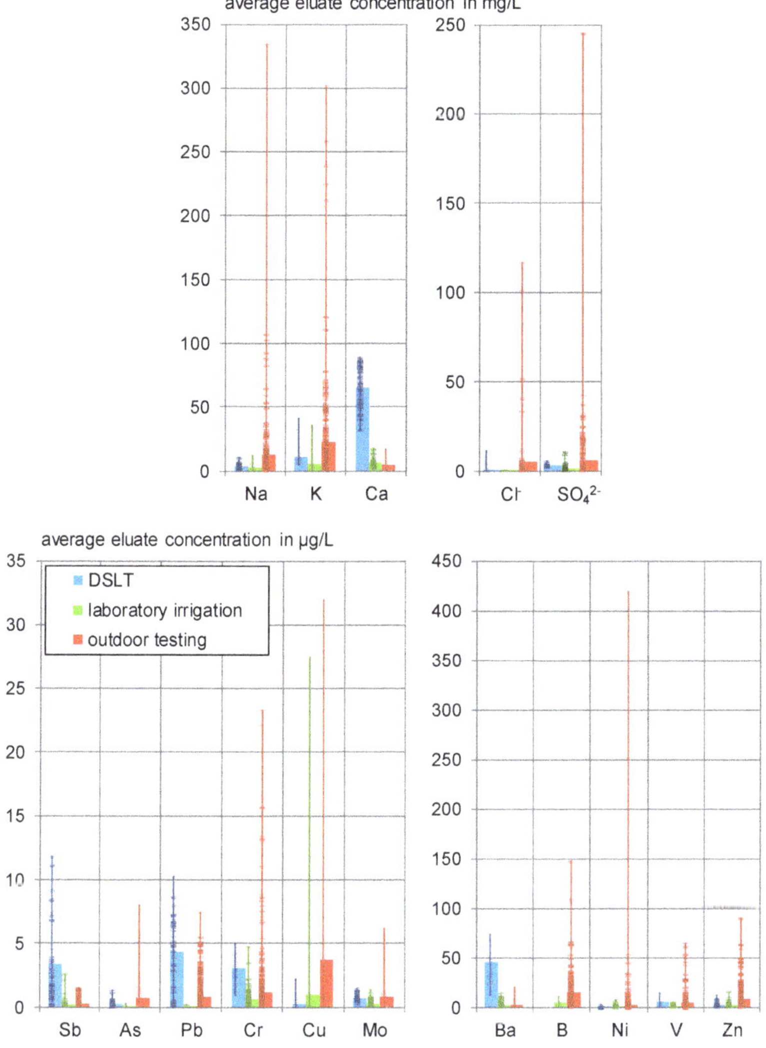

Figure 13. Average eluate concentrations and value distribution of DSLT, laboratory irrigation, and outdoor testing.

The substances named in the last two points are therefore main subjects to further interpretation.

Neither the number of reinforcement layers nor the thickness of the concrete cover showed significant influence on the leaching behavior [27,35]. Figure 10 and the figures for the discussion therefore show the averaged substance release of the different test specimens.

3.3. Laboratory Irrigation

During the laboratory irrigation, only slight changes in the test specimens' surface were observed. In the first week of the testing, a droplet pattern developed, lightening up the surface when converging over time. Figure 14 exemplarily shows the smooth surface of specimen L 1-20 A before the testing and the surface after 4 weeks of irrigation, which led to a slightly rougher surface and a few calcite-filled pores.

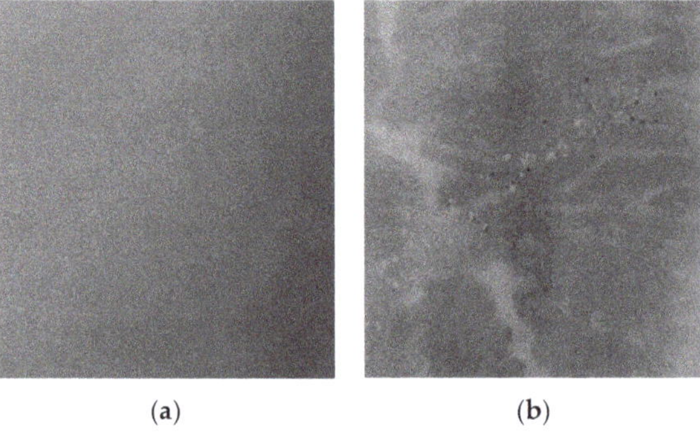

(**a**)   (**b**)

Figure 14. (**a**) Surface of test specimen L1-20 A before the laboratory irrigation; (**b**) surface of test specimen L 1-20 A after 4 weeks of laboratory irrigation.

The eluate pH development differed slightly between the test specimens as can be seen in Figure 15. All tests show a decreasing pH over the testing period. Most eluates had a starting pH of 10 to 11, two started at around pH 9 and one at only pH 8. This might be a consequence of inaccurate sealing and therefore pre carbonation of the test specimens' surface. The different ages of the samples had no systematic effect on the pH. It is noticeable that the specimens starting at a lower pH show a more stable and constantly decreasing development than the specimens coming from a higher pH. All eluates level at a pH around 7.5 after three weeks of irrigation, which also continued in the subsequent irrigation after six months.

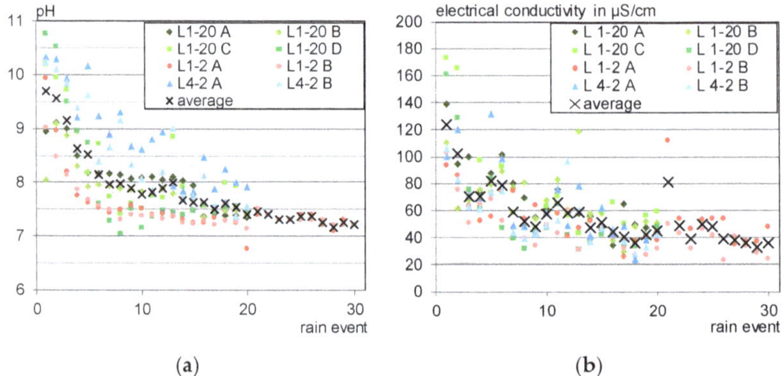

(**a**)   (**b**)

Figure 15. (**a**) Development of the eluate pH in the laboratory irrigation experiment; (**b**) development of the electrical conductivity in the laboratory irrigation experiment.

The conductivity of all eluates varied according to the amount of water applied in one step and showed a decreasing tendency from 100 to 180 µS/cm in the first week to 30 to 60 µS/cm in the fourth week of testing.

Figure 16 exemplarily shows the course of the incremental concentrations measured in the eluates for sodium, calcium, arsenic and vanadium.

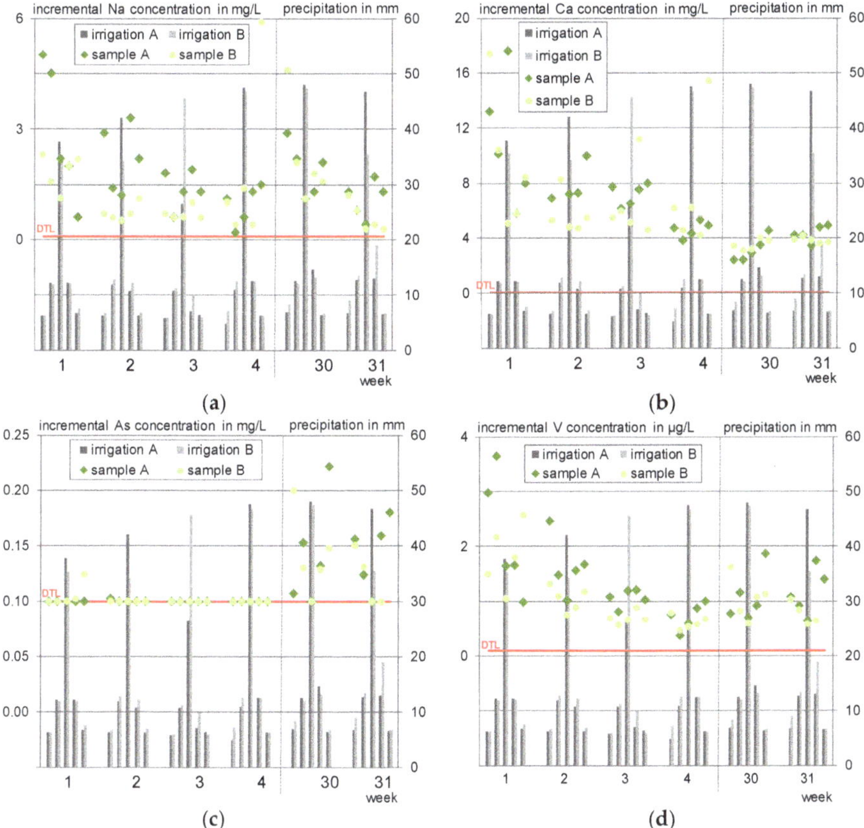

Figure 16. Incremental eluate concentrations of the test specimens L 1-2. (**a**) sodium; (**b**) calcium; (**c**) arsenic; (**d**) vanadium.

A decreasing tendency in the eluate concentrations can be observed for all substances over the initial testing period of four weeks. Leaching of arsenic could not be detected during these four cycles. The concentrations of all substances are lower when more water is applied which shows dilution and is hinting at a diffusion-controlled leaching. However, calcium appears to level at concentrations around 4 mg/L after three weeks.

Contrary to the conception expectation that drying phases will increase the following eluate concentrations, the scheduled weekly irrigation breaks do not lead to changed, higher, concentrations compared to the rain event with the same conditions of the previous cycle. Only the long drying phase leads to increased concentrations for all shown substances except for calcium, suggesting that the conception of the drying phases is too short or rather not forceful enough to depict effects that are usually evoked by intermittent wetting.

A change in pH by further advanced carbonation processes can be excluded as a cause for the course changes from week 30. Figure 15 pictures the very similar pH of 7.5 before the break, showing that the samples are already carbonated, dropping to minimum pH 7.2 after the break. Moreover, this slight difference would not affect the examined substances even if they were leached pH dependent and close to their maximum solubility.

3.4. Outdoor Testing

All outdoor test specimens developed the same optical changes over the time of exposure, pictured exemplarily for test specimen F1A in Figure 17.

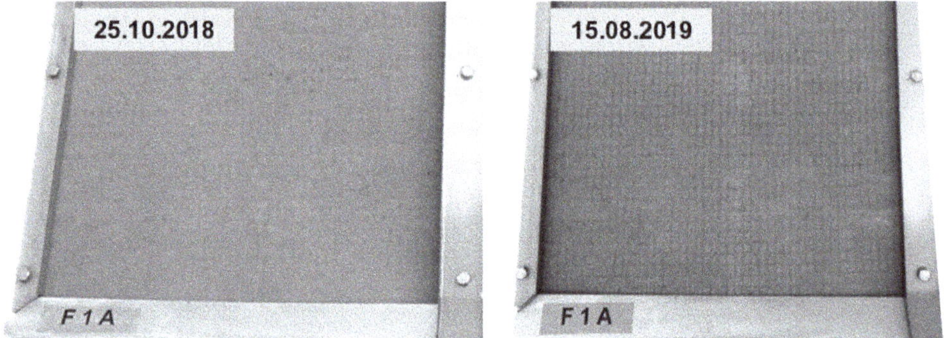

Figure 17. Surface of outdoor test specimen F1A after two and 44 weeks of exposure.

After one year of weathering, the surface appeared more porous and the reinforcement which was covered by only 2 mm of concrete became clearly visible. Some pores were filled with calcite. The dark lines above the textile are attributed to water flow along the reinforcement due to less suction depth of the concrete and probably cavities next to the textile.

Apparently, the crack width of the cracked surfaces remained constant over the testing time. It was measured in week 4, 8 and 44 with a crack measuring gauge. Nevertheless, light microscope images taken from cross sections after the test revealed that micro cracks, positioned transversely towards the main cracks, and also pores were closed over time as can be seen in Figure 18.

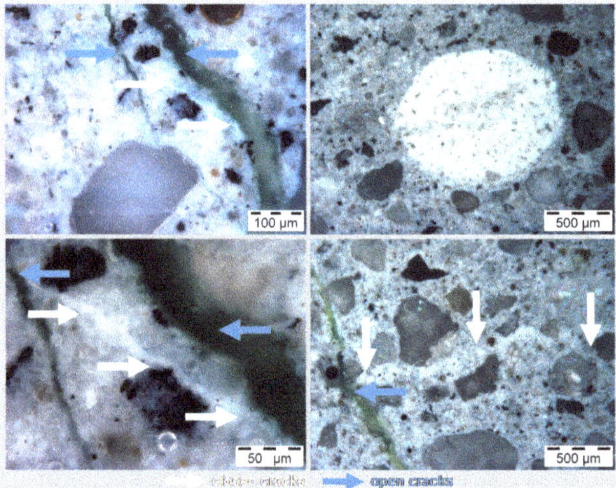

Figure 18. Cross section of tests specimen F1gB after one year of exposure; closed cracks and pores. Blue: main cracks, open; white: transverse cracks, closed.

The eluate pH, imaged in Figure 19a, developed the same way for all test specimens. It started at around pH 10 and dropped to pH 8 within the first 5 weeks, after a drying phase it increased to pH 9.5 again but levelled at pH 7 ± 0.75.

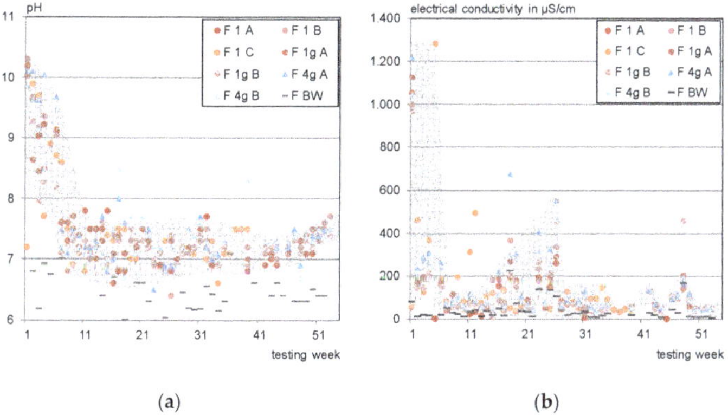

(a)   (b)

Figure 19. (**a**) Eluate pH in the outdoor testing; (**b**) electrical conductivity in the outdoor testing.

The conductivity of the eluates varied as per Figure 19b. Except from the first rain events (1000 to 1200 µS/cm dropped to 200 to 400 µS/cm), no decreasing tendency was observed.

Figure 20 exemplarily shows the individual concentrations of sodium, calcium, arsenic and vanadium for the outdoor test specimen F 1g A and F 1g B. Sodium, arsenic and vanadium developed a nearly identical concentration pattern. Sodium shows a slightly higher response to dry or semi-dry phases, which is attributed to its lower bonding to the concrete matrix.

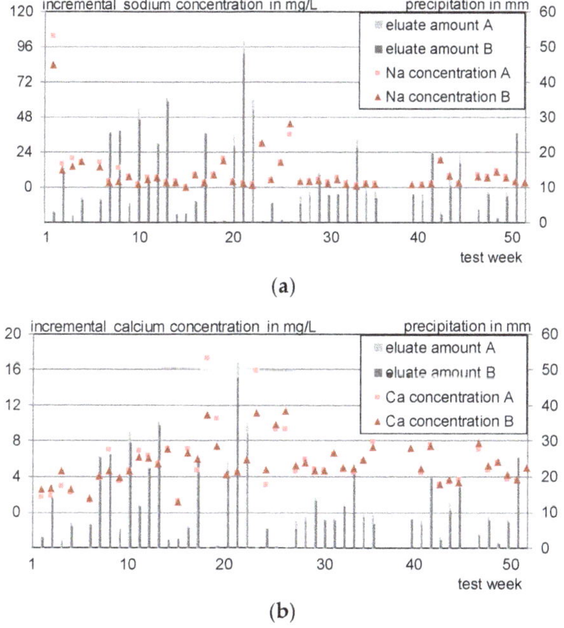

Figure 20. *Cont.*

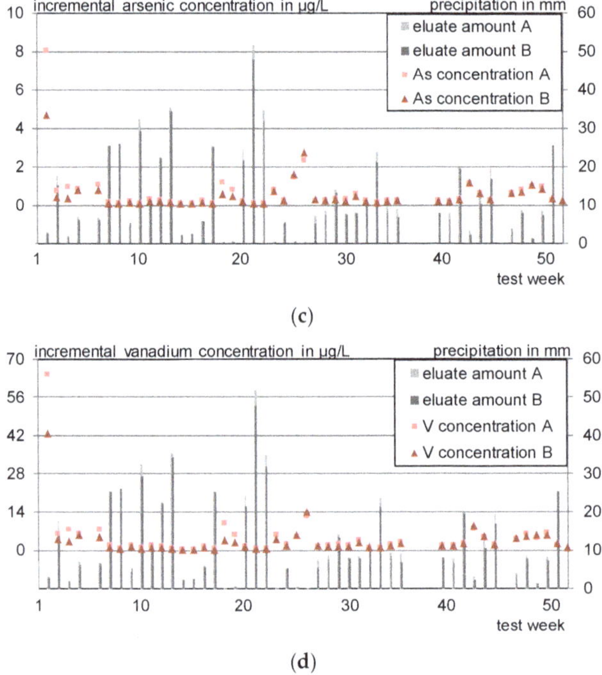

(c)

(d)

Figure 20. Incremental leachate concentrations of the test specimens F 1g. (**a**) Sodium; (**b**) calcium; (**c**) arsenic; (**d**) vanadium.

Calcium concentrations are increasing in the beginning and level, similar to the laboratory irrigation, between 4–8 mg/L. (Higher concentrations in cases of very low rainfall can be assigned to high blank values.)

3.5. Standardized Leaching Tests

In the DSLTs, the eluate pH for all test specimens was stable at 11.2 to 11.8 with a maximal difference of 0.3 between the specimens of one composition.

The established DSLT is based on a water-to-surface ratio of 80 L/m^2. The amount of water applied in the tests described in this work was lower (D 1-2 47.6 L/m^2, D 4-2 42.1 L/m^2), which is irrelevant in this tests, as long as the release is controlled by diffusion. However, if the pH is significantly different or concentrations reach equilibrium, the release will be changed.

To determine the leaching mechanism of the evaluated substances under standardized laboratory conditions, the evaluation method described in [26] was used. If the leaching process of the DSLT is diffusion controlled, granted that the diffusion coefficient does not change over the process, the release is in a linear relation to the root of time. The defined leaching steps lead to eluate concentrations that allow for distinguishing the leaching mechanism, mainly diffusion or solubility.

For the evaluation, a normalized concentration is calculated. The eluate concentrations of each step are divided by the averaged concentration of all steps and plotted as a column diagram. A diffusion-controlled process will picture a three-step diagram (see Figure 21c), while a solubility-controlled process shows a continuous normalized concentration of one (see Figure 21b) as it is constantly leached in case of a stable pH.

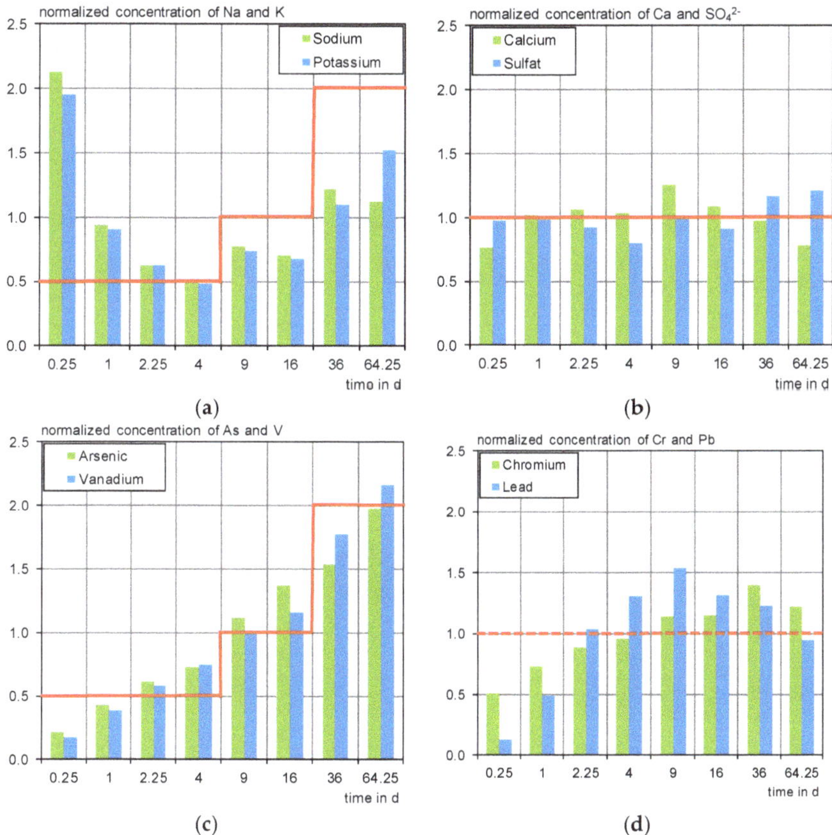

Figure 21. Leaching mechanism, three-step diagram after CEN/TS 16637-2. (**a**) Sodium and potassium, diffusion controlled with initial wash-off effect; (**b**) calcium and sulfate, solubility controlled; (**c**) arsenic and vanadium, diffusion controlled; (**d**) chromium and lead, no mechanism determined.

It is obvious that the described pattern is never exactly met and some mechanism overly each other. This is a known phenomenon; the criteria for a diffusion-controlled process are therefore defined as valid for the root of the mean square error ($\sqrt{MSE}$) below 0.4 for each eluate concentration [26].

For the evaluation of the DSLT, the results of D 1-2 and D 4-2 were averaged and subsequently analyzed following Annex B of CEN/TS 16637-2 [26]. Sodium, potassium (Figure 21a), arsenic and vanadium (Figure 21c) were determined as diffusion controlled, whereas the alkalis seem superimposed by an initial wash-off effect. Consistent with the findings of [40], calcium and sulfate leach mainly solubility controlled (Figure 21b). In this work, the standard deviations between the single eluate concentrations were >0.25; the leaching mechanism is therefore not calculative affirmed within the meanings of [26]. For chromium and lead (Figure 21d) no leaching mechanism can be identified according to [26]. In [40], chromium is found to be diffusion controlled, lead is depleting over the duration of the tank test. Looking at the normalized concentrations, it is assumed that chromium and lead leach diffusion controlled followed by depletion or by equilibrium concentrations.

One approach to check if concentrations are approaching equilibrium is to compare them to the concentrations reached in the pH$_{stat}$ test. The pH$_{stat}$ test (2.3.1) indicates the solubility of the respective elements at different pH values. Figure 22 pictures the results of the pH dependence test compared to the concentrations of the other leaching experiments of this work.

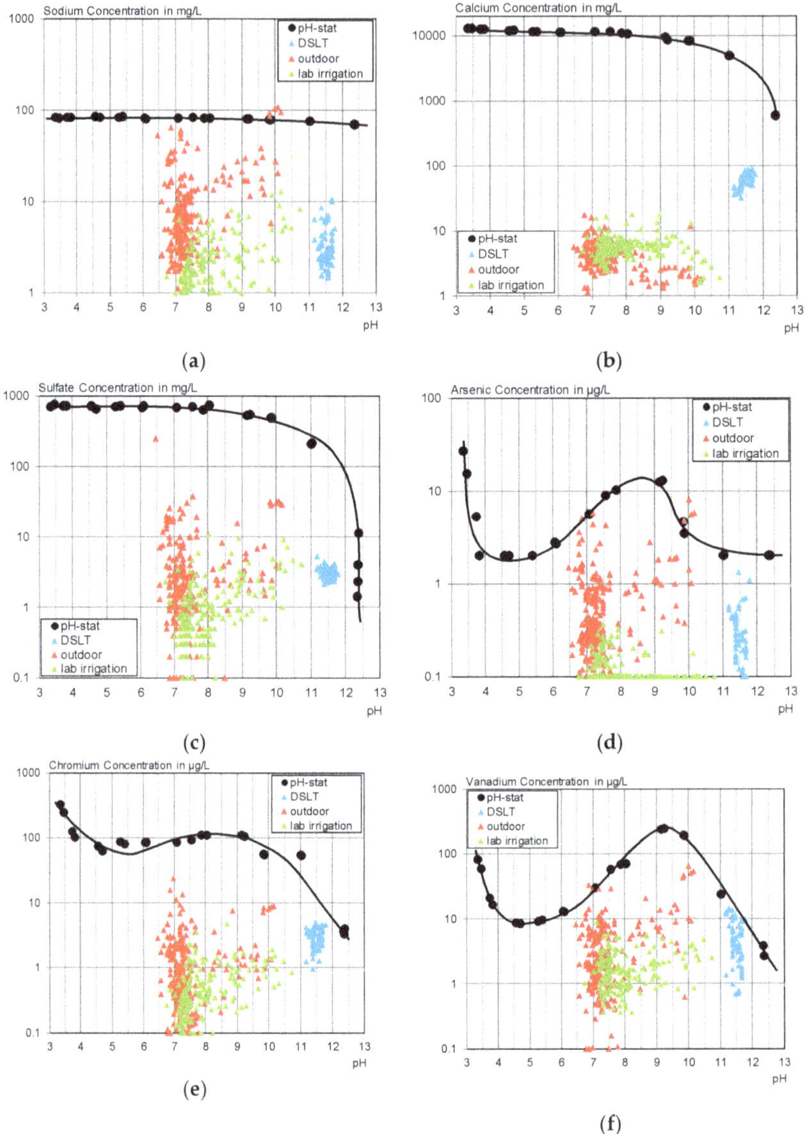

Figure 22. pH-dependent leaching of selected substances from C^3 compared to the leachate concentrations of the DSLT, the laboratory irrigation and the outdoor testing. (**a**) Sodium; (**b**) calcium; (**c**) sulfate; (**d**) arsenic; (**e**) chromium; (**f**) vanadium.

For permanent water contact, it becomes visible that the concentrations of the solubility-controlled substances calcium and sulfate remain in a closer range compared to the concentrations of the other substances. However, they stay clearly below their possible maximum concentrations. This might be an effect of leaching steps that are not long enough to reach equilibrium concentrations combined with decreasing concentration gradients due to physical barriers through the concrete.

Only vanadium, and, considering analytical and computational uncertainties, possibly chromium, reached the determined pH$_{stat}$ concentrations in the last two time steps of some DSLTs. A higher water to surface ratio would probably lead to a higher total release at the same or lower concentrations in this case.

During intermittent wetting in the laboratory, a concentration gradient seems to be kept for all substances. The concentrations are mainly below or in the range of the DSLT.

Outdoors, all concentrations except for calcium exceed both laboratory experiments' concentrations, which is partially a result of the rainwater contamination. However, the net values between the test specimens' concentration and the glass panel results show that the outdoor leaching sometimes leads to significantly increased leaching compared to the laboratory testing. This can be a result of harsher conditions as the material experiences longer drying phases, wind is carrying moisture from the surface, and temperature changes contribute to faster drying. However, it also has to be considered that the outdoor specimens were larger than the laboratory ones, which might extend the contact time for each droplet. For diffusion-controlled leaching, a longer time span results in higher concentrations.

- Sodium leaches strongest under outdoor conditions, which confirms the findings concerning the influence of dry phases from the laboratory irrigation. It is assumed that capillary transport mechanisms transport dissolved substances to the surface as described in [11,13] and therefore cause an increased availability of sodium on the test specimens' surface.
- Sulphate also shows higher outdoor concentrations but considering the blank, the net concentrations of sulphate were mainly below the DSLT and in the range of the lab data.
- Calcium concentrations show a good concordance between laboratory and outdoor irrigation.
- Arsenic, chromium and vanadium leach more strongly in the outdoor irrigation compared to the laboratory, especially for higher pH values.

There is no general correlation between the laboratory leaching data and outdoor emissions. A direct prediction of the outdoor emissions seems only possible for solubility-controlled substances with low background concentration in the eluent.

4. Discussion

To classify the measured concentrations and allow a statement concerning the environmental compatibility of C^3, the results are compared to reference values. Figure 23 shows the maximum concentrations reached in the irrigation experiments related to current threshold values of Groundwater in Germany [41] and for surface water bodies in Europe [6]. The concentrations of the DSLT eluates are pictured comparatively.

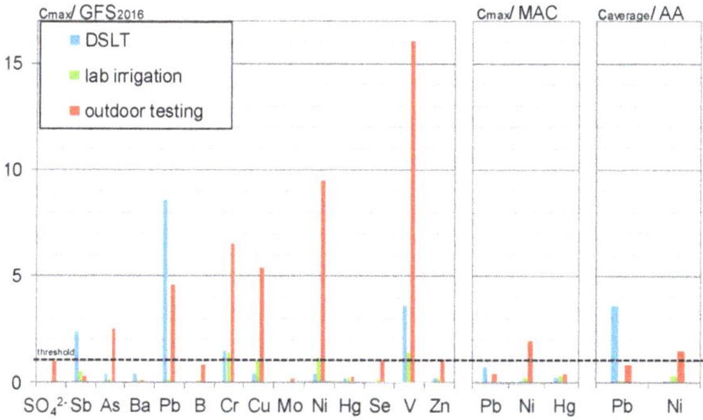

Figure 23. Maximum leachate concentrations related to the de-minimis thresholds after [6,41].

The MACs (maximum allowable concentrations) and AAs (annual averages) after Article 3, Point 1, together with Part A of Annex I of the European directive on environmental quality standards for surface water qualities [6,42] only applicable for lead, nickel, and mercury, are only exceeded for nickel during the outdoor investigation. The MAC of 34 µg/L was exceeded in 3% of the time, the annual average is at 110% of the threshold of 4 µg/L. Considering the average background concentration of 1.3 µg/L in the rainwater, this exceedance is not only attributable to the C^3. A direct discharge to a surface water compartment can be considered without negative effects.

As pictured in Figure 20, during the laboratory irrigation, only chromium (1.25% of 160 eluates), nickel (1%), and vanadium (5%) occasionally exceeded the de-Minimis threshold for groundwater of the German Working Group on water issues of the Federal States and the Federal Government represented by the Federal Environment Ministry (LAWA) [41]. Since the experiment was conducted using aggressive deionized water and furthermore the thresholds are designed for groundwater not seepage water, a direct infiltration in the soil is considered to be environmentally harmless.

Outdoor, arsenic (4%), lead (25%), copper (17%), nickel (14%), vanadium (32%), and zinc (1%) were measured in higher concentrations, whereby lead, copper, and zinc concentrations originated only from the rain water and were even adsorbed by the concrete. These findings are similar to a study of Schiopu et al., where leachate concentrations from irrigated concrete slabs were only higher than the rainwater concentrations in 47%, respectively, 21% of the eluates for copper and zinc [24].

Arsenic, nickel and vanadium are leached from the C^3 and contribute to the rainwater discharge concentrations. It has to be taken into account that exceeding the threshold values after [41] means, that a direct entry into the groundwater in high amounts could possibly affect the environment. The retention by the covering soil layers and the dilution with unaffected rainwater leads to decreasing concentrations.

Generally speaking, it is a known problem that municipal rainwater discharge contains, amongst others, too much heavy metals and trace elements. The study of [3], for example, showed that all heavy metals except from vanadium and nickel exceed the MACs and AAs in the rainwater runoff in Berlin.

Furthermore, it has to be mentioned that it is an internationally discussed topic for currently more than 10 years to except concrete from further testing on the release of regulated dangerous substances. Schiopu et al., for example, studied irrigated concrete slabs and only found leachate concentrations close to the rainwater concentrations for sodium, potassium, calcium, sulfate, copper, boron and zinc [24]. Table 8 shows the results of this work for the critical substances chromium and vanadium compared to the concentrations measured in the studies of Scherer and Vollpracht [15,21,43], to a study of Schiopu [44] and to the collection of concrete and mortar DSLT-Data by Dijkstra and v. d. Sloot from the project ECRICEM [45]. The dossier [45] was intended as a proposal for the without further testing (WFT) qualification of concrete whereby the DSLT values should confirm these properties. C^3 is clearly below the maximum values of [45] and shows low concentrations compared to [21,43].

Table 8. Leachate concentrations of chromium and vanadium for different mineral building materials.

Source	Examined Material	Eluate/Leachate Concentration in µg/L					
		Cr			V		
		Outdoor	Lab. Irr.	DSLT	Outdoor	Lab. Irr.	DSLT
[15,21]	reinforcement fiber plaster	<0.08–68.9	<0.5–26.4	<0.5–2.8	0.86–29.4	<0.5–8.3	<0.5–6.4
	lime-cement plaster	<0.08–141	<0.5–30.1	<0.5–5	<0.08–8.81	0.6–47.6	<0.5–5.1
	face masonry mortar	<0.08–93.2	<0.5–42.4	<0.5–7.8	<0.08–35.0	0.8–44.3	<0.5–8.4
[4]	concrete new	16–39	-	-	-	-	-
	concrete old	<1–6	-	-	-	-	-
[43]	unspecified mineral building materials	-	0.1–150	<1–49	-	<0.2–132	1–128
[44]	concrete slabs	<2–104	-	<2 *	-	-	-
[45]	ECRICEM mortar statistics	-	-	0.45–14	-	-	0.15–50
this work	C^3	<0.1–13.4	<0.1–4.71	0.96–5.02	<0.1–64.2	0.358–5.55	0.7–14.5

* results of a multi batch test similar to the DSLT (64 days, 50 L/(m^2 ·water renewal), eluate collection at every renewal after 2, 4, 8 h and 1, 2, 4, 9, 16, 36 and 64 days).

Considering the outdoor data of Table 8, a contribution from C^3 to the total load might only be assumed for vanadium. Looking at the raw materials' vanadium content (Table 4), the vanadium is most likely to be leached from the cement not the reinforcement.

After the comparison to general and actual rainwater background concentrations, German and European ground and surface water thresholds, as well as other mineral building products, it can be concluded that the concentrations, measured for all leaching experiments conducted, showed no environmental critical results. However, it also has to be mentioned that any emission from any material might be of concern as it is contributing to the total contaminant load and fostering potentially harmful concentrations in the rainwater discharge. The reduction and prevention of substances release must stay a main objective.

5. Summary and Conclusions

The use of composite building materials is currently increasing, which leads to the use of new materials or material combinations with unknown leaching behavior. Mutual influences or reactions of combined substances can change emissions despite an unchanged content level. To avoid a possibly harmful release of substances from the outset, it is necessary to prove their environmental compatibility.

In this research, the leaching behavior of C^3, especially for the application as irrigated façade elements, was investigated. Laboratory and outdoor exposure tests were run to determine and assess the heavy metal and trace element emissions.

A laboratory irrigation stand and an outdoor experimental setup with reproducible result production were introduced. Investigations through this stand on C^3, compared to each other, to the standardized laboratory test DSLT, and to reference values, indicated that this innovative composite material as environmentally compatible concerning its heavy metal and trace elements leaching behavior. Further research on organic trace substances, as, e.g., bisphenol A used as a raw material or amines used as hardeners in epoxy resins, would complement this classification.

The concentration thresholds for German groundwater [41] and the EU surface waters [6,41] were exceeded by arsenic, nickel and vanadium at times; only vanadium leached more strongly than an average concrete. In comparison to leachate concentrations of other irrigated mineral building materials, C^3 is neglectable. Moreover, zinc, lead, and copper were found not to be leached but rather to be adsorbed from concrete construction elements. As these substances are the ones with the comparably highest concentrations in rain waters, building materials other than concrete, for example, metal sheets or renders, must be considered as a source of the emissions from façades described in [3]. Considering the data of [45] for the WFT qualification of concrete reinforces this assumption; although it has to be kept in mind that this study is related to DSLT results while intermittent wetting causes altered leaching behaviors.

The non-traditional reinforcement showed no discernible influence on the leaching behavior of the substances observed. However, different leaching patterns, dependent on outdoor factors, could be identified for different elements, providing a foundation for further assessment method development.

In addition to the leaching behavior of C^3, the validity of laboratory experiments and the transferability to the outdoor behavior were investigated. No general correlation between laboratory leaching data and outdoor emissions were found. The results depend on the examined substance and the method used.

For further research regarding the laboratory irrigation, the wetting cycles have to be optimized to represent outdoor influencing factors. Although the test delivered reproducible results, the outdoor behavior was not sufficiently reproduced. The full irrigation cycle as a time- and effort-saving prediction method is not practicable and does not seem expedient. However, it could be used to vary single factors and better understand outdoor leaching behavior. In [46], a 24 h weathering cycle for the leaching of organic materials is introduced by Bandow et al. As inorganic constituents are partially incorporated into the leached material and generally show another emission behavior than organics, an adaption on concrete would have to be evaluated.

A transfer function or model to predict outdoor emissions of irrigated building materials from laboratory data would be desirable for a fast and reasonable assessment of materials that are of concern for the environment. In part 2 of this research, this topic will be debated. A first step towards an assessment method would be to check existing assessment concepts on their suitability by using the data collected for this work. If there is no direct application possibility, the investigation of the transferability, for example, by transfer functions, would be a further approach. Lastly, if necessary, a new approach on how to estimate outdoor emissions based on known leaching mechanisms and present data has to be developed.

Supplementary Materials: The following are available online at http://www.mdpi.com/1996-1944/13/19/4405/s1, Table S1: Laboratory irrigation results of the test specimen L 1-2 A: leachate amount, pH value, electrical conductivity and concentrations of sodium, potassium, calcium, chloride, and sulfate, Table S2: Laboratory irrigation results of the test specimen L 1-2 A: heavy metal and trace element concentrations, part 1, Table S3: Laboratory irrigation results of the test specimen L 1-2 A: heavy metal and trace element concentrations, part 2, Table S4: Laboratory irrigation results of the test specimen L 1-2 B: leachate amount, pH value, electrical conductivity and concentrations of sodium, potassium, calcium, chloride, and sulfate, Table S5: Laboratory irrigation results of the test specimen L 1-2 B: heavy metal and trace element concentrations, part 1, Table S6: Laboratory irrigation results of the test specimen L 1-2 B: heavy metal and trace element concentrations, part 2, Table S7: Laboratory irrigation results of the test specimen L 4-2 A: leachate amount, pH value, electrical conductivity and concentrations of sodium, potassium, calcium, chloride, and sulfate, Table S8: Laboratory irrigation results of the test specimen L 4-2 A: heavy metal and trace element concentrations, part 1, Table S9: Laboratory irrigation results of the test specimen L 4-2 A: heavy metal and trace element concentrations, part 2, Table S10: Laboratory irrigation results of the test specimen L 4-2 B: leachate amount, pH value, electrical conductivity and concentrations of sodium, potassium, calcium, chloride, and sulfate, Table S11: Laboratory irrigation results of the test specimen L 4-2 B: heavy metal and trace element concentrations, part 1, Table S12: Laboratory irrigation results of the test specimen L 4-2 B: heavy metal and trace element concentrations, part 2, Table S13: Laboratory irrigation results of the test specimen L 1-20 A: leachate amount, pH value, electrical conductivity and concentrations of sodium, potassium, calcium, chloride, and sulfate, Table S14: Laboratory irrigation results of the test specimen L 1-20 A: heavy metal and trace element concentrations, part 1, Table S15: Laboratory irrigation results of the test specimen L 1-20 A: heavy metal and trace element concentrations, part 2, Table S16: Laboratory irrigation results of the test specimen L 1-20 B: leachate amount, pH value, electrical conductivity and concentrations of sodium, potassium, calcium, chloride, and sulfate, Table S17: Laboratory irrigation results of the test specimen L 1-20 B: heavy metal and trace element concentrations, part 1, Table S18: Laboratory irrigation results of the test specimen L 1-20 B: heavy metal and trace element concentrations, part 2, Table S19: Laboratory irrigation results of the test specimen L 1-20 C: leachate amount, pH value, electrical conductivity and concentrations of sodium, potassium, calcium, chloride, and sulfate, Table S20: Laboratory irrigation results of the test specimen L 1-20 C: heavy metal and trace element concentrations, part 1, Table S21: Laboratory irrigation results of the test specimen L 1-20 C: heavy metal and trace element concentrations, part 2, Table S22: Laboratory irrigation results of the test specimen L 1-20 D: leachate amount, pH value, electrical conductivity and concentrations of sodium, potassium, calcium, chloride, and sulfate, Table S23: Laboratory irrigation results of the test specimen L 1-20 D: heavy metal and trace element concentrations, part 1, Table S24: Laboratory irrigation results of the test specimen L 1-20 D: heavy metal and trace element concentrations, part 2, Table S25: Outdoor irrigation results of the test specimen F BW: leachate amount, pH value, electrical conductivity and concentrations of sodium, potassium, calcium, chloride, and sulfate, Table S26: Outdoor irrigation results of the test specimen F BW: heavy metal and trace element concentrations, part 1, Table S27: Outdoor irrigation results of the test specimen F BW: heavy metal and trace element concentrations, part 2, Table S28: Outdoor irrigation results of the test specimen F 1 A: leachate amount, pH value, electrical conductivity and concentrations of sodium, potassium, calcium, chloride, and sulfate, Table S29: Outdoor irrigation results of the test specimen F 1 A: heavy metal and trace element concentrations, part 1, Table S30: Outdoor irrigation results of the test specimen F 1 A: heavy metal and trace element concentrations, part 2, Table S31: Outdoor irrigation results of the test specimen F 1 B: leachate amount, pH value, electrical conductivity and concentrations of sodium, potassium, calcium, chloride, and sulfate, Table S32: Outdoor irrigation results of the test specimen F 1 B: heavy metal and trace element concentrations, part 1, Table S33: Outdoor irrigation results of the test specimen F 1 B: heavy metal and trace element concentrations, part 2, Table S34: Outdoor irrigation results of the test specimen F 1 C: leachate amount, pH value, electrical conductivity and concentrations of sodium, potassium, calcium, chloride, and sulfate, Table S35: Outdoor irrigation results of the test specimen F 1 C: heavy metal and trace element concentrations, part 1, Table S36: Outdoor irrigation results of the test specimen F 1 C: heavy metal and trace element concentrations, part 2, Table S37: Outdoor irrigation results of the test specimen F 1g A: leachate amount, pH value, electrical conductivity and concentrations of sodium, potassium, calcium, chloride, and sulfate, Table S38: Outdoor irrigation results of the test specimen F 1g A: heavy metal and trace element concentrations, part 1, Table S39: Outdoor irrigation results of the test specimen F 1g A: heavy metal and trace element concentrations, part 2, Table S40: Outdoor irrigation results of the test specimen F 1g B: leachate amount, pH value, electrical conductivity and concentrations of sodium, potassium, calcium, chloride, and sulfate, Table S41: Outdoor irrigation results of the test specimen F 1g B: heavy metal and trace element concentrations, part 1, Table S42: Outdoor irrigation results of the test specimen

F 1g B: heavy metal and trace element concentrations, part 2, Table S43: Outdoor irrigation results of the test specimen F 4g A: leachate amount, pH value, electrical conductivity, and concentrations of sodium, potassium, calcium, chloride, and sulfate, Table S44: Outdoor irrigation results of the test specimen F 4g A: heavy metal and trace element concentrations, part 1, Table S45: Outdoor irrigation results of the test specimen F 4g A: heavy metal and trace element concentrations, part 2, Table S46: Outdoor irrigation results of the test specimen F 4g B: leachate amount, pH value, electrical conductivity, and concentrations of sodium, potassium, calcium, chloride, and sulfate, Table S47: Outdoor irrigation results of the test specimen F 4g B: heavy metal and trace element concentrations, part 1, Table S48: Outdoor irrigation results of the test specimen F 4g B: heavy metal and trace element concentrations, part 2.

Author Contributions: Conceptualization, methodology, validation, formal analysis, investigation, data curation, visualization, writing—original draft preparation: L.W.; writing—review and editing, funding acquisition, supervision A.V. All authors have read and agreed to the published version of the manuscript.

Funding: This research was funded by the German Federal Ministry for Education and Research (BMBF), grant number 03ZZ0328B. The Author L. Weiler was partially funded by the Ministry for Culture and Research of the state of North Rhine-Westphalia (MKW) under the funding scheme "Forschungskollegs NRW".

Acknowledgments: The authors would like to thank the VDZ gGmbH, especially G. Spanka, for generating and providing the DSLT data, Hentschke Bau GmbH, F. Jesse, for developing and providing the test specimen and pictures, and the Geographical Institute of the RWTH Aachen University, C. Schneider, G. Ketzler, for providing the weather data.

Conflicts of Interest: The authors declare no conflict of interest. The funders had no role in the design of the study; in the collection, analyses, or interpretation of data; in the writing of the manuscript, or in the decision to publish the results.

References

1.	Lupsea, M.; Tiruta-Barna, L.; Schiopu, N. Leaching of hazardous substances from a composite construction product—An experimental and modelling approach for fibre-cement sheets. *J. Hazard. Mater.* **2014**, *264*, 236–245. [CrossRef] [PubMed]

2.	Schiopu, N.; Jayr, E.; Méhu, J.; Barna, L.; Moszkowicz, P. Horizontal environmental assessment of building products in relation to the construction products directive (CPD). *Waste Manag. (N. Y.)* **2007**, *27*, 1436–1443. [CrossRef]

3.	Wicke, D.; Matzinger, A.; Rouault, P. *Relevanz Organischer Spurenstoffe im Regenwasserabfluss Berlins—OgRe. (Relevance of Organic Trace Substances in the Rainwater Discharge of Berlin)*; Kompetenzzentrum Wasser Berlin GmbH: Berlin, Germany, 2017.

4.	Persson, D.; Kucera, V. Release of Metals from Buildings, Constructions and Products during Atmospheric Exposure in Stockholm. *Water Air Soil Pollut. Focus* **2001**, *1*, 133–150. [CrossRef]

5.	Björklund, K. Substance flow analyses of phthalates and nonylphenols in stormwater. *Water Sci. Technol. A J. Int. Assoc. Water Pollut. Res.* **2010**, *62*, 1154–1160. [CrossRef]

6.	European Parliament. *Directive 2013/39/EU of the European Parliament and of the Council on Environmental Quality Standards in the Field of Water Policy*; European Parliament: Strasbourg, France, 2013.

7.	Clara, M.; Ertl, T.; Giselbrecht, G.; Gruber, G.; Hofer, T.F.; Humer, F.; Kretschmer, F.; Kolla, L.; Scheffknecht, C.; Weiß, S.; et al. *SCHTURM—Spurenstoffemissionen Aus Siedlungsgebieten Und Von Verkehrsflächen: Studie Im Auftrag Des Bundesministeriums Für Land- Und Forstwirtschaft, Umwelt Und Wasserwirtschaft. (Trace Substances Emissions from Residential and Traffic Areas)*; BMLFUW: Wien, Austria, 2014.

8.	Gasperi, J.; Sebastian, C.; Ruban, V.; Delamain, M.; Percot, S.; Wiest, L.; Mirande, C.; Caupos, E.; Demare, D.; Kessoo, M.D.K.; et al. Micropollutants in urban stormwater: Occurrence, concentrations, and atmospheric contributions for a wide range of contaminants in three French catchments. *Environ. Sci. Pollut. Res. Int.* **2014**, *21*, 5267–5281. [CrossRef]

9.	Vollpracht, A.; Brameshuber, W. Investigations on the leaching behaviour of irrigated construction elements. *Environ. Sci. Pollut. Res. Int.* **2010**, *17*, 1177–1182. [CrossRef]

10.	CEN European Committee for Standardization. *EN 16105 Paints and Varnishes—Laboratory Method for Determination of Release of Substances from Coatings in Intermittent Contact with Water*; CEN European Committee for Standardization: Brussels, Belgium, 2011.

11.	Hecht, M. Quellstärke ausgewählter Betone in Kontakt mit Sickerwasser. (Source Strength of Selected Concretes in Contact with Leachates). *Beton Und Stahlbetonbau* **2005**, *100*, 85–88. [CrossRef]

12. Burkhardt, M.; Junghans, M.; Zuleeg, S.; Boller, M.; Schoknecht, U.; Lamani, X.; Bester, K.; Vonbank, R.; Simmler, H. Biozide in Gebäudefassaden—Ökotoxikologische Effekte, Auswaschung und Belastungsabschätzung für Gewässer. (Biocides in facades of buildings—Ecotoxicologic effects, wash off and pollution assessment for water bodies). *Environ. Sci. Eur.* **2009**, *21*, 36–47. [CrossRef]

13. Schoknecht, U.; Gruycheva, J.; Mathies, H.; Bergmann, H.; Burkhardt, M. Leaching of biocides used in façade coatings under laboratory test conditions. *Environ. Sci. Technol.* **2009**, *43*, 9321–9328. [CrossRef]

14. Bruntsch, S.; Nebel, H. Untersuchung Des Auslaugverhaltens Von Metallischen Werkstoffen. Bachelor's Thesis, RWTH Aachen University, Aachen, 2011.

15. Nebel, H.; Vollpracht, A.; Brameshuber, W. Auslaugverhalten von Putzen und Mörteln. (Leaching behaviour of renderings and mortars). *Mauerwerk* **2012**, *16*, 2–9. [CrossRef]

16. Schoknecht, U.; Töpfer, A.; Uhlig, S.; Baldauf, H. *Characterisation of Leaching of Biocidal Active Substances of Main Group 2 'Preservatives' from Different Materials under Weathering Conditions*; UBA: Dessau-Roßlau, Germany, 2012; Project No. (FKZ) 3708 65 404, Report No. (UBA-FB) 001641/E), UBA-Texte No. 6.

17. Nebel, H.; Scherer, C.; Schwitalla, C.; Schwerd, R.; Vollpracht, A.; Brameshuber, W. Leaching Behaviour of Renderings and Mortars. In Proceedings of the Eighth International Masonry Conference, MASONRY 11, Dresden, Germany, 4–7 July 2010; Jäger, W., Haseltine, B., Fried, A., Eds.; International Masonry Society: Dresden, Germany, 2010; pp. 1801–1808.

18. Rougelot, T.; Burlion, N.; Bernard, D.; Skoczylas, F. About microcracking due to leaching in cementitious composites: X-ray microtomography description and numerical Approach. *Cem. Concr. Res.* **2010**, *40*, 271–283. [CrossRef]

19. Garrabrants, A.; Sanchez, F.; Gervais, C.; Moszkowicz, P.; Kosson, D. The effect of storage in an inert atmosphere on the release of inorganic constituents during intermittent wetting of a cement-based material. *J. Hazard. Mater.* **2002**, *91*, 159–185. [CrossRef]

20. Hendriks, I.C.F.; Raad, J.S. Report—Principles and background of the building materials decree in the Netherlands. *Mat. Struct.* **1997**, *30*, 3–10. [CrossRef]

21. Scherer, C. *Umwelteigenschaften Mineralischer Werkmörtel. (Environmental Properties of Mineral Mortars) Forschungsergebnisse Aus Der Bauphysik 12*; Fraunhofer Verlag: Stuttgart, Germany, 2013.

22. van Gerven, T.; van Baelen, D.; Dutré, V.; Vandecasteele, C. Influence of carbonation and carbonation methods on leaching of metals from mortars. *Cem. Concr. Res.* **2004**, *34*, 149–156. [CrossRef]

23. Vega-Garcia, P.; Schwerd, R.; Scherer, C.; Schwitalla, C.; Johann, S.; Rommel, S.H.; Helmreich, B. Influence of façade orientation on the leaching of biocides from building façades covered with mortars and plasters. *Sci. Total Environ.* **2020**, *734*, 139465. [CrossRef] [PubMed]

24. Schiopu, N.; Jayr, E.; Méhu, J.; Moszkowicz, P. Environmental impact of construction products: Leaching Behaviour During Service Life. In Proceedings of the 2008 World Sustainable Building Conference, World SB08 Melbourne, Melbourne Convention Centre, Melbourne, Australia, 21–25 September 2008; Foliente, G., Lützkendorf, T., Newtron, P., Paevere, P., Eds.; ASN Events: Balnarring, Australia, 2008.

25. Schneider, K.; Butler, M.; Mechtcherine, V. Carbon Concrete Composites C 3—Nachhaltige Bindemittel und Betone für die Zukunft. (Carbon Concrete Composites C 3—Sustainable binders and concretes for the future). *Beton und Stahlbetonbau* **2017**, *112*, 784–794. [CrossRef]

26. DIN e.V. *Construction Products—Assessment of Release of Dangerous Substances—Part. 2: Horizontal Dynamic Surface Leaching Test*; German Version; Beuth Verlag: Berlin, Germany, 2014; (DIN CEN/TS 16637-2).

27. Weiler, L.; Vollpracht, A.; Jesse, F.; Müller, C.; Reiners, J.; Spanka, G. Umweltverträglichkeit von Carbonbeton: Ergebnisse eines Forschungsvorhabens. (Environmental Compatibility of Carbon Reinforced Concrete—Results of a Research Project). *Beton* **2020**, *70*, 166–175.

28. VDZ e.V. *Tätigkeitsbericht 2003–2005: VI Umweltverträglichkeit Von Zement Und Beton. (Activity Report of the Association of German Cement Factories for the Years of 2003 to 2005))*; Verlag Bau+Technik GmbH: Düsseldorf, Germany, 2005.

29. Spanka, G. *DSLT Analysedaten Email [Excel Dataset]*; Düsseldorf, Germany, 2018.

30. Jesse, F. *Fertigungsstand Mit Frischbetonkennwerten, Email [PDF]*; Bautzen, Germany, 2018.

31. DIN e.V. *DIN EN 14429:2015, Characterization of Waste –Leaching Behaviour Test—Influence of PH on Leaching with Initial Acid/Base Addition German Version*; Beuth: Berlin, Germany, 2015; (DIN EN 14429:2015).

32. Baumgartner, A.; Liebscher, H.-J. *Lehrbuch Der Hydrologie: Quantitative Hydrologie. (Textbook of Hydrology—Quantitative Hydrology)*; Borntraeger: Stuttgart, Germany, 1996.

33. Marshall, J.S.; Palmer, W.M.K. The distribution of raindrops with size. *J. Meteor.* **1948**, *5*, 165–166. [CrossRef]

34. German Meteorological Service. Wetterlexikon Des DWD: Weather Encyclopaedia of the German Meteorological Service. Available online: https://www.dwd.de/DE/service/lexikon/Functions/glossar.html?lv2=101812&lv3=101906 (accessed on 1 October 2020).

35. Weiler, L.; Vollpracht, A. Environmental Compatibility of Carbon Reinforced Concrete: Irrigated Construction Elements. *KEM* **2019**, *809*, 314–319. [CrossRef]

36. Schneider, C.; Ketzler, G. *Climate Data Logger Aachen-Hörn—Data of Annual Report 2018*; Datensatz (Dataset); RWTH Aachen University: Aachen, Germany, 2018; ISSN of summarized version 1861-3993.

37. Schneider, C.; Ketzler, G. *Climate Data Logger Aachen-Hörn—Data of Annual Report 2019*; Datensatz (Dataset); RWTH Aachen University: Aachen, Germany, 2019; ISSN of summarized version 1861-3993.

38. ISO International Organization for Standardization. *Hygrothermal Performance of Buildings—Calculation and Presentation of Climatic Data—Part. 3: Calculation of a Driving Rain Index for Vertical Surfaces from Hourly Wind and Rain Data*; ISO International Organization for Standardization: Genf, Switzerland, 2009; (ISO 15927-3:2009).

39. Davis, J.R. *Stainless Steels, ASM Specialty Handbook*; ASM International: Materials Park, OH, USA, 2010.

40. Vollpracht, A.; Brameshuber, W. Binding and leaching of trace elements in Portland cement pastes. *Cem. Concr. Res.* **2016**, *79*, 76–92. [CrossRef]

41. Bund-/Länderarbeitsgemeinschaft Wasser (LAWA); Ministerium für Umwelt; Klima und Energiewirtschaft Baden-Württemberg. *Ableitung Von Geringfügigkeitsschwellenwerten Für Das Grundwasser: Aktualisierte und überarbeitete Fassung 2016*; Bund-/Länderarbeitsgemeinschaft Wasser (LAWA): Stuttgart, Germany, 2017.

42. Wenzel, A.; Schlich, K.; Shemotyuk, L.; Nendza, M. *Revision der Umweltqualitätsnormen der Bundes-Oberflächengewässerverordnung: Nach Ende der Übergangsfrist für Richtlinie 2006/11/EG und Fortschreibung der europäischen Umweltqualitätsziele für prioritäre Stoffe. (Revision of the Environmental Quality Standards of the Federal German Surface Water Body Directive)*; UBA: Berlin, Germany, 2014; Volume 2015.

43. Brameshuber, W.; Vollpracht, A. Umweltverträglichkeit von mineralischen Baustoffen. (Environmental compatibility of mineral building materials). *Mauerwerk* **2009**, *13*, 190–194. [CrossRef]

44. CONE SCHIOPU, N. Caracterisation des Emssions dans l'eau des Produits de Construction Pendant Leur vie en Oeuvre. Ph.D. Thesis, L'Institut National des Sciences Appliquées de Lyon, Lyon, France, 2007.

45. Dijkstra, J.J.; van der Sloot, H. *Dossier of Information Justifying That Concrete Qualifies for the Potential Release of Regulated Dangerous Substances without-Further-Testing (WFT) by the Producer)*; ECN: Petten, Netherlands, 2013.

46. Bandow, N.; Aitken, M.D.; Geburtig, A.; Kalbe, U.; Piechotta, C.; Schoknecht, U.; Simon, F.-G.; Stephan, I. Using Environmental Simulations to Test the Release of Hazardous Substances from Polymer-Based Products: Are Realism and Pragmatism Mutually Exclusive Objectives? *Materials* **2020**, *13*, 2709. [CrossRef] [PubMed]

 *materials*

Article

Leaching of Carbon Reinforced Concrete—Part 2: Discussion of Evaluation Concepts and Modelling

Lia Weiler *, Anya Vollpracht and Thomas Matschei

Institute of Building Materials Research, RWTH Aachen University, 52062 Aachen, Germany; vollpracht@ibac.rwth-aachen.de (A.V.); matschei@ibac.rwth-aachen.de (T.M.)
* Correspondence: weiler@ibac.rwth-aachen.de; Tel.: +49-241-809-5134

Received: 9 October 2020; Accepted: 30 October 2020; Published: 3 November 2020

Abstract: Possible threats on the environment and human health by the leaching of new building materials and composites in contact to water should be prevented from the outset. It is therefore necessary to assess and ensure their environmental compatibility. For irrigated construction elements this is a challenging task, as there is no general correlation between known testing methods and outdoor emissions. A feasible assessment concept is needed for these conditions. In this work the German assessment method for permanently wet building materials is applied on different carbon reinforced concrete (C^3) leaching data. Furthermore, emission prediction approaches of the Dutch building Materials Decree and the software COMLEAM are tested. The established methods are not yet suitable to determine the complex long term outdoor emissions of irrigated C^3. In order to achieve realistic results in time saving testing methods and to define reasonable release limits, it is necessary to determine and verify the relevant influencing parameters on leaching through intermittent water contact. This research works out leaching patterns and correlations between inorganic substances. It is shown that the input parameters time of exposure, contact time, air temperature, air humidity, runoff and background concentration should be considered to predict the leaching processes from irrigated concrete phenomenologically.

Keywords: leaching; irrigated construction elements; environmental compatibility; irrigated building materials; environmental assessment; evaluation concepts

1. Introduction

Following the construction products regulation (CPR), directive No 305/2011 of the European Parliament, building products must not harm the user or the environment throughout their whole life cycle [1]. The directive has to be executed directly, but member states can enact additional regulations. This theoretically leads to high requirements concerning the environmental compatibility of building products.

Substances that are harmful to the environment can be emitted during processing, service life or recycling and disposal of building products. Emissions occur in the form of particles or by outgassing into the air and by leaching into the soil and/or ground- and surface water bodies.

To protect these compartments and also human health, but also to secure future recyclability and therefore resource efficiency, contaminants in building products should remain in low ranges. However, a maximum allowable content is regulated only for a few potentially harmful substances. Furthermore, the total content is often not applicable to assess possible emissions. It is well known that material composition and leaching conditions rather than the total substances content determine the release of substances from different materials to the environment, e.g., [2–7].

Even established building materials are not to be neglected for further research on their environmental behavior. Due to the continual developing states of knowledge, precision in analytics, and legal regulations, as for example the EU POP- [8] and REACH regulations [9], commonly used products can turn out to be hazardous to human health and the environment [2,10]. Well-known cases are the carcinogenic asbestos and, more recent, the flame retardant Hexabromocyclododecane (HBCD), bearing PBT characteristics (persistent, bio accumulative, toxic) [11].

To tackle this issue, the European Commission is, based on the CPR, required to release harmonized testing standards and assessment methods. This is partially realized for some cases by EN 16516 for Volatile Organic Compounds [12], CEN/TS 16637, Part 1 to 3, for the leaching of hazardous substances from building products [13], or EN 16105:2011 for the leaching of paints and varnishes [14]. The assessment methods of the harmonized testing standards are desired to be harmonized, too, but are to date implemented nationally so that different European countries, as for example the Netherlands or Germany, developed different regulations for the release of defined substances from building products.

In Germany, an assessment concept for the leaching behavior of construction products and materials from the "Centre of Competence for Construction" (DIBt) is used in the context of technical approvals for new building materials [15]. The concept earmarks leaching tests for different materials in different application scenarios. Monolithic building materials in permanent contact with water are tested with the so-called Dynamic Surface Leaching Test (DSLT) regulated by the harmonized European technical specification CEN/TS 16637-2 [13]. The released amount of relevant substances is then compared to specific limits [16], derived from threshold values for groundwater of the German Working Group on water issues of the Federal States and the Federal Government represented by the Federal Environment Ministry (LAWA) [17].

The concept applies for construction elements in direct contact to soil. Irrigated construction elements, as for example roofs or façades, are not considered in the DIBt concept so far [18,19]; though the relevance of runoff emissions is, inter alia, shown by Wicke et al. [20], Gasperi et al. [21], Clara et al. [22] and Scherer [6] and the issue has been discussed by an expert group of the DIBt and different studies [18,23].

Emissions from intermittently wetted construction elements are difficult to predict and therefore also to assess as these materials experience a permanent wet–dry stress, which causes a deviating leaching behavior. Dry phases may lead to faster capillary transport and therefore an increased availability for leaching or wash-off in the following rain [24]. Also increased or lower release depending on the respective substance, chosen point in time, and reference value as, e.g., contact time or amount of water applied [18,25], and changing release patterns compared to permanent wet components [6,25] can be observed.

However, the Netherlands assess irrigated construction elements by using the DSLT and a transfer function that considers the reduced water contact time with a factor of 0.1 [26–28]. This method might not cover the worst case leaching conditions and therefore underestimates the actual emissions. Hecht and Schoknecht et al. for example showed that not only the duration of water contact, but also the transport during drying phases determines the emissions [24,29], the DSLT does not achieve realistic exposure conditions [30]. Still it is desirable to find or develop a horizontal and therefore universally applicable and easily adaptable concept for the assessment of the environmental compatibility of irrigated construction components. This concept could be used to recognize possible threats from the outset, and ensure a sustainable application of new building materials and composites.

In order to achieve realistic results in time saving testing methods and to define reasonable release limits, it is necessary to determine and verify the relevant influencing parameters on leaching through intermittent water contact and other relevant environmental factors on the respective material and therefore also to create a wider database [5,25,29].

In part 1 of this study [25] the leaching behavior of carbon textile reinforced concrete (C^3) was investigated under two established laboratory tests: DSLT and pH-dependent leaching, an artificial indoor irrigation, and under outdoor conditions. The respective eluate concentrations were measured and investigated on their environmental relevance with respect to currently allowed threshold values for similar cases and leachate data of previous studies on mineral building materials. The material

C^3 showed a low leaching in all cases and was found to be environmentally compatible. However, outdoor concentrations often exceeded values measured in the laboratory tests and show a significantly different release pattern which leaves the question whether current assessment concepts are applicable for intermittently wetted construction products. This question becomes more relevant in case of more critical emissions.

In order to evaluate this problem, this work examines different approaches on their suitability and adaptability on the emission prediction of C^3. This way an attempt was made to calculate the long-time emissions of the examined material, and to relate the DSLT leaching data to the actual irrigation data. As not many concepts regarding the emission prediction of irrigated concrete exist, the conventional German method for building components in permanent water contact [15,16], the approach of the Netherlands, applying to intermittent wetted cementitious materials [27], and the modelling program COMLEAM [31,32], designed for the release of organic substances from construction products, are tested.

Subsequently, as for the evaluation or development of an appropriate assessment method, the leaching mechanisms and especially the emission determining factors should be known to provide for a short term test or a transfer model meeting real conditions, a comprehensive data analysis is conducted.

2. Methods

2.1. Experimental Setup

To determine the leaching behavior of C^3 a DSLT according to [13], a laboratory irrigation, and outdoor exposure were conducted on test specimens with different sizes, conditions and concrete covering. Samples were collected from the respective runoff or leachates and analyzed on several heavy metal and trace elements concentrations.

The detailed experimental setups and methods used for data acquisition are described in [25], the measured data are to be found in the electronical annex of [25].

2.2. Cumulative Release

Following [13] and based on the determined concentrations of [25], the emissions and the cumulated release were calculated using Equations (1) and (2). For the outdoor experiments the concentrations of the blind test were subtracted from the concentrations of the eluates from the test specimens to compensate for the background concentrations in the rainwater and potential contamination from deposited particles.

$$r_i = \frac{V_i}{A} \cdot c_i \tag{1}$$

$$R_n = \sum_{i=1}^{n} r_i \tag{2}$$

where r_i = release during the interval i in mg/m², V_i = volume of eluate applied in interval i in L, A = surface of the sample in m², c_i = concentration of element in the eluate of interval i in mg/L, R_n = cumulative release including the intervals 1 to n in mg/m².

2.3. Contact Time

The contact time for the laboratory experiments was determined by the duration of the test (in case of the DSLT) respectively the time span of each irrigation period. For outdoor exposure the weather data from the nearby weather station "Aachen Hörn" [33,34], positioned on a roof within 800 m distance from the testing site, with 10 min resolution were used. All intervals with recorded precipitation were counted and assumed as continuously wet. This approach was chosen based on the high data resolution, which allows minor deviations per increment. Moreover, the assumption was made that the error of semi-dry intervals accounted for as totally wet is averaging with the error of test specimens remaining wet for a certain period of time after actual precipitation.

2.4. Outlier Identification

The complex interaction of the influencing factors resulted in a wide data distribution, exemplarily pictured in Figure 1 for potassium and chromium. It was therefore difficult to define criteria for a statistical outlier model.

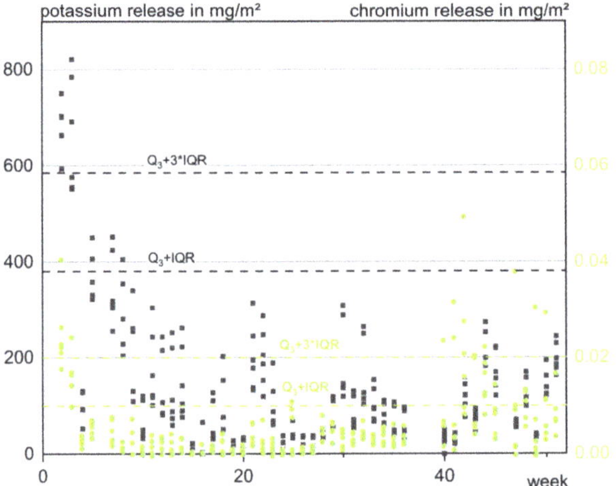

Figure 1. Scatter plot of the incremental releases of potassium (black) and chromium (green) over the testing period of 1a.

The usually robust outlier test build on a multiple of the interquartile range (IQR) [35] identified the data of practically all first rain events and also plausible values after drying periods as outliers when using Equation (3).

$$Q_1 - 3 \times IQR > x > Q_3 + 3 \times IQR \tag{3}$$

where Q_1 = first quartile, 25% of the data; Q_3 = third quartile, 75% of the data; and IQR = interquartile range, defined as $Q_3 - Q_1$.

As a consequence the obvious outliers (e.g., an incremental molybdenum release of 2 mg/m² in comparison to an average of 0.037 mg/m²) were sorted out manually by assessment of the incremental release values (calculated after 2.2). The release instead of the concentrations was used to eliminate the impact of background concentrations and the runoff amount on the measured values. Besides the absolute value of a specific incremental release, the environmental factors and replicate samples were considered to estimate the plausibility of the value. In conclusion 1.02% of the data points were screened out or, if applicable, replaced by the value of the replicate. Replacement was favored before discarding because missing values lead to constant instead of increasing release. In situations of heavy rainfall even one missing value may cause improbable release developments and high deviations in the final cumulative release.

2.5. Transfer Functions and Modelling

2.5.1. Approach of the Soil Quality Decree

The Dutch soil quality decree defines a transfer function (see Equation (4)) to predict the long term leaching behavior by using data obtained by the DSLT. Building components are categorized in two categories: A and B of which B is for irrigated, partially wet components and defined as wetted during 10% of exposure time.

$$I_{soil} = E_{material} = E_{64d} \cdot f_{ext-V}(h,\ x\%,\ D_e) \cdot f_{temp} \tag{4}$$

where I_{soil} = immission into the ground in one year, respectively 100 years in mg/m^2, $E_{material}$ = emission from the building component in one year respectively 100 years in mg/m^2, E_{64d} = result of the DSLT after 64 days in mg/m^2, f_{ext-V} (h, x%, D_e) = extrapolation factor from 64 days to years, f_{temp} = temperature correction factor from laboratory to outdoor.

The soil quality decree sets the factor f_{temp} to 0.7. The factor f_{ext-V} is calculated considering the thickness, the wetting time and the diffusion coefficient of common building materials. It was agreed on using f_{ext-V} = 5 in order to calculate the 100 year cumulated emission and f_{ext-V} = 0.8 to determine the one year cumulated emission of irrigated construction elements [27].

This function was used to predict the 1 year and 100 year cumulated emissions based on the data obtained from the DSLT.

2.5.2. Modelling with the Software COMLEAM

The non-commercial software COMLEAM (version 2.0) [31], developed and provided by the HSR, University of Applied Sciences Rapperswil (Rapperswil-Jona, Switzerland), and financed by the German Environment Agency (UBA), is a tool to assess the leaching of organic substances from building components exposed to wind and rainfall on a macroscopic scale. As the model has been successfully used for the assessment of the aquatic risk by potential harmful substances [32,36], e.g., biocides and organic additives, the software is tested for its suitability for inorganic elements using the leaching data of this work. It has to be considered, that most likely processes on a microscopic scale determine the release of the investigated elements but since the processes are induced by external factors, an adaption might be possible.

The software allows defining building geometries, weather data, and surface materials to calculate the runoff and resulting emissions from buildings by using customized emission functions. Moreover, the concentration course resulting from the modelled emissions can be calculated for different environmental compartments, e.g., surface water classes. To describe the runoff emission correlation, the different functions have to be provided with coefficients derived from experimental studies.

Input Data

The software uses Equation (5), taken from DIN EN ISO 15927:2009-08, to calculate the runoff from provided weather data [32],

$$r_{SR} = \alpha \cdot r^{0.88} \cdot w \cdot \cos(\gamma) \tag{5}$$

where r_{SR} = wind driven rain in mm, α = location factor (dimensionless), r = amount of precipitation in mm, w = wind speed in m/s, γ = angle between building exposition and wind direction in °.

Since the total amount of rain that hit the laboratory test specimens was directly measured, no wind driven rain was calculated.

For outdoor simulations the hourly averages for wind direction, wind speed and precipitation of the actual weather data of the weather station Aachen Hörn [33,34] and the runoff coefficient for uncoated concrete of 85% were used. Because of the exposed position with low obstruction possibilities and the comparably small test specimens size, the location factor α was set to 1.

Regarding the geometry data COMLEAM does only distinguish between facades (90°) and horizontal components (0°) to calculate the amount of wind driven rain [31]. To test the sensitivity and take the 45° ground angle of the test specimens into account, additional calculations were made with the two imaginary building components pictured schematically in Figure 2. The surface for normal rain and the surface for wind driven rain were defined as two parts of one building in the geometries section. The geometry data used for the simulations are summarized in Table 1.

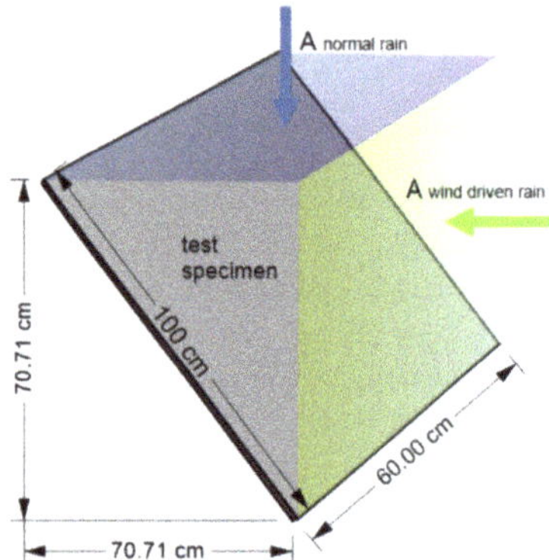

Figure 2. Scheme of the fictitious building surface used for wind driven rain calculations.

Table 1. Geometry Data used for COMLEAM.

Scenario	Object ID	Building ID	Facade					Mineral 1 (Concrete) ID
			Width [m]	Height [m]	Area [m²]	Exposition [°]	Ground Angle [°]	
Laboratory	1	1	0.30	0.40	0.12		45	
Outdoor	1	1	0.60	1.00	0.60	270	45	106
Outdoor-split	1	1	0.60	0.71	0.42		90	
	2	1	0.71	0.60	0.42		0	

Emission Functions

The runoff to emission correlation in COMLEAM can be described by five different functions, which were chosen on the basis of fitting existing mathematical descriptions with the prerequisite of a constantly decreasing slope. These functions are not necessarily describing the real physical processes. They are called: "Logarithmic function", "emission function for limited growth", "Langmuir emission function and Michaelis–Menten emission function", "double logarithmic emission function", and "diffusion controlled emission function" [32]. These functions were adjusted and parameterized to match the emission processes of irrigated building components. Another option to describe the runoff to emission correlation in COMLEAM is, to implement a dataset of measured runoff and corresponding cumulated emissions.

The logarithmic emission function (Equation (6)) turned out to describe the emission course of organics from irrigated construction elements in the most appropriate way [32]. Since inorganic trace element emissions from concrete with permanent water contact are mainly solubility and diffusion controlled, both functions and a measured dataset are tested in this work to model the laboratory irrigation experiment. Equation (7) shows the function used for diffusion controlled release in COMLEAM. It becomes apparent, that this modified function does not describe a real diffusion controlled process, as it considers only the square root of the amount of water applied, but no time factor, which is a basic parameter of diffusion.

$$E_{cum} = a \cdot \ln\left(1 + b \cdot q_{c,cum}\right) \tag{6}$$

$$E_{cum} = k \cdot \sqrt{q_{c,cum}} \tag{7}$$

where E_{cum} = cumulated release in mg/m², $q_{c,cum}$ = cumulated runoff in L, a = proportional factor "characteristic substance percentage" (dimensionless), b = proportional factor (not defined) in m²/L and k = "diffusion coefficient" in m/ $\sqrt{L}$.

The following input parameters were used for this work; the parameters listed in Table 2 were derived from the laboratory irrigation data by regression with the least square method.

Table 2. Emission function input parameters for COMLEAM derived from the laboratory irrigation data.

Element	Logarithmic Function Parameters		Diffusion Parameter
	a	b in m²/L	k in m/ √L
Ba	0.271	0.0194	0.0299
V	0.261	0.0109	0.0208

For the implementation of full datasets the runoff–emission correlations from the DSLT, the laboratory irrigation and the outdoor data were used.

2.6. Spearman Correlation

To evaluate the correlation between the weather data and the observed substance emissions as well as the mutual correlations between the substances, nonparametric correlation estimations on monotonic relationship after Charles Spearman [37] were conducted using the software SPSS version 25.0 and Origin 2019b. Due to the lack of normal distribution and the wide data range a parametric test (e.g., Pearson) was no option.

The Spearman correlation coefficient (r_s) is based on the ranked values for each variable and thereby considering only the order instead of the total value of the raw data. The coefficient r_s is basically calculated after Equation (8), ties are considered in an extended equation by the number of their incidence [38].

$$r_s = 1 - \frac{6 \cdot \sum_{i=1}^{n} (r_i - s_i)^2}{n^3 - n} \tag{8}$$

where r_s = Spearman rank coefficient, r_i = rank of variable X of data pair i; s_i rank of variable Y of data pair i; and n = number of data pairs.

The test is considered as robust against wide ranges and outliers [39,40]. Nevertheless, the impact of the first months of exposure (see also Section 2.3) on the correlation factor r_s was tested by running a second calculation leaving the first four weeks apart. An improved Spearman correlation value (r_s) for 13% of the data was observed but 49% were downgraded. The influence on strong correlations was expectedly low so the full data set was used for further assessment.

To describe the strength of the correlation factor r_s a rough classification was done in reference to Kendall [38] and Cohen [41]. In this work r_s is referred to as weak for $|r_s| \leq 0.25$; moderate for $0.25 < |r_s| \leq 0.50$; strong for $0.50 < |r_s| \leq 0.75$ and very strong for $0.75 < |r_s| \leq 1.00$. The *p*-value was also calculated and the significance level set to $p = 0.05$.

To recognize relationships other than monotonic (e.g., parabolic), scatterplots were created additionally and inspected on their course.

2.7. Multiple Regression

To calculate linear multiple regressions Minitab® 19 Statistical Software was used. Key assumptions for this kind of regressions are:

- A linear relationship between input parameters and the outcome variable;
- No multicollinearity of input variables (The software excludes strongly correlating variables by regression of one predictor on another one. Moreover, collinear input parameters were partially excluded by knowledge based selection in advance.);
- and homoscedasticity of residuals, ratable by the software's residual plots.

The interactions between the twelve possibly determining, partially correlating, variables:

- Time of exposure in days (t_{ex});
- contact time in hours (t_{con});
- air temperature in °C (T);
- normal rain in mm (NR);
- wind driven rain in mm (WDR);
- total rain in mm (TR);
- rain intensity in mm/h (I);
- runoff in L/m^2 (runoff);
- wind speed in m/s (v);
- wind direction in ° (α);
- air humidity in % (RH); and
- rain water pH/background concentration.

As well as their respective contribution to the emission value were examined. To take different slopes into account also transformed data (e.g., logarithmized or to the power of −1) were used.

Depending on their integrity, $n = 302$ to 334 datasets were used to fit a function using stepwise backward elimination. In doing so, quadratic equations and terms with twofold interactions between the parameters were allowed. The elimination method starts calculating with all potential terms in the model and removes the least significant terms. The α value for removal was set to 0.1 in the first step. Terms with $p < 0.05$ or with contributions of lower than 0.01% were directly removed from the models as well. The quality of the models is rated by the distribution of residuals and the models R^2, adjusted R^2 and predicted R^2 (see Table A1). If not stated otherwise the adjusted R^2 is used in the results and discussions section.

3. Results

3.1. Assessment of Cumulated Release of C^3 Using the Concepts for Permanent Water Contact

In [26] the eluate concentrations of irrigated C^3 were evaluated and found to be uncritical. For a long term assessment, also the cumulative total release has to be considered. The conventional assessment methods for building materials in contact to water in Germany and in the Netherlands are therefore applied to the results of the DSLT and for comparison to the other tests conducted.

Table 3 shows the cumulative releases compared to the threshold values after [16] and to the threshold values for soil and groundwater protection of the Netherlands [42], which both apply to the DSLT results. As the German values are based on the assumption of a direct release into the groundwater and refer to the emission from the material whereas the Dutch consider soil retention and are set as immissions values, the Dutch thresholds are less rigorous and probably more suitable for irrigated construction elements.

Table 3. Maximum cumulative release of substances after 64 days (Dynamic Surface Leaching Test (DSLT)), 28 days (laboratory irrigation) and 365 days (outdoor exposure) of testing in comparison to groundwater protection values after [16] and [42], (green: <1% of German threshold, blue: >15% of German threshold).

Substance	DSLT [mg/m^2]	Laboratory [mg/m^2]	Outdoor [mg/m^2]	Threshold D [mg/m^2]	Threshold NL [mg/m^2]
SO$_4$$^{2-}$	1170	393	233	264,495	165,000
Sb	1.34	0.049	-	5.5	8.7
Ba	17.5	0.869	−0.27	375	1500
Cr	1.16	0.371	0.53	7.7	120
Cu	0.209	0.271	−0.10	15.4	98
Mo	0.324	0.083	0.10	38.6	144
Ni	0.204	0.816	3.64	15.4	81
V	2.18	0.750	1.43	4.4 [1]	320
Zn	1.23	0.829	−3.61	63.9	800

[1] Currently suspended.

It becomes apparent that firstly, most substances except for nickel and arsenic show lower releases during the irrigation cases compared to permanent water contact and secondly, even the German thresholds are not reached by any substance in any test. The element closest to the threshold would be vanadium with a release of 2.18 mg/m^2 in the DSLT and a threshold of 4.4 mg/m^2; however this threshold is currently suspended. Next would be antimony with 1.34 mg/m^2 released compared to a threshold of 5.5 mg/m^2.

A tendency of the 4-layer, cracked surface specimens towards a higher release can be assumed for sodium, potassium, arsenic, copper, molybdenum, selenium, and vanadium from the outdoor results (see also Figure A1). Nevertheless this is not a significant difference and might also be a result of the multiply cracked specimens and thus an extended surface. This can be confirmed by the lab results from intact surfaces and also the differences in the total release of most substances from the specimens F 1gA with 2–3 cracks and F 1gB with only one crack. An influence from the carbon reinforcement and its SBR coating on the release of heavy metals and trace elements is therefore improbable.

Also when looking at the median releases of cementitious materials collected in an in-house database from the Institute for Building Materials Research (134 DSLTs) and comparing it to the average releases determined in this study (see Figure 3), it is revealed, that the overall cumulative release of all substances, except for antimony and molybdenum, is lower than the median release observed in the previous DSLTs. Since the C^3 consists of a fine grained concrete with a low water binder ratio and a dense matrix, this is an expected effect. It also shows once more that very likely no matrix-reinforcement interactions are influencing the leaching of heavy metals.

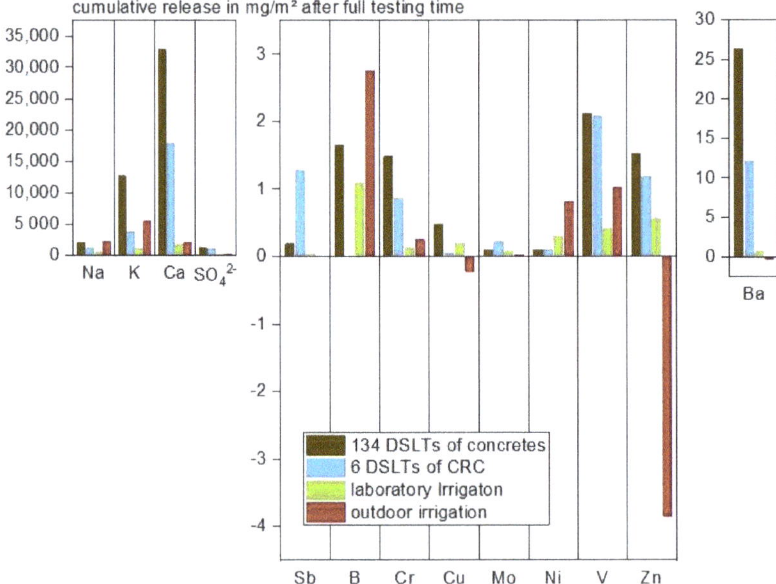

Figure 3. Cumulative release of selected substances from concretes after 64 days of DSLT, 28 days under laboratory irrigation and 365 days of outdoor exposure; comparison of the median of 134 DSLT data sets to C^3.

The direct comparison of the cumulative releases to the threshold values and to similar materials allows, in accordance with the findings of [25], a positive evaluation of the emission behavior of C^3 in terms of leaching.

However, it has to be mentioned that the applied assessment method and therefore the threshold values can only be seen as a benchmark. As the concepts are designed for the case of direct contact between concrete and groundwater, including an immediate dilution, lower allowable emission values may have to be applied on irrigated construction elements. In order to define new limit values the point of compliance has to be agreed on first. If the leachate is considered to infiltrate in a soil compartment and the assessment takes place in a certain soil depth or even in the groundwater, interactions of the leachate with the soil and dilution with pure rainwater or groundwater can be considered as diminishing factors. In that case lower thresholds than the ones of [16] can be expected. These aspects are not focused on further in this paper as it only deals with the prediction of the source term.

3.2. Transfer and Modelling

Since no direct correlation between the laboratory results and the outdoor leaching behavior can be determined, subsequent transfer options are investigated on their suitability. The results of the approaches specified in 2.5 are summarized in the following. Due to their consistent but distinct leaching behavior, the elements vanadium and barium were chosen exemplarily for the discussions; results are illustrated using the test specimens and rain intensities of L 1–2 A respectively F 1 A.

3.2.1. COMLEAM

Laboratory Irrigation

Using the software COMLEAM with the input data described in chapter 2.5.2, it was possible to reproduce the experimental irrigation data. Consistent with [32], the logarithmic function performed best (see also Figure 6) even compared to the original dataset. Therefore, further modelling was done using this function. The modelled effect on environmental compartments, in which the rainwater run-off infiltrates, is not further analyzed in this paper.

The software was not applicable to use the compiled laboratory data of this work for further prognosis and assessment. The DSLT was developed to match diffusion controlled processes, therefore the same amount of water is applied in different time steps. Since all available functions, even the "diffusion function", are neglecting the factor time, it is not possible to describe the runoff–emission relation well.

To illustrate the problem of the emission data related to contact time or runoff, Figures 4 and 5 show the relation of the averaged cumulated releases of both irrigation investigations of all specimens' types to the DSLT. The evaluation is done at specific contact times (230 and 800 h) and specific water amounts (200 and 400 L/m^2).

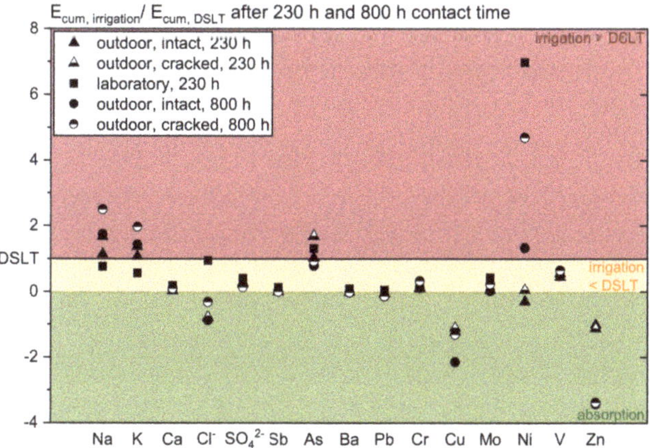

Figure 4. Relation of the averaged cumulated release of the irrigation scenarios related to the DSLT after 230 and 800 h of contact time; 800 h $\hat{=}$ precipitation time during one year.

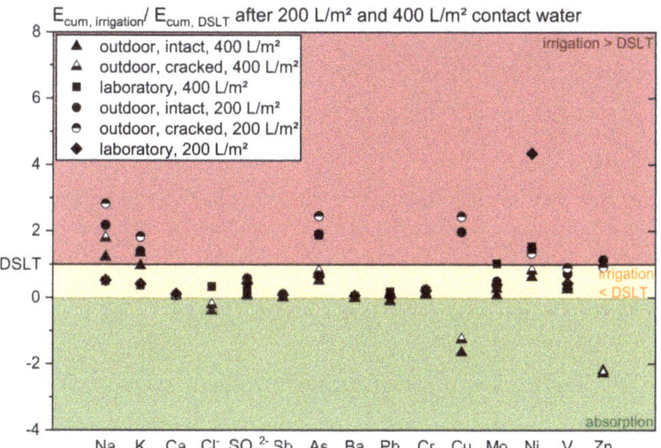

Figure 5. Relation of the averaged cumulated release of the irrigation scenarios related to the DSLT after a runoff of 200 and 400 L/m^2.

The different progresses of the leaching and the significance of the influencing factors become visible. In many cases (for Na, SO$_4^{2-}$, As, Cu, Mo, V, and Zn) the emissions at the considered contact times show a slightly better agreement than at the two cumulated amounts of water applied. However, a systematic relation between the tests cannot be observed. Depending on the substance the ratios vary widely. It is clearly revealed that the amount of water and the contact time are not the only main factors determining the release. Hence, the complexity of concrete leaching cannot be described in one function.

Figure 6 illustrates this problem. The input function derived from experimental laboratory irrigation data of four weeks is compared to further experimental data. The second irrigation sequence was carried out six month after the end of the first irrigation period; during the storage time the test specimens were exposed to the laboratory environment (20 ± 4 °C, 60 ± 15% RH). It is visible, for barium more strongly than for vanadium, that the logarithmic function based on the first two sequences describes the first four weeks of irrigation well but cannot be used as a prognosis for the last two weeks. Even at low concentrations, as measured in this work, and at the controlled ambient conditions in the laboratory, longer phases in which the sample dries show an unneglectable influence. Since the leaching behavior changes over time, due to an altering microstructure and decreasing pH due to carbonation, it might be helpful to include other functions derived from either experimental data and/or geochemical modelling in this simulation, in case that a precise modelling of inorganic substances is targeted.

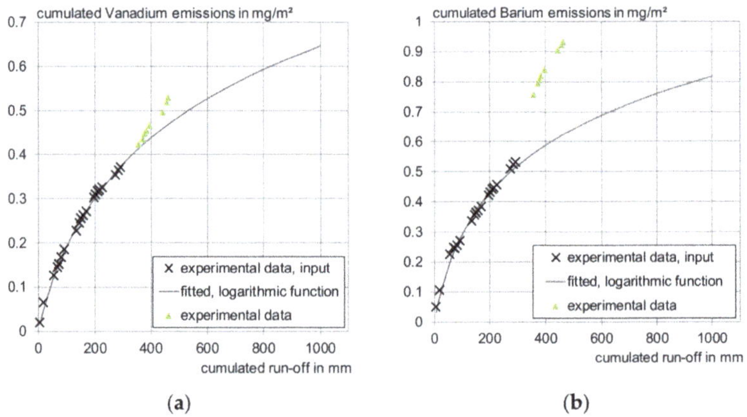

Figure 6. Fitted emission functions compared to experimental data: (**a**) Vanadium, (**b**) barium.

Field Tests

The outdoor experiment was modelled for the sample F 1A. The log-function describes a continuously decreasing slope so that it could obviously not picture the process of the outdoor simulation (compare Figures 7 and A1), modelling was therefore conducted using the full dataset of the runoff–emission correlation, the actual weather data, and the runoff coefficient of 85%. In this case, the program calculates the run-off from the weather data and shows the corresponding emissions. As for the laboratory experiments it was possible to picture the actual process. The simulation led to a runoff of 567 L/m^2 (collected: 524 L/m^2) and cumulated vanadium emissions of 0.96 mg/m^2 (measured: 0.9 mg/m^2) after one year.

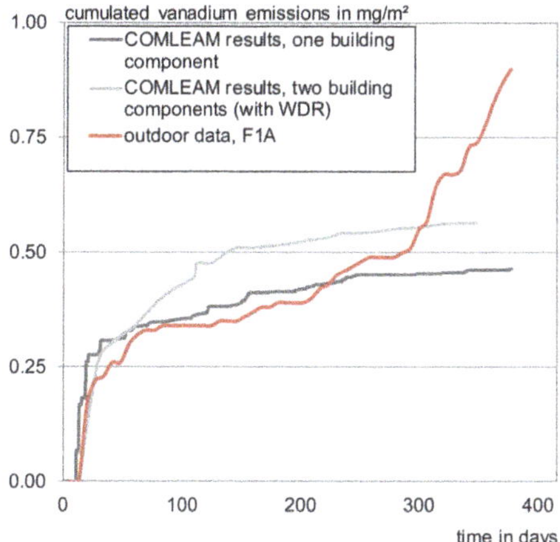

Figure 7. Cumulated emissions of vanadium from F 1 A after one year of exposure; actual release compared to COMLEAM calculations.

Predictions for the even more complex outdoor leaching behavior of concrete cannot be described by the software yet. Figure 7 exemplarily shows the approach of calculating the second half of the testing year for vanadium by implementing the first half as input data and using the same factors as the reproduction simulation. With 0.48 mg/m^2 only approximately half of the amount released in the outdoor experiment is predicted as cumulated release after one year.

3.2.2. Transfer Functions

For constantly wet construction elements the Dutch approach [27] provides formulae to extrapolate cumulative emission data from the DSLT to outdoor exposure after certain time periods. For the case of partially moistened elements [27] it suggests the factors 2.4 (1 year) and 15 (100 years) using Equation (4) described in paragraph 2.5.1. This turns out to just be a factor of 1/3 to calculate from constantly wet to irrigated components.

CEN/TS 16637-2, Annex b.7.4 [13] provides a formula based on the diffusion function to estimate cumulative emissions for permanently wet components on the basis of the DSLT results.

Figure 8 shows the different prognoses for vanadium by using:

- The Dutch approach, Equation (4), for intermittent moistened materials;
- the extrapolation after CEN/TS 16637-2 for permanently wet materials and applying the factor of 1/3 from the Dutch approach on this results;
- the COMLEAM calculations with an input dataset of the DSLT; and
- the COMLEAM calculations with an input logarithmic function derived from lab irrigation.

The calculation results are compared to the actual outdoor release.

Using the DSLT data as an input for the software COMLEAM leads to a very high overestimation. The data extrapolated from the DSLT by the Dutch and the combination of German/Dutch standards can be seen as similar after one year, but the actual emissions are underestimated by a factor of 4.

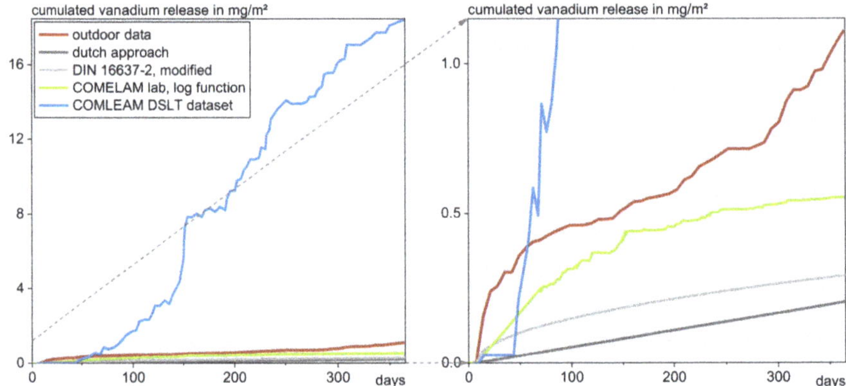

Figure 8. Prognosis of the cumulated vanadium release after 1 year of irrigation, calculated with different approaches.

It is revealed that every calculation method provides a different estimation. Even the same method delivers different tendencies for different substances.

Table 4 shows the prognoses after the Dutch approach [27] leading to extremely inconsistent over- and underestimations.

An assessment concept will have to consider more environmental factors and also distinguish between the substance characteristics.

Table 4. Emission prognosis after the Dutch soil quality decree [27] compared to measured emission values.

Substance	Substance Release After 1a in mg/m^2		Deviation from the Measured Value in %
	Calculated After [28]	Outdoor	
Na	650	2578	−297
K	2086	6393	−207
Ca	14328	2157	85
Cl$^-$	30.4	−163	638
SO$_4$$^{2-}$	669	180	73
As	0.0395	0.108	−174
Ba	9.03	−0.286	103
Pb	0.482	−0.240	150
Cr	0.631	0.288	54
Cu	0.176	−0.188	207
Mo	0.191	0.0573	70
Ni	0.159	0.511	−221
V	1.27	1.11	12
Zn	0.651	−3.99	712

4. Discussion

All known available methods to assess irrigated construction elements are based on laboratory tests and only refer to the factors time and/or amount of water applied. They are not yet applicable to estimate and therefore assess the emissions of irrigated concrete.

4.1. Leaching Patterns in the Field Experiments

Different influencing factors that are not considered in laboratory tests lead to different outdoor leaching patterns. Characteristic release graphs of the substances of the same leaching behavior are grouped and summarized in Table 5. For comparison the actual release curves of sulfate, calcium, barium, chromium, vanadium, and zinc over the testing period of one year are exemplarily shown in Appendix A, Figure A1.

Table 5. Categorization of leaching patterns, outdoor testing compared to the blank.

Substance	Blank (Glass Plate)		Emission from C³	
	Development of Concentrations	Cumulated Amount in mg/m² after One Year	Cumulated Release in mg/m² after One Year	Schematic Release Graph
Ca	Independent from season and weather conditions	220	2157	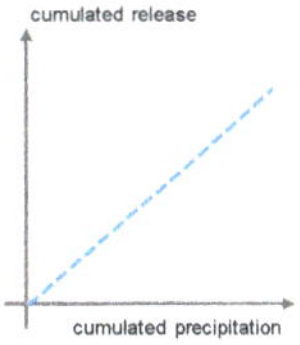 Graph 1 (cumulated release vs. cumulated precipitation)
Cl⁻	Unstable, consistently in the range of eluate concentrations	870	−163	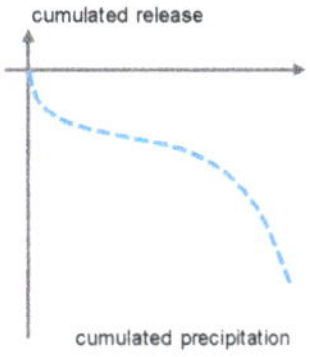 Graph 2 (cumulated release vs. cumulated precipitation)
Ba	Independent from season and weather conditions	1.11	−0.286	
Pb		0.464	−0.240	
Zn	Higher concentrations after dry phases, probably due to particle deposit	6.09	−3.99	
Cu		0.910	−0.188	
SO₄²⁻	Unstable	609	180	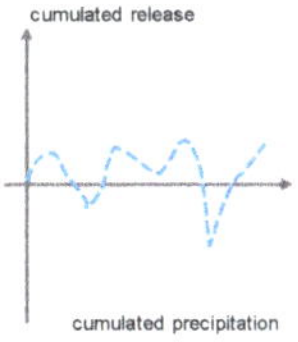 Graph 3 *2 (cumulated release vs. cumulated precipitation)
Sb		0.081	0.000	
Mo *1	Independent from season and weather conditions, but consistently in the range of eluate concentrations	0.227	0.213	
B	Independent from season and weather conditions	1.87	3.21	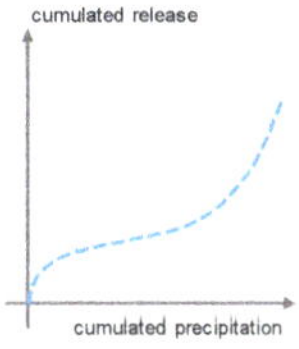 Graph 4a (cumulated release vs. cumulated precipitation)
As		0.071	0.108	
Cr		0.130	0.288	
V	Stable, slight increase from march to august	0.143	1.11	
Na		597	2578	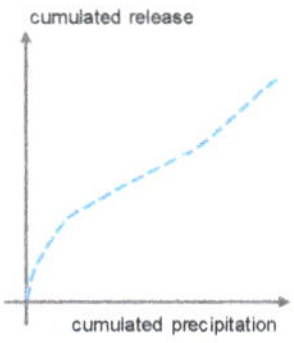 Graph 4b (cumulated release vs. cumulated precipitation)
K	Independent from season and weather conditions	71.6	6393	

*1 Molybdenum shows a tendency to graph 4. *2 Very low emission to blank ratio: High impact of allocation, possible contaminations or analytical errors.

It is noticeable that the oxyanion forming metals form one pattern group (4a) which would point at a pH-dependent leaching or a change in the chemical structure, e.g., decomposition of cement hydrates such as ettringite. As the cations show a similar pattern (4b) it is assumed that external factors, determining the transport mechanisms, are responsible for a part of the emissions.

In [6] Scherer observed similar patterns for the leaching of the elements boron and antimony from renders as found for arsenic, boron, chromium, and vanadium. Contrary to the findings of this work, vanadium and chromium were assorted to a more linear leaching course. However, the slope changes observed for the leaching of C^3 can be, especially in the case of chromium, identified for the mortars of [6] as well (see pp. 104–111 of [6]). The leaching behavior therefore seems reproducible for cementitious materials. In [6] it was assumed, that particle deposition or surface damage by weather, for example hail, led to the increase phases of group 4 emissions. Both can be excluded as dominant factors for this work, as, apart from the mentioned deposits, the blind test showed no irregularities at the respective events, hail was not observed and frost (at around 180–200 mm cumulative runoff) had no significant influence on the subsequent release.

4.2. Influencing Factors on Outdoor Leaching

4.2.1. Spearman Correlation

To calculate the effects on the outdoor leaching behavior, the main effects on the release have to be considered. The data of this work were examined concerning their correlation between weather data and emissions by calculating the Spearman correlation coefficient. The results are presented as a heat map in Figure 9, where strong positive correlations are pictured in dark green and strong negative correlations in dark red. All correlations $|r_s| > 0.3$ are significant on a level of $p = 0.05$. Seven insignificant moderate correlations ($0.25 < |r_s| < 0.3$) were calculated: pH—contact time, pH—chromium release, air temperature—sulfate release, EC—arsenic release, normal rain—boron release, vanadium release—calcium release, and zinc release—arsenic release.

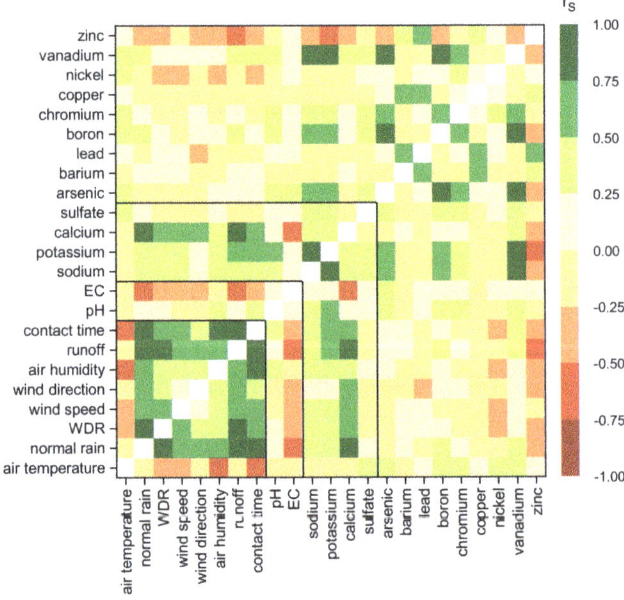

Figure 9. Spearman correlation of weather parameters and substance leaching.

It is revealed that calcium and zinc are the only substances showing direct, strong correlations to single weather parameters. Figure 10 shows that under outdoor conditions calcium leaching can be seen as exclusively dependent on the amount of water applied. The emissions correlate very strong to the amount of normal rain/run-off ($r_s = 0.84$, $p < 0.0001$ and $r_s = 0.88$, $p < 0.0001$) and strong for wind driven rain ($r_s = 0.63$, $p < 0.0001$) respectively weather parameters influencing WDR. Zinc uptake correlates strong with runoff ($r_s = -0.51$, $p < 0.0001$) and moderate with contact time ($rs = -0.43$, $p < 0.0001$).

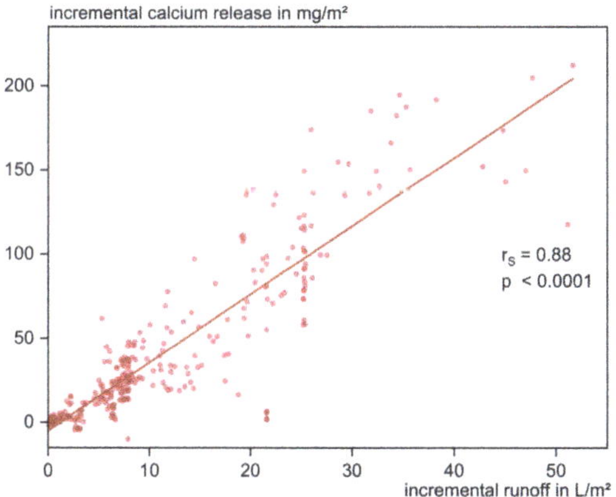

Figure 10. Incremental calcium leaching related to amount of runoff.

Air temperature and air humidity show a strong mutual correlation but different impact on the substances. Mainly diffusion controlled heavy metals are more sensitive to temperature. Some moderate correlations to air temperature but not to humidity were determined for the leached substances sulfate ($r_s = 0.25$), vanadium ($r_s = 0.27$), nickel ($r_s = 0.37$), chromium ($r_s = 0.50$), boron ($r_s = 0.40$), and arsenic ($r_s = 0.33$) while low air humidity seems to accelerate the capillary transport of the more soluble substances like sodium, potassium and calcium independently from temperature.

Contrary to the other heavy metals boron and vanadium additionally correlate moderately ($r_s = 0.30$ and 0.40) to the amount of runoff.

Furthermore, the pattern groups formed and presented in Table 5 are verified. Figure 11 underlines that very strong ($r_S > 0.75$) to strong ($r_S > 0.5$) linear correlations occur between the substances of group 4 with ratios of 10:1 for vanadium to chromium and arsenic, and 1:2 for vanadium to boron. Because of its chemical similarity and the present moderate correlations to sodium ($r_s = 0.35$), potassium ($r_s = 0.31$), arsenic ($r_s = 0.41$), boron ($r_s = 0.50$), chromium ($r_s = 0.34$), and vanadium ($r_s = 0.35$) sulfate is most probably a part of this group as well.

All correlations determined for the grouped substances are most likely not causal and therefore suggesting that the release of the groups is determined by the same parameters. The grouping probably could be used to divide the observed parameters into three to four transfer groups for an assessment concept.

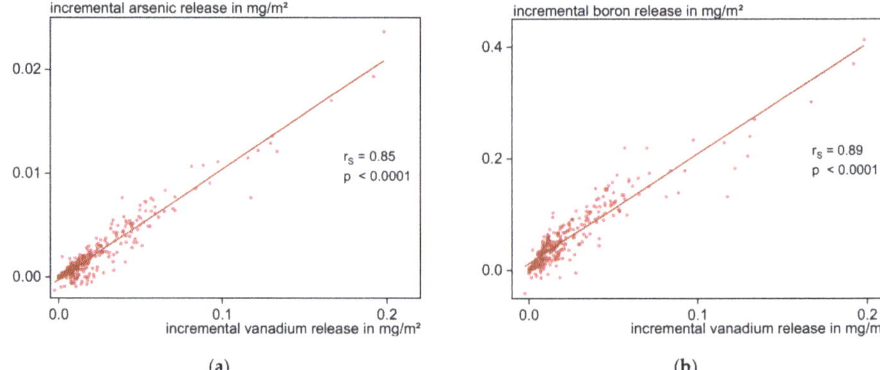

Figure 11. Incremental release of (**a**) arsenic and (**b**) boron in relation to vanadium release (group 4 of Table 5).

4.2.2. Influence of Dry Phases

Since the correlation calculation was not able to straightly prove single influencing parameters on most substances' leaching behavior, the influence of dry phases, which is known to be significant (compare also Figure 6) is examined by qualitatively relating the release curves to weather events.

Figure 12 shows the cumulated release of calcium and the elements of group No. 4 as a percentage of their total release related to contact time and weather. The first change in the leaching process for all substances correlates also to the first drying phase in week 6, suggesting a physical change in the pore structure. Most likely this traces back to an accelerated carbonation as carbon dioxide can enter the concrete better and dissolves in the pore water by a partial drying of water filled pores. This may cause a densification of the cement matrix or at least a covering of the respective, substance incorporating, phases. This theory is supported by the observed pH drop from ~pH 9 to pH 7 ± 0.5, increased calcium leaching, and decreased leaching of sulfate, arsenic, boron, chromium, molybdenum, selenium, and vanadium. Lead is absorbed to a higher extent [25], probably due to elevated background concentration, and might be replacing calcium [43]. Only sodium and potassium as easily soluble constituents are only indirectly influenced by carbonation, due to changes in the pore structure.

The next obvious changes can be observed at week 20 after two and from week 28 after four weeks of low precipitation and therefore phases of probably half wetted concrete. After week 20 calcium, sodium, and potassium stay unaffected while arsenic, boron, chromium, and vanadium releases increase. Since this effect is only visible shortly after the semidry phase, it is attributed to capillary transport. The next semidry phase results in a further increase of the anionic substances pictured in Figure 12. An explanation for this phenomenon would be an advanced carbonation process, leading to a degradation of ettringite and C–S–H and therefore the release of incorporated substances [44].

The total dry phases rise heavy metal leaching but show no effect on sodium and potassium. Calcium is leached but does not seem to be influenced immediately by the drying cycles.

As a conclusion to the simultaneous changes in the leaching process, it can be assumed, that matrix change is the prevalent factor affecting the leaching behavior of many investigated substances, often leading to the characteristic curves No. 2 and 4. It is assumed that heavy changes occur especially after semidry phases as a result of accelerated carbonation and precipitation of dissolved substances and after dry phases as a result of increased capillary transport by the greater humidity gradient.

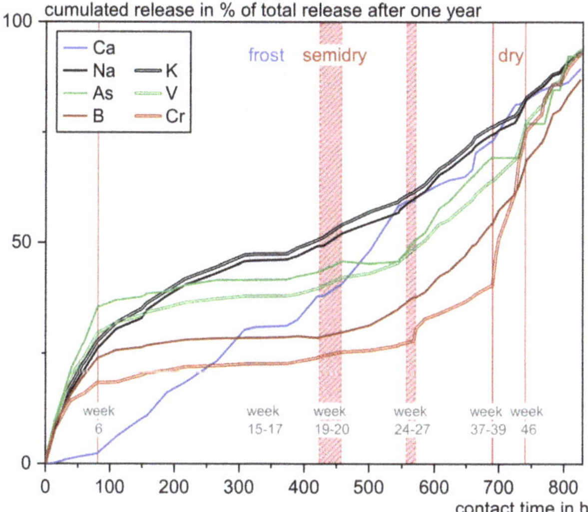

Figure 12. Cumulated release of selected elements as a percentage of their total release related to contact time and weather.

4.2.3. Multiple Regressions

The influence of interactions of the aging matrix and weather conditions on the substance leaching are not to be shown by the spearman correlation and can be assumed but not quantified by interpreting the release curves. A multiple regression was therefore conducted to formulate a model which can describe mutual interactions and quantify the observed emission behavior.

Six input parameters were found to be decisive for most of the examined substances: Time of exposure (t_{exp}), contact time (t_{con}), air temperature (T), air humidity (RH), and runoff; for the leachate pH and zinc uptake additionally the respective background concentrations. Using these factors, it was possible to calculate the incremental releases of all parameters with correlation factors of $R^2 > 0.7$. A consideration of the individual runoffs mostly led to an improvement of ~2%, showing a better adaption to the spread between the respective test specimens. A categorization into the three categories following a dry (Y), semidry (S), and wet phase (N) led to an estimation improvement of 0.1 to maximum 4%; however, at a disproportionate increase of function complexity and partially seeming to result into overfitting.

Figures 13–16 show the fitted data for pH, calcium, vanadium, and zinc. A third to half of the collected data was used to fit a function and predict the remaining part of the test period. Table A2 summarizes the contributions of single terms and therefore parameters to the whole model.

pH

As can be concluded from Figure 13, 82% of the measured leachate pH can be described by terms considering the time of exposure (40%) and the background pH (42%) of the rainwater. Including RH or t_{con}^2 into the function will improve the R^2 by around 2% but does not change the general course. Thus the leachate pH value is mainly resulting from the decreasing materials pH and the background pH of the rain water.

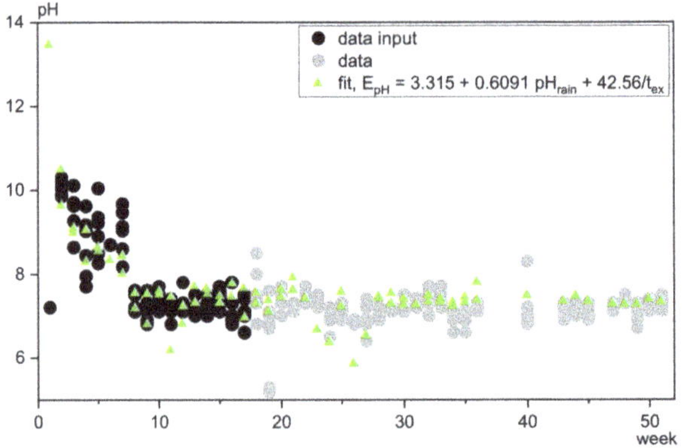

Figure 13. Fit and resulting prognosis of incremental pH values in comparison to original data.

Calcium

The leaching of calcium is primary determined by the amount of water applied. A good approximation can therefore already be achieved, as indicated by the Spearman correlation factor, by calculating the calcium emission either as a function of NR ($R^2 = 0.78$) or runoff ($R^2 = 0.86$). Figure 14 shows the function and fit derived from the first 17 weeks of exposure. Adding a time or temperature factor improves the correlation factor by 1% (t_{ex}) to 2% (T, t_{con}^2) and results into a slightly more uniform distribution of residuals.

It becomes visible that the regression function allows a good prediction of the future emissions. However, it has to be mentioned, that the validity has to be tested for more than one concrete material and that the factors are only applicable for the observed case of C^3. For a long term prognosis t_{ex} will have to be examined for its relevance and contribution again.

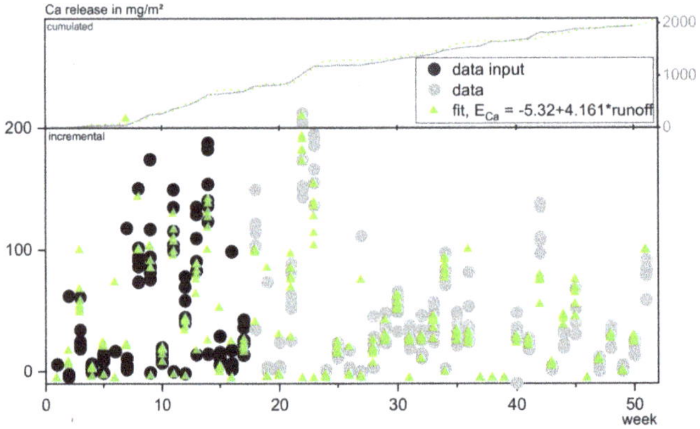

Figure 14. Fit and resulting prognosis of calcium release in comparison to original data.

Group 4, Vanadium

The release of the substances of group 4 can be approximated and predicted by a function derived from half of the data with a combination of the input parameters time of exposure, contact time, air temperature, and air humidity. A factor for the amount of water applied is not immediately considered. As contact time is already covering the amount of rain and runoff to a certain extent, the amount of water shows a minor impact on the release of this substances. Figure 15 exemplarily shows the fit for vanadium. In cases of heavy rainfall (see weeks 6 and 44) the release is underestimated while for low runoffs (weeks 24 and 49) the reduced emissions, probably due to low concentration gradients, cannot be met. However, intensity as an input variable neither was calculated as significant nor led to improved fitting results.

Air temperature and time of exposure are the main influencing parameters, explaining ~70% of the vanadium emissions. This again reinforces the assumption that matrix changes play a major role in the release of group 4 substances, while the diffusion process is subordinate.

As was estimated on the basis of the Spearman correlation results, using a multiple of the vanadium function will depict the main release pattern of arsenic and boron sufficiently, too. Since chromium shows too heavy deflections after the dry period, it could not be modelled well using the available factors.

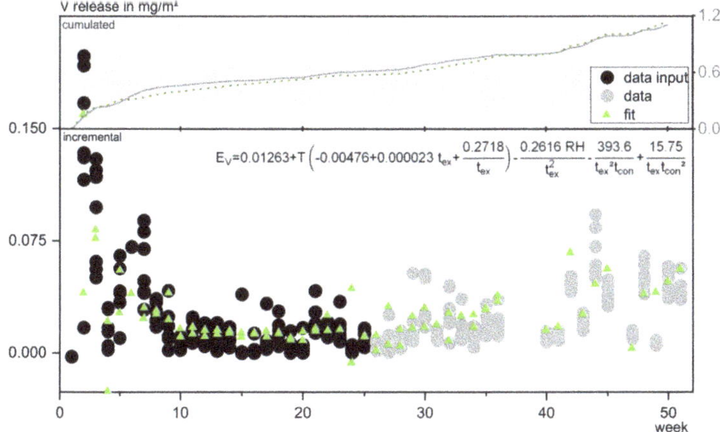

$$E_V = 0.01263 + T\left(-0.00476 + 0.000023\, t_{ex} + \frac{0.2718}{t_{ex}}\right) - \frac{0.2616\, RH}{t_{ex}^2} - \frac{393.6}{t_{ex}^2 t_{con}} + \frac{15.75}{t_{ex} t_{con}^2}$$

Figure 15. Fit and resulting prognosis of vanadium release in comparison to original data.

Zinc

Seventy percent of the zinc uptake can be explained using factors based on the rainwater background concentration (Zn_b) and amount of runoff combined with contact time. Including temperature and RH into the calculation improves the correlation. This is suggesting that zinc uptake in this case is a combination of capillary transport and diffusion into the test specimen, and a temperature dependent chemical process like complexation and incorporation, e.g., as hyroxides or carbonates.

It has to be mentioned, that the previously described, derived functions, especially the coefficients, should thereby not be seen as universal rules and are not describing the underlying physical-chemical process. Pre carbonated samples for instance will certainly show another slope for the exposure time (t_{ex}) influence. However, they clearly show the influence of the mentioned parameters on the leaching behavior. Despite the complex interactions it seems possible to predict the leaching behavior of concrete by using the factors responsible for matrix changes and capillary transport instead of modelling the actual process. Considering the uncertainties of an outdoor experiment, the overall low emissions, and possible contaminations or analytical errors, an adequately precise fit was achieved for the examined indicators.

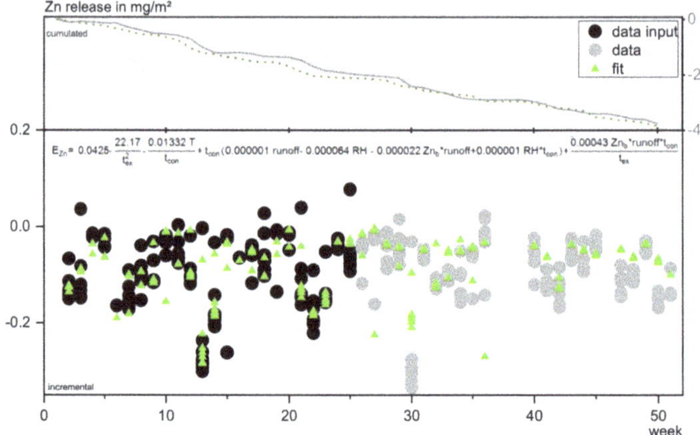

$$E_{Zn} = 0.0425 \cdot \frac{22.17}{t_{Ex}^2} - \frac{0.01332\,T}{t_{con}} + t_{con}(0.000001\,runoff - 0.000064\,RH - 0.000022\,Zn_b \cdot runoff + 0.000001\,RH \cdot t_{con}) + \frac{0.00043\,Zn_b \cdot runoff \cdot t_{con}}{t_{Ex}}$$

Figure 16. Fit and resulting prognosis of zinc release in comparison to original data.

5. Summary and Conclusions

- Investigations on the inorganic leaching behavior of carbon textile reinforced concrete confirmed the findings of Part 1 of this study: No environmentally harmful leaching was observed.

- Different approaches were tested to predict outdoor leaching behavior from laboratory data. Calculation models provided by Dutch and German standards mostly underestimated the total release. The tested modelling software COMLEAM does not suit the difficult case of heavy metal and trace elements leaching from cementitious materials, as it is designed for prediction on a macroscopic scale. A different modelling concept is needed for inorganics released from concrete.

- The investigated elements were divided into four groups, characterized by their respective leaching pattern formed through external factors. The influencing factors were determined using a Spearman correlation calculation, whereby most substances show only moderate correlations to single weather parameters.

- The influence of combined weather conditions was calculated. Considering six external factors is sufficient to describe and predict the leaching processes phenomenologically. The main leaching mechanisms (solution and diffusion) remain important but are significantly superimposed by outdoor influences with different impact on the particular substances.

- More research is necessary to develop a matching concept on transfer functions for irrigated building components. For an improved transferability to other cementitious materials, the underlying physicochemical processes should be identified, e.g., using geochemical modelling. The findings of this work concerning pattern groups and influencing parameters are providing a foundation for further assessment method development and definition of physicochemical relations.

Author Contributions: Conceptualization, methodology, validation, formal analysis, investigation, data curation, visualization, writing—original draft preparation: L.W. Writing—review and editing, funding acquisition, supervision and advice A.V. Supervision: T.M. All authors have read and agreed to the published version of the manuscript.

Funding: The Author L. Weiler was funded by the Ministry for Culture and Research of the state of North Rhine-Westphalia (MKW) under the funding scheme "Forschungskollegs NRW".

Acknowledgments: The authors would like to thank M. Burkhardt and M. Rohr of the TH Rapperswil for providing and supporting with the software COMLEAM (https://www.comleam.ch/de).

Conflicts of Interest: The authors declare no conflict of interest.

Appendix A

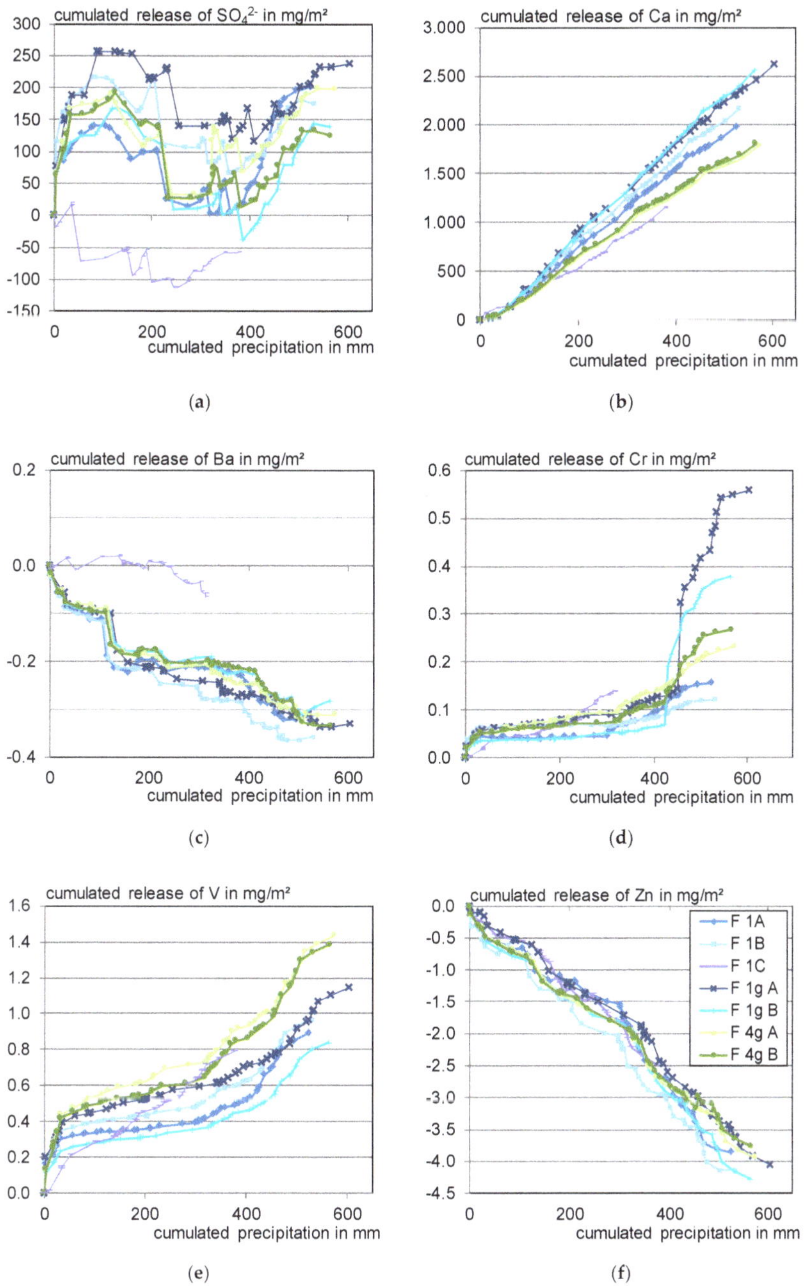

Figure A1. Cumulative release of (**a**) sulfate, (**b**) calcium, (**c**) barium, (**d**) chromium, (**e**) vanadium, and (**f**) zinc. Negative values occur due to adsorption of substances contained in the rain water.

Appendix B

Table A1. Comparison of fit qualities for predictor variations.

Substance	R^2 in %	R^2 adjusted in %	R^2 predicted in %
pH	81.76	81.38	80.30
Ca	85.76	85.71	85.42
V	75.20	74.50	71.30
Zn	71.23	69.90	67.09

Table A2. Excerpt of the multiple regressions variance analyses.

Variable	Contribution in %	p Value
pH regression	81.76	0.000
pH_{rain}	41.95	0.000
$1/t_{ex}$	39.80	0.000
calcium regression	85.76	0.000
runoff	85.76	0.000
zinc regression	71.23	0.000
t_{ex}^{-2}	0.85	0.000
T/t_{con}	14.43	0.002
$RH * t_{con}$	43.86	0.000
$Zn_b * runoff * t_{con}$	4.23	0.000
$runoff^2 * t_{con}$	4.25	0.000
$RH * t_{con}^2$	0.13	0.011
$(Zn_b * runoff * t_{con})/tex$	3.48	0.000
vanadium regression	80.80	0.000
T	2.12	0.000
$t_{ex} * T$	37.73	0.000
T/t_{ex}	30.02	0.000
t_{ex}^2/t_{con}	1.30	0.000
RH/t_{ex}^2	0.93	0.000
$1/t_{ex}^2 t_{con}$	6.44	0.000
$1/t_{ex} t_{con}^2$	2.27	0.000

References

1. European Parliament. *Regulation (EU) No 305/2011 of the European Parliament and of the Council of 9 March 2011 Laying Down Harmonised Conditions for the Marketing of Construction Products and Repealing Council Directive 89/106/EEC (EUV 305/2011)*; European Parliament: Strasbourg, France, 2011.
2. Ilvonen, O.; Dürkop, J.; Horn, W.; Englert, N.; Plehn, W.; Bolland, T.; Däumling, C.; Dorer, C.; Fischer, J.; Kirschbaum, B.; et al. *Umwelt-und Gesundheitsverträgliche Bauprodukte: Ratgeber für Architekten, Bauherren, Planer. (Environmental-and Health Compatible Building Products)*; UBA: Dessau-Roßlau, Germany, 2015.
3. Van der Sloot, H.; Van Zomeren, A.; Stenger, R.; Schneider, M.; Spanka, G.; Stoltenberg-Hansson, E.; Dath, P. *Environmental CRIteria for CEMent—Based Products (ECRICEM) Phase I: Ordinary Portland Cements and Phase II: Blended Cements, Executive Summary (ECV-E-11-020)*; Netherlands Energy Research Foundation: Petten, The Netherlands, 2008.
4. Van der Sloot, H. Characterization of the leaching behaviour of concrete mortars and of cement–stabilized wastes with different waste loading for long term environmental assessment. *Waste Manag.* **2002**, *22*, 181–186. [CrossRef]

5. Schiopu, N.; Jayr, E.; Méhu, J.; Barna, L.; Moszkowicz, P. Horizontal environmental assessment of building products in relation to the construction products directive (CPD). *Waste Manag.* **2007**, *27*, 1436–1443. [CrossRef] [PubMed]

6. Scherer, C. *Umwelteigenschaften Mineralischer Werkmörtel (Environmental Properties of Mineral Mortars)*; Fraunhofer Verlag: Stuttgart, Germany, 2013.

7. Wachtendorf, V.; Kalbe, U.; Krüger, O.; Bandow, N. Influence of weathering on the leaching behaviour of zinc and PAH from synthetic sports surfaces. *Polym. Test.* **2017**, *63*, 621–631. [CrossRef]

8. European Parliament. *REGULATION (EC) No 850/2004 OF THE EUROPEAN PARLIAMENT AND OF THE COUNCIL of 29 April 2004 on Persistent Organic Pollutants and Amending Directive 79/117/EEC: POP*; European Parliament: Strasbourg, France, 2004.

9. European Parliament. *REGULATION (EC) No 1907/2006 OF THE EUROPEAN PARLIAMENT AND OF THE COUNCIL of 18 December 2006 Concerning the Registration, Evaluation, Authorisation and Restriction of Chemicals (REACH), Establishing a European Chemicals Agency, AMENDING Directive 1999/45/EC and Repealing Council Regulation (EEC) No 793/93 and Commission Regulation (EC) no 1488/94 as well as Council Directive 76/769/EEC and Commission Directives 91/155/EEC, 93/67/EEC, 93/105/EC and 2000/21/EC: REACH*; European Parliament: Strasbourg, France, 2006.

10. Neubrand, H. Unerkannte Schadstoffrisiken Bei Vorhandenen und Neuen Baustoffen (Undetected Pollution Risks in Conventional and new Building Materials). In *Aachener Bausachverständigentage 2016*; Eckrich, W., Halstenberg, M., Herold, C., Liebert, G., Mohrmann, M., Moriske, H.-J., Neubrand, H., Oswald, M., Pohlenz, R., Seibel, M., et al., Eds.; Springer Fachmedien Wiesbaden: Wiesbaden, Germany, 2016; pp. 161–167.

11. Koch, C.; Schmidt-Kötters, T.; Rupp, R.; Sures, B. Review of Hexabromocyclododecane (HBCD) with a focus on legislation and recent publications concerning toxicokinetics and—Dynamics. *Environ. Pollut.* **2015**, *199*, 26–34. [CrossRef] [PubMed]

12. DIN German Institute for Standardization. *Construction Products—Assessment of Release of Dangerous Substances—Determination of Emissions into Indoor Air (EN 16516:2017)*; Beuth Verlag GmbH: Berlin, Germany, 2017.

13. Din, E.V. *Construction Products—Assessment of Release of Dangerous Substances—Part 2: Horizontal Dynamic Surface Leaching Test*; German Version (DIN CEN/TS 16637-2); Beuth Verlag GmbH: Berlin, Germany, 2014.

14. CEN European Committee for Standardization. *Paints and Varnishes—Laboratory Method for Determination of Release of Substances from Coatings in Intermittent Contact with Water (EN 16105:2011)*; CEN: Brussels, Belgium, 2011.

15. DIBt. *Grundsätze zur Bewertung der Auswirkungen von Bauprodukten auf Boden und Grundwasser (Principles for the Assessment of the Influence of Building Products on Soil and Groundwater Bodies)*; DIBt: Berlin, Germany, 2011.

16. DIBt. *Muster-Verwaltungsvorschrift Technische Baubestimmungen (MVV TB)*; Deutsches Institut für Bautechnik: Berlin, Germany, 2019.

17. Bund-/Länderarbeitsgemeinschaft Wasser (LAWA), Ministerium für Umwelt, Klima und Energiewirtschaft Baden-Württemberg. *Ableitung von Geringfügigkeitsschwellenwerten für das Grundwasser: Aktualisierte und Überarbeitete Fassung 2016*; LAWA: Stuttgart, Germany, 2017.

18. Vollpracht, A.; Brameshuber, W. Investigations on the leaching behaviour of irrigated construction elements. *Environ. Sci. Pollut. Res. Int.* **2010**, *17*, 1177–1182. [CrossRef] [PubMed]

19. Weiler, L.; Vollpracht, A. Environmental compatibility of carbon reinforced concrete: Irrigated construction elements. *KEM* **2019**, *809*, 314–319. [CrossRef]

20. Wicke, D.; Matzinger, A.; Rouault, P. *Relevanz organischer Spurenstoffe im Regenwasserabfluss Berlins—OgRe (Relevance of Organic Trace Substances in the Rainwater Discharge of Berlin)*; Kompetenzzentrum Wasser Berlin GmbH: Berlin, Germany, 2017.

21. Gasperi, J.; Sebastian, C.; Ruban, V.; Delamain, M.; Percot, S.; Wiest, L.; Mirande, C.; Caupos, E.; Demare, D.; Kessoo, M.D.K.; et al. Micropollutants in urban stormwater: Occurrence, concentrations, and atmospheric contributions for a wide range of contaminants in three French catchments. *Environ. Sci. Pollut. Res. Int.* **2014**, *21*, 5267–5281. [CrossRef] [PubMed]

22. Clara, M.; Ertl, T.; Giselbrecht, G.; Gruber, G.; Hofer, T.F.; Humer, F.; Kretschmer, F.; Kolla, L.; Scheffknecht, C.; Weiß, S.; et al. *SCHTURM—Spurenstoffemissionen aus Siedlungsgebieten und von Verkehrsflächen: Studie im Auftrag des Bundesministeriums für Land- und Forstwirtschaft, Umwelt und Wasserwirtschaft. (Trace Substances Emissions from Residential and Traffic Areas)*; BMLFUW: Wien, Austria, 2014.

23. Dijkstra, J.J.; Van der Sloot, H.A.; Spanka, G.; Thilen, G. *How to Judge Release of Dangerous Substances from Construction Products to Soil and Groundwater: CPD Topic 1. Soil and Groundwater Impact*; ECN-C-05-045; ECN: Petten, The Netherlands, 2005.

24. Hecht, M. Quellstärke ausgewählter Betone in Kontakt mit Sickerwasser (Source Strength of Selected Concretes in Contact with Leachates). *Beton Stahlbetonbau* **2005**, *100*, 85–88. [CrossRef]

25. Weiler, L.; Vollpracht, A. Leaching of Carbon Reinforced Concrete—Part 1: Experimental Investigations. *Materials* **2020**, *13*, 4405. [CrossRef] [PubMed]

26. Eikelboom, R.T.; Ruwiel, E.; Goumans, J. The building materials decree: An example of a Dutch regulation based in the potential impact of materials on the environment. *Waste Manag.* **2001**, *21*, 295–302. [CrossRef]

27. Verschoor, A.J.; Lijzen, J.; van den Broek, H.H.; Cleven, R.; Comans, R.; Dijkstra, J.J.; Vermij, P. *Kritische Emissiewaarden voor Bouwstoffen: Milieuhygiënische Onderbouwing en Consequenties voor Bouwmaterialen*; RIZA-Rapport nr. 2006.029; RIVM: Bilthoven, The Netherlands, 2006.

28. Hendriks, I.C.F.; Raad, J.S. Report—Principles and background of the building materials decree in the Netherlands. *Mat. Struct.* **1997**, *30*, 3–10. [CrossRef]

29. Schoknecht, U.; Gruycheva, J.; Mathies, H.; Bergmann, H.; Burkhardt, M. Leaching of biocides used in façade coatings under laboratory test conditions. *Environ. Sci. Technol.* **2009**, *43*, 9321–9328. [CrossRef] [PubMed]

30. Bandow, N.; Aitken, M.D.; Geburtig, A.; Kalbe, U.; Piechotta, C.; Schoknecht, U.; Simon, F.-G.; Stephan, I. Using environmental simulations to test the release of hazardous substances from polymer-based products: Are realism and pragmatism mutually exclusive objectives? *Materials* **2020**, *13*, 2709. [CrossRef] [PubMed]

31. Burkhardt, M.; Engelke, D.; Gehrig, S.; Hochstrasser, F.; Rohr, M.; Tietje, O. *Introduction and Application of the Software COMLEAM:—Manual, Version 1.0*; HSR University of Applied Sciences: Rapperswil, Switzerland, 2018.

32. Tietje, O.; Burkhardt, M.; Rohr, M.; Borho, N.; Schoknecht, U. *Emissions—und Übertragungsfunktionen für die Modellierung der Auslaugung von Bauprodukten. (Emission and Transfer Functions for the Modelling of the Leaching of Building Materials)*; UBA: Dessau-Roßlau, Germany, 2018.

33. Schneider, C.; Ketzler, G. *Climate Data Logger Aachen-Hörn—Data of Annual Report 2018. Datensatz (Dataset)*; RWTH Aachen University: Aachen, Germany, 2018; ISSN 1861-3993.

34. Schneider, C.; Ketzler, G. *Climate Data Logger Aachen-Hörn—Data of Annual Report 2019. Datensatz (Dataset)*; RWTH Aachen University: Aachen, Germany, 2019; ISSN 1861-3993.

35. Hedderich, J.; Sachs, L. *14. Überarb. und erg. Aufl. Angewandte Statistik: Methodensammlung mit R*; Springer: Berlin/Heidelberg, Germany, 2012.

36. Burkhardt, M.; Klingler, M.; Savi, D.; Rohr, M.; Tietje, O.; Junghans, M. *Entwicklung Einer Emissionsbasierten Bauproduktebewertung—Anwendung des Konzepts für Dachbahnen und Fassadenputze*; BAFU: Basel, Switzerland, 2020.

37. Spearman, C. The proof and measurement of association between two things. *Am. J. Psychol.* **1987**, *100*, 441–471. [CrossRef] [PubMed]

38. Kendall, M.G. *Rank Correlation Methods*, 4th ed.; Oxford University Press: New York, NY, USA, 1975.

39. Shevlyakov, G.L.; Oja, H. *Robust Correlation: Theory and Applications*; Wiley Series in Probability and Statistics; Wiley: Chichester, UK, 2016.

40. Croux, C.; Dehon, C. Influence functions of the Spearman and Kendall correlation measures. *Stat. Methods Appl.* **2010**, *19*, 497–515. [CrossRef]

41. Cohen, J. *Statistical Power Analysis for the Behavioral Sciences*, 2nd ed.; L. Erlbaum Associates: Hillsdale, NJ, USA, 1988.

42. Ministerie van Volkshuisvesting, Ruimtelijke Ordening en Milieubeheer. Regeling Bodemkwaliteit (nr. DJZ2007124397) per 1 January 2008. In *Staatscourant*; Ministerie van Binnenlandse Zaken en Koninkrijksrelaties: Den Haag, The Netherlands, 2007; p. 67.

43. Vollpracht, A.; Brameshuber, W. Binding and leaching of trace elements in Portland cement pastes. *Cem. Concr. Res.* **2016**, *79*, 76–92. [CrossRef]
44. Mulugeta, M.; Engelsen, C.J.; Wibetoe, G.; Lund, W. Charge-based fractionation of oxyanion-forming metals and metalloids leached from recycled concrete aggregates of different degrees of carbonation: A comparison of laboratory and field leaching tests. *Waste Manag.* **2011**, *31*, 253–258. [CrossRef] [PubMed]

materials

Article

Testing of Eluates from Waterproof Building Materials for Potential Environmental Effects Due to the Behavior of *Enchytraeus albidus*

Marya Anne von Wolff [1,2] and Dietmar Stephan [1,*]

[1] Group of Building Materials and Construction Chemistry, Department of Civil Engineering, Technische Universität Berlin, Gustav-Meyer-Allee 25, 13B, 13555 Berlin, Germany; vonwolff@tu-berlin.de
[2] Joint Laboratory of Applied Ecotoxicology, Environmental Safety Group, Korea Institute of Science and Technology Europe (KIST Europe), Stuhlsatzenhausweg 97, 66123 Saarbrücken, Germany
* Correspondence: stephan@tu-berlin.de; Tel.: +49-30-314-72101

Abstract: In order to determine the potential environmental impact of construction products, it is necessary to evaluate their influence on organisms exposed to them or their eluates under environmental conditions. The behavior of the white worm *Enchytraeus albidus* is a useful tool for assessing the potential environmental impact of construction products in contact with water and soil. This study investigates the environmental effects of eluates from two construction products, a reactive waterproofing product, and an injection resin, on the reproduction and avoidance behavior of *E. albidus*. The eluates were prepared according to existing guidelines. The soil used for the tests was moistened with the eluates of the construction products. The reproduction results of the worms were collected after six weeks of exposure. Offsprings were counted under the microscope and statistically analyzed. Results from the avoidance behavior were collected after 48 h of exposure, and results were compared with the reproduction results. The eluates from both construction products induced significant changes in the reproduction behavior of *E. albidus*. Undiluted or only slightly diluted eluates of the injection resin drastically reduced the reproduction of the worms, whereas the leaches of the reactive waterproofing product only had a minor effect. The avoidance results for the injection resin indicates that its presence in the habitat is clearly detrimental to the survival of *E. albidus*, while the avoidance results for the waterproofing resin showed an initial avoidance of the eluates, but no harmful effects were observed. The avoidance test is a way of rapid toxicity screening of environmental samples when time is a critical parameter to measure possible environmental effects. This study shows that ecotoxicological tests using *Enchytraeids* are a valuable and important tool for understanding the mode of action of eluates from construction products in the environment.

Keywords: *Enchytraeids*; waterproof building materials; ecotoxicology; biotest

Citation: von Wolff, M.A.; Stephan, D. Testing of Eluates from Waterproof Building Materials for Potential Environmental Effects Due to the Behavior of *Enchytraeus albidus*. *Materials* **2021**, 14, 294. https://doi.org/10.3390/ma14020294

Received: 16 November 2020
Accepted: 6 January 2021
Published: 8 January 2021

Publisher's Note: MDPI stays neutral with regard to jurisdictional claims in published maps and institutional affiliations.

1. Introduction

The environmental impacts of existing construction materials are important to consider while developing new products. Once the construction products are exposed to weathering, they could potentially be leached, and the resulting eluates could have a negative impact on organisms in the environment [1]. The chemical composition of construction materials and their leaching behavior are crucial for the environmental compatibility of a product [2].

In addition to this prospect, other stages of the building material also play a role in its environmental compatibility. Even during the construction phase, dismantling, recycling, and disposal [3,4], water-soluble substances can be released, especially from fine-grained materials, which have an impact on the environment [5]. As all building materials have a limited life span and weather over time, the assessment of leaching behavior and the evaluation of the environmental relevance of the resulting eluates are of great importance for the certification of building materials in the European market [6].

Dynamic surface leaching test (DSLT), according to CEN/TS 16637-2:2016 [7], is widely applied in Europe to determine the leaching behavior of construction materials and is recommended by the German Environmental Agency (Umweltbundesamt, UBA- Berlin, Germany) for their environmental impact assessment [8]. The DSLT is one of the standard methods to evaluate the release of dangerous substances from building products. The test determines the release of inorganic and non-volatile organic substances through contact with leaching agents per unit area of the construction product under investigation as a function of time. During the leaching process, 8 eluates are generated under specified test conditions, which are then examined for chemical and ecotoxicological parameters. The end of the standard experiment is the 64th day of the experiment, and the release of substances related to the specific surface is determined.

Several studies have investigated the release behavior of inorganic ions such as Na^+, Al^{3+}, Ca^{2+}, Si^{4+}, and Cu^{2+} and organic substances (e.g., superplasticizers) from concrete materials and construction waste [8–11]. However, only a few are taking the results of an ecotoxicological analysis into account [12–14].

Currently, the most used toxicological bioassays for testing products are made using species as *Daphnia sp.*, Algae, and *Danio rerio*. Although those species are very representative for the aquatic environments, the scenarios of construction products are often in the terrestrial environment, turning those species not always the ideal match for the necessary bioassays. Ecological risk assessment from construction materials can be examined by aquatic and terrestrial biomarkers [15], among others, the ecotoxicological tests using the Oligochaeta species of genus *Enchytraeidae*. Although ecotoxicological tests using these species have been developed and standardized only within the last two decades, it became an important indicator organism for the determination of impacts on soil ecosystem due to their sensitivity to a broad spectrum of xenobiotics, ease of maintenance in the laboratory [15], and widely representation worldwide [16–18].

Between the possible ecotoxicological tests available, two are specifically used: the reproduction and the avoidance behaviors. The reproduction test using Oligochaetas is very commonly used, but it takes much more work and is very time-consuming since the complete ecotoxicological test takes up to 6 weeks of work and constant measurements. The avoidance behavior test is much simpler compared to the reproduction test and is also less time-consuming since an avoidance test takes only 48 h to get results. As shown in previous studies, avoidance behavior can be used as a first indication of the occurrence of damage in *Enchytraeids* when exposed to nanoparticles [19], giving a very fast result when compared to the traditional studies using reproduction tests.

However, although faster results can be obtained using test methods as avoidance behavior, the reproduction tests with *Enchytraeids* seem to be the most accurate in looking at the long term exposure scenario [20] and is likewise the best way to associate with real field exposures of chemicals from anthropogenic activities [21].

Tests performed in this study intended to fill the information gap about exposures of Oligochaetes in the presence of eluates from construction product particles. This study aims to verify the possible effects of waterproof building materials in the environment using the reproduction and avoidance behavior as ecotoxicological tools [22] to elucidate the effects of leaches from construction materials prepared with different types of water for the environment.

2. Materials and Methods

2.1. Construction Products

This research selected two examples of construction products that are applied directly in contact with soil or water.

The first construction product used was a waterproof material to protect walls from water ingress, for example. The non-commercial material consisted of 2 components, mixed by hand with a spatula in the mass proportion of 1:1 in a beaker. The resulting paste was placed in silicone forms and cured for 46 days until the product was completely dry.

The second construction product is a very fast-reacting silicate injection resin. The injection resin also consists of 2 components, which have to be mixed within a few seconds. These were filled in two-chamber cartridges. A static mixer was screwed to the outlet of the cartridges, and the cartridge was placed on a compressed air gun. The components react immediately, and the mixture was extruded with the air-gun in silicone molds, where it hardens within a few minutes. Although the material cures very quickly due to the reaction heat, in this study, we waited for 7 days for complete stabilization of the product after the reaction before submerging in the water baths for leaching. Due to a long time for leaching and constantly washing the pieces on the baths, all the eluates were tested without further dilution.

Since the two products cover very different areas in terms of their structural application, the results are not intended to be compared either. At this point, it must be pointed out again that special laboratory formulations with known ingredients were deliberately worked on and not on ecotoxicologically tested and marketable products, so that effects can be achieved in the comparative investigations with a high probability.

2.2. Leaching Method

Leaching of the two construction materials was performed according to CEN/TS 16637-2 [7]. The decision to prepare the eluates with two different types of water was taken to search for the best conditions for survival and thus for the reproduction of the *Enchytraeids*. The eluates were prepared using tap water and distilled water as the medium for leaching the components from the construction products. The pieces of the construction products previously prepared according to item 2.1 were submerged in baths of distilled water or tap water. The dimensions of the container used for leaching was $31 \times 23 \times 9 \text{ cm}^3$ (l × w × h) and has a capacity of 3.5 L. Eluates from both construction products were collected after exposure times of 6 h and 1, 2, 4, 9, 16, 36, and 64 days, and vacuum filtered using a micropore filter of 0.45 µm. All eluates were characterized for pH, electric conductivity, and the following inorganic components: Al^{3+}; Ca^{2+}; Cu^{2+}; K^+; Na^+; Si^{4*}; NO_3^-; SO_4^{2-}; Cl^-. The leaching procedure was performed in triplicates, and results are presented further with the mean values obtained for the parameters.

2.3. Test Organisms

This study used the species *Enchytraeus albidus* as the test organism [22]. Worms were cultured for many years in the ecotoxicological laboratory of the Technische Universität Berlin using a bio garden soil, kept at a controlled temperature of 10 °C and fed at libitum with bio rolled oats, autoclaved and finely grounded. Organisms were cleaned from soil particles and acclimatized before starting the test procedure.

2.4. Soil

Standard soil (LUFA 2.2) was used to perform the avoidance and reproduction tests. The soil was commercially acquired from Landwirtschaftliche Untersuchungsund Forschungsanstalt (LUFA) Speyer, Germany [23]. The characteristics of the soil are: soil type: loamy sand; dry matter of the soil: 94.8 wt.%; water content 5.4 g water/100 g soil; maximum water holding capacity (WHC): 44.8 ± 2.9 wt.%; pH: 5.6 ± 0.4; cation exchange capacity: 9.2 cmol/kg ± 1.4.

2.5. Experimental Design

2.5.1. Reproduction Test

Rounded glass vessels with 100 mL were used for all the reproduction tests. In each vessel, 50 g of soil was placed inside and moistened using the eluate until the maximum water holding capacity of 46 wt.%. As control sets, the same conditions were placed using three different types of water: tap water, distilled water, and reconstituted freshwater. All the control vessels were completely free of leaching from the construction products. For each condition tested, four replicates were placed [21].

Ten adult *Enchytraeids* individuals with a well-developed clitellum were placed in each vessel. All vessels were covered with a lid containing small holes in order to avoid escaping the worms. The worms in each vessel were fed with 0.2 mg of bio rolled oats per week, distributed equally. Vessels were weighed and kept at a controlled temperature of 20 °C, and water content was replaced when evaporation occurred. The experimental plan ran for six weeks. After the first four weeks of exposure, the adult *Enchytraeids* were removed, and vessels were kept at the same conditions in order to wait for the hatching of the cocoons. After additional two weeks, organisms were fixed and colored using the extraction method of staining with Bengal red according to ISO 16387:2014 [21] and counted under the microscope. The results obtained from the tests were compared to the control results.

2.5.2. Avoidance Test

To perform the avoidance tests, once again, rounded glass vessels with 100 mL were used. The tests were performed according to ISO 17512 [19]. The vessels were divided into two sections using a removable wall. On one side, was placed 25 g of LUFA soil 2.2, moistened until the maximum water holding capacity of the soil was reached using the eluate of the construction product. On the opposite side of the vessel, LUFA soil moistened with the same corresponding type of water was placed. That means, if the eluate from the construction product was previously prepared with tap water, the soil on the control side was also moistened with tap water, but without any previous contact with the construction product. This was an attempt to prove the avoidance behavior by the possible components leached and not just by the type of water.

After placing both soils on the vessel, the wall was removed, and 10 *Enchytraieds* per vessel with a well-developed clitellum were introduced in the fine line that divides both soils. The vessels were covered with a lid containing small holes to permit air exchange. Four replicates per treatment were prepared, and vessels were left at a temperature of 20 °C and period light control (16/8—light/dark) for 48 h without food.

After 48 h, a removable wall was again introduced in the division of the soils and both sides were searched individually for the worms. Worms were counted, and results were compared. Figure 1 shows the schematic representation of the avoidance test, with all steps.

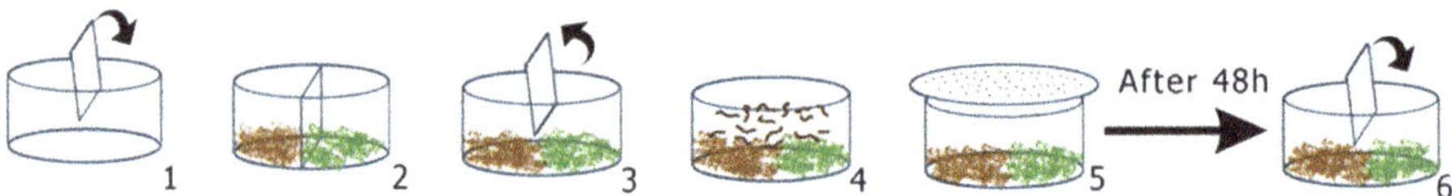

Figure 1. Schematic representation of the experimental procedure of the *Enchytraeid* avoidance test: (**1**) inserting the movable wall into the center of the test vessel; (**2**) introduction of the soils to be tested; (**3**) the movable wall is removed; (**4**) placing the *Enchytraeids* in the middle of the soil; (**5**) covering the test vessel with a lid (perforated); (**6**) reintroduce the wall to separate the floors and count the organisms present on each side.

Negative control was also placed where both sides of the vessel contained the same type of control to evidence the non-avoidance behavior when both sides contain the same component.

3. Results and Discussion

3.1. Eluate Characteristics

The values measured for pH and electric conductivity of the eluates are represented in Figure 2. The chart represents the mean values obtained from the triplicates made for each day of measurement for the following eluates: Silicate resin product distilled water (DW) and silicate resin product tap water (TW) and also for the waterproof product distilled water (DW) and waterproof product tap water (TW).

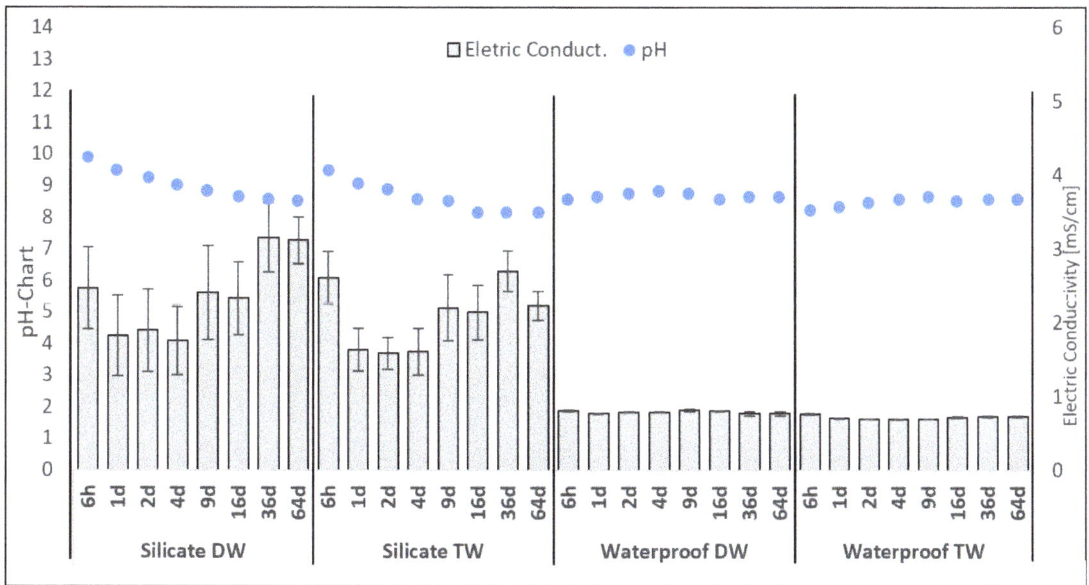

Figure 2. Graphic representation of pH and electric conductivity. Values are measured for all eluates of silicate and waterproof products.

The variation of electric conductivity of the silicate injection resin for DW presented a variation between 1.5 and 3.5 mS/cm. This variation can be explained due to the chemical behavior of the silicate building material when exposed to constant contact with water [24]. The range variation of the electric conductivity of eluates from the waterproof material with TW and DW was very stable, reaching a maximum value of 1 mS/cm during the total time of leaching.

The eluates with TW and DW of the silicate injection resin showed very similar behavior in the variation of the pH, starting with values of pH 10 and slowly stabilizing until pH 8. The pH of the eluates from the waterproof material in TW and DW presented a small variation of results along the 64 days of leaching alternating between pH 8 and 9.

Table 1 presents the mean values of inorganic content analyzed for the eluates of the construction products and the blank samples. All samples were analyzed in triplicates by inductively coupled plasma atomic emission spectroscopy ICP-AES method.

Table 1. Inorganic characterization of the eluates: mean values for leachings and blanks in mg/L.

Parameter	Silicate TW	Silicate DW	Waterproof TW	Waterproof DW	Blank TW	Blank DW
Al^{3+}	0.34	0.14	0.5	0.1	0.01	0.001
Ca^{2+}	50.2	47.6	123.5	53.8	1.6	0.9
Cu^{2+}	0.6	0.5	0.9	0.1	0.1	0.1
K^+	11.0	10.1	32.7	7.9	1.1	0.2
Na^+	1193.6	771.8	232.5	58.6	181.1	9.1
Si^{4+}	53.8	48.6	8.2	6.8	6.2	0.2
$NO_3{}^-$	11.3	9.3	6.4	1.2	5.7	0.3
$SO_4{}^{2-}$	167.5	111.4	136.5	14.6	102.4	3.4
Cl^-	66.8	57.8	64.8	53.7	49.8	2.2

The highest variations in analogy with the correspondent blank sample occurred for the silicate product TW, indicating that the constant baths of the pieces stimulated the increase of components as Al^{3+}; Ca^{2+}, Na^+, Si^{4+}, and NO_3^-. The same occurred for the silicate samples leached with DW; however, the increase was proportional to the initial presence of inorganic components. For the waterproof product in TW and DW, inorganic components as Ca^{2+} and Na^+ were leached out.

The DSLT has already been used in many studies. Brameshuber et al. [25] applied this to the mortar to investigate the release of organic constituents from concrete under practice-relevant conditions, whereby the effects of organic substances in the eluates could be classified as minor. However, contamination with sodium, sulfates, aluminum, and some heavy metals could be proven while using contaminated concrete blocks. As part of the selection of the processes suitable for the ecotoxicological assessment of building materials [13], the DSLT was carried out on 37 representative building products that contained mobilizable organic substances. The point of criticism of the DSLT was that some substances could no longer be identified due to their biodegradability over the duration of the DSLT of 64 d. This argument is also put forward by Bandow et al. [1] because the conversion of organic substances cannot be excluded, especially under real conditions. The transferability of the test results from the DSLT to real environmental and practical conditions also appears problematic. However, in its principles for evaluating the effects of construction products on soil and groundwater [26], the Deutsche Institut für Bautechnik already states how the laboratory results from the horizontal leaching test can be transferred to real conditions through model considerations. Scherer described the DSLT as "the authoritative and recognized test procedure for evaluating the environmental impact" [27]. Due to the frequent use of tests, test variants modified by the DIN standard are already available. Märkl et al. [12] adapted the DSLT for plastic products and used it to leach polyurethane resin during the curing phase. Within the research project, the DSLT was used both for reactive sealing and for a silicate injection resin. As described in DIN CEN/TS16637-2, the DSLT applies to evaluating the surface-dependent release for monolithic, plate-like, and film-like products. The type of construction product determines the implementation conditions and the dimensions of the test specimen. The waterproof product was classified as a plate-like product, while the silicate injection resin was classified as a monolithic product.

3.2. Reproduction Results

In order to determine the best control design for the reproduction behavior and the subsequent comparison of the results with leachings of the construction products, this study collected data of three different scenarios of control sets using tap water, distilled water, and reconstituted freshwater as watering medium of the LUFA soil. The results presented in Figure 3 evidence the highest number of offsprings for the vessels moistened with reconstituted freshwater, while a very similar result was achieved using tap water. Based on these results, which were combined to simulate real construction scenario situations, this study carried out the leaching of the construction components with tap water and distilled water.

After the total exposure time of 6 weeks, the offsprings of *E. albidus* were counted, and the results are shown in Figures 4 and 5. The control represents the mean number of *Enchytraeids* on the water reproduction test, where no eluates of construction products were in contact with this group.

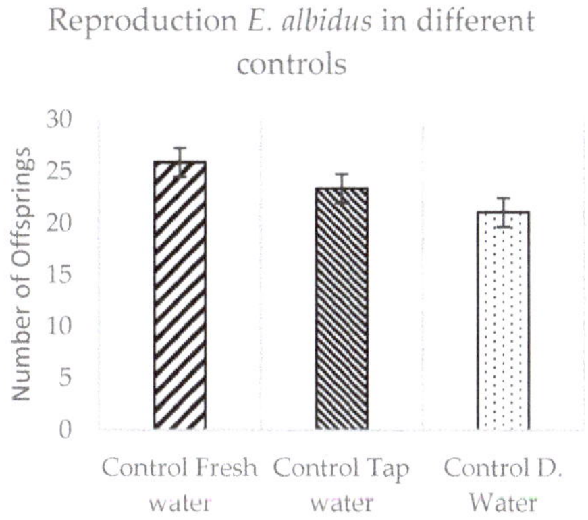

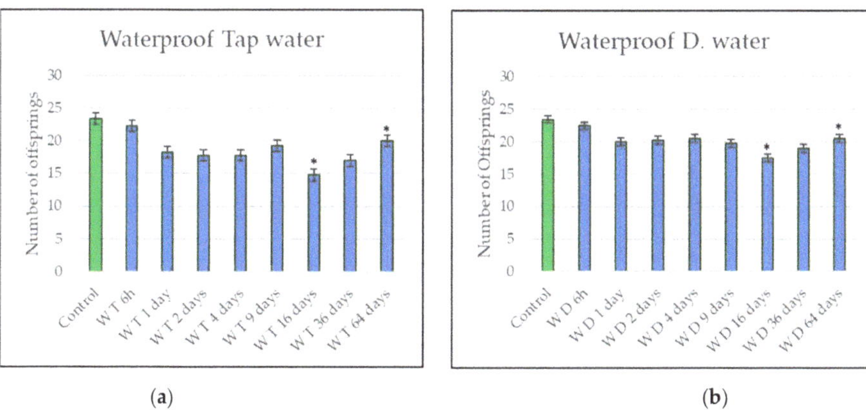

Figure 3. Reproduction behavior of *E. albidus* comparing three different types of water.

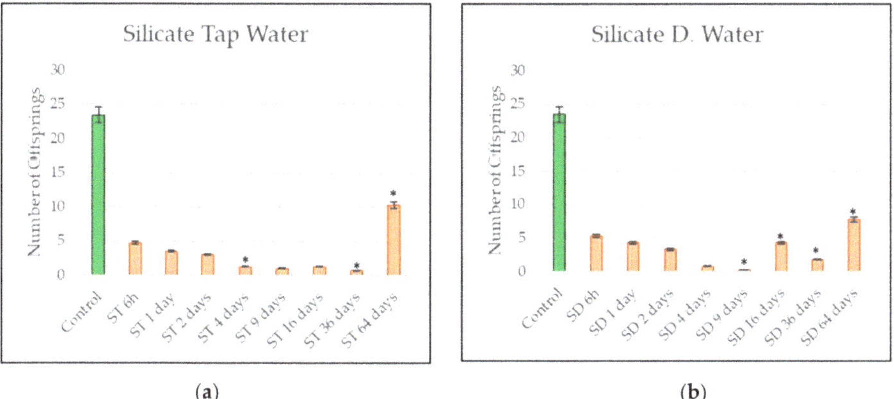

(**a**)   (**b**)

Figure 4. (**a**) *Enchytraeid* reproduction results for eluates from the waterproof material using tap water. (**b**) *Enchytraeid* reproduction results for eluates from the waterproof material using distilled water. Significant results are marked with an (*).

(**a**)   (**b**)

Figure 5. (**a**) *Enchytraeid* reproduction results for the silicate leach using tap water. (**b**) *Enchytraeid* reproduction results for the silicate leach using distilled water. Significant results are marked with an (*).

An analysis of Figure 4a shows a low decrease in *Enchytraeids* reproduction behavior. Nevertheless, after 16 days of leaching, an apparent decrease in the number of offsprings can be seen, with less than 15 worms counted. The same decrease was perceived for the eluates of the waterproof material with DW after 16 days, (Figure 4b), however with a total amount of 17 *Enchytraeids*. The most significant results are emphasized with an (*) point.

The reproduction tests were performed for all the eluates. When observing the reproduction results for both eluates, after 64 days, a stabilization of the reproduction behavior was noticed at the end of the test, indicating a reduction of the toxicity and decrease in the presence of active ingredients from the waterproof material when exposed to water for a longer time [28]. The drop in the number of offsprings was not very alarming for eluates of the waterproof material. However, effects on the *Enchytraeid* populations can be noticed. An alarming situation would be configurated when the component tested affects the reproduction behavior in an exposed concentration that affects fifty percent of the population. The different results over time represent typical leaching behavior. In the beginning, there was a first wash-off followed by reduced diffusion-controlled release.

Both types of eluates (TW and DW) of the waterproof material contained a low variation in the quantities of ions when compared to the corresponding blank samples. Ions of Na^+ and Ca^{2+} increased with the leaching. However, it is already proven that sodium and calcium have a small influence on toxicological effects for *Enchytraeids* [29], and the presence of these minerals on the leachings can help to clarify the reduction of offsprings in the results.

Analyzing the results of the eluates of the silicate resin are plotted in Figure 5a,b. A high impact of the eluates from the silicate product on the reproduction behavior of the *Enchytraeids* can be seen. In all phases of the leaching process, small numbers of offsprings were counted. The silicate eluates prepared with distilled water presented the highest impact on the population, bringing special attention to the reproduction test from day 9 silicate DW, where a mean number close to 0 was achieved. For both reproduction tests of silicate leachings from day 64, the *Enchytraeids* presented a tendency to recover the population of worms, and a higher number of offsprings was counted; however, the number of offsprings is still under fifty percent of the population when compared to the control group. The most significant results are emphasized with an (*) point.

The silicate leaching for DW on day 16 presented a slight recovery of the results compared to the previous reading on the same test. This effect can sometimes occur once the *Enchytraeids* are biological indicators, and the sensibility can change according to the exposure [30]. Few other details were also visible during the reproduction test days, e.g., some of the silicate eluates revealed the presence of a few eggs, but these were too weak to hatch and did not hatch until the end of the test. In other situations, the initially introduced adult worms used died at the beginning of the test or even one week before the end of the test. Besides, the absence of adult worms had a direct effect on the number of offsprings.

Besides the increased concentration of sodium, chlorides, and sulfates in the eluates from the silicate component, it is impossible to confirm with certainty what causes the strong toxicity influence from the eluates from the silicate resin for the *Enchytraeids*. The very small number of offsprings, in one factor, can be explained by the sensitive sensors on the body of the *Enchytraeids*, which perceive limited survival conditions early on and will avoid habitats [31].

3.3. Avoidance Results

The same eluates were tested for the avoidance behavior of the *Enchytraeids*, and results are represented in the following graphics. Figure 6 represents the avoidance behavior of the *Enchytraeids* using the same treatment on both sides (negative control).

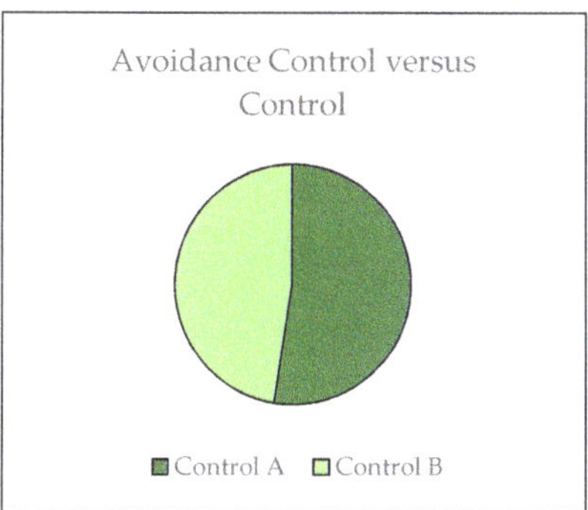

Figure 6. *Enchytraeid* avoidance distribution for both sides with control soil.

Figure 6 shows the graphic proportion of worms that prefer to stay on each side of the soil. On both sides, the same control conditions were used, and it was possible to see that when both sides contain a control soil free of toxicants, the worms do not show one preferred side. The same average distribution was observed for both sides, being one side with 49% of the population and the other side with 51% of the population of *Enchytraeids*. According to the ISO 17512 (2008), the control is considered valid if the proportions stay between 40–60% in the distribution hack.

The avoidance results for the waterproof leachings are plotted in Figure 7a,b. The green line represents the number of *Enchytraeids* that chose the control side, while the red line represents the number of *Enchytraeids* that preferred the eluate side.

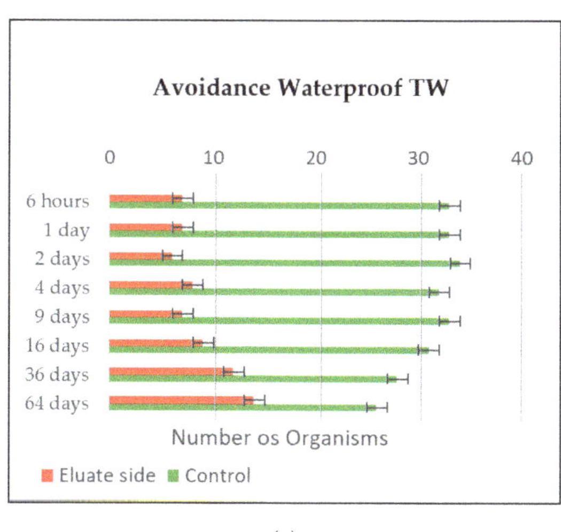

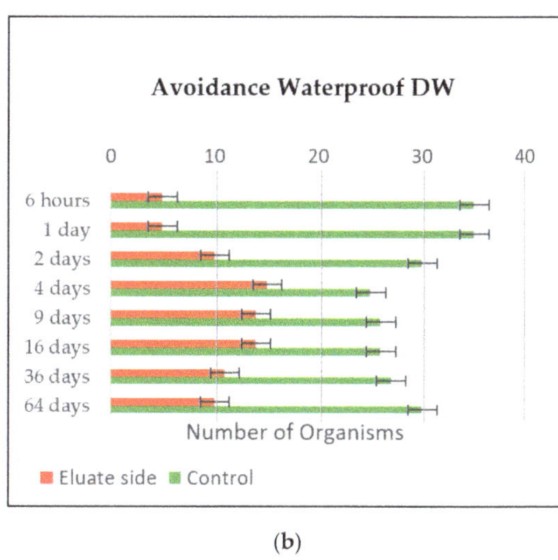

(a)   (b)

Figure 7. (a) *Enchytraeid* avoidance results for the waterproof leach using tap water. **(b)** *Enchytraeid* avoidance results for the waterproof leach using distilled water.

Analyzing the avoidance behavior of the *Enchytraeids* testing the waterproof leaches, it was possible to see that the worms tend to prefer mostly the side of the control, but for few leaches, the difference in the number of worms in each side is small. A less poor scenario in the number of organisms on each side is achieved in the vessels of waterproof TW 64 days. The statistical avoidance percentage was calculated as represented in Table 2.

Table 2. Avoidance results percentages (%) for *E. albidus*, calculated for all eluates of waterproof and silicate.

Time	Waterproof TW	Waterproof DW	Silicate TW	Silicate DW
6 h	65	75	85	85
1 day	65	75	85	85
2 days	70	50	55	70
4 days	60	25	70	50
9 days	65	30	85	80
16 days	55	30	40	40
36 days	40	40	40	45
64 days	30	50	30	50

Figure 8a,b represents the avoidance behavior of the *Enchytraeids* for the eluates of the silicate. The result indicates that one more time, the *Enchytraeids* tend to move mostly to the side of the control soil, where no contaminations of chemicals are detected. The avoidance was similar in both tests, with few variations in numbers for 9 and 64 days.

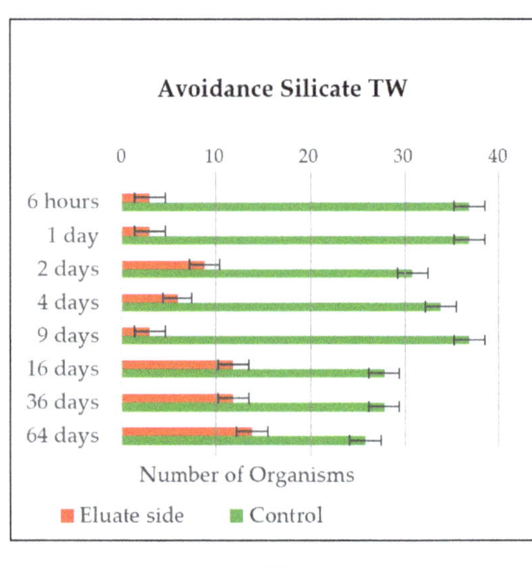

(a)

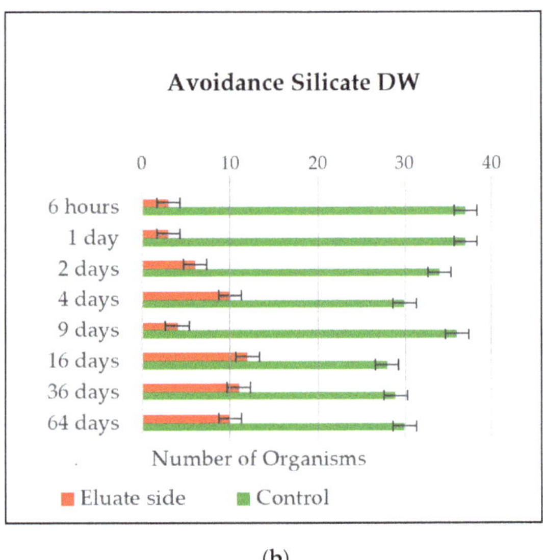

(b)

Figure 8. (**a**) *Enchytraeid* avoidance results for the silicate leach using tap water. (**b**) *Enchytraeid* avoidance results for the silicate leach using distilled water.

A statistical analysis was carried out to calculate the percentage of worms affected. Avoidance was calculated according to the following Equation (1):

$$x = \left(\frac{n_c - n_t}{N} \right) \times 100 \tag{1}$$

where:

x is avoidance, expressed as a percentage;
n_c is the number of worms in the control soil (either per vessel or in the control soil of all replicates);
n_t is the number of worms in the test soil (either per vessel or in the test soil of all replicates);
N is the total number of worms (usually 10; either per vessel or in the control soil of all replicates).

Using the results collected on the avoidance tests, the percentage of avoidance was calculated and the results are represented in Table 2.

According to ISO 17512, when the avoidance results are $\geq 80\%$, a limited survival habitat function is configurated. If an attraction of $>80\%$ by the test soil is observed, the presence of chemical substances cannot be excluded. The result indicates an impact on the behavior of the organisms.

The limited survival habitat was determined on the following eluates: silicate TW (6 h, 1 day, 9 days) and silicate DW (6 h, 1 day, 9 days). Although eluates from waterproof did not represent a limited habitat function, in few eluates, the avoidance percentage reaches 75% being very close to this value and indicating the possible presence of chemicals.

The concentration of harmful substances consequently seems to be significantly increased in the first leaching steps since many components that can be leached out at the beginning of a DSLT can be released through washing. With the length of the leaching times, the number of Enchytraeids that migrate to the eluate side increases, indicating a decreased release of harmful substances with the time of leaching.

A better distribution of the number of organisms on both sides of the vessels for all leachings is noted in leaches of 64 days when an avoidance rate of 30% and 50% is reached.

3.4. Evaluation of Enchytraeid Results—Reproduction Versus Avoidance

Comparing the results of the reproduction tests with the results of the avoidance tests, a certain similarity can be observed in the sensitivity of both test results.

The initial eluates of the silicate product caused an impacting reduction on the number of *Enchytraeids* in the reproduction test, while in the avoidance test, the same eluates indicated an avoidance of 85% of the population.

Besides, the eluates from waterproof for reproduction tests did not present high toxicity for the organisms, but a high avoidance percentage was observed for the same components. It is important to remember that the reproduction test conditions supply constant contact with the soil + eluate for a chronic exposure time, allowing the leaches to stabilize for a longer period of time [32].

For all eluates of 64 days, in the reproduction test, an increase in the number of offsprings was observed. At the same time, the avoidance test also shows better survival habitat conditions with smaller avoidance behavior, being both test results an indicator of stabilization of chemicals in the soil.

The reproduction test is applied to detect effects resulting from sublethal concentrations and long term scenario exposures. The avoidance test exists to investigate the habitat function of soil with earthworms as representatives of the soil biocenosis. The endpoints are determined to obtain information on the environmental effects. The reproduction test is very labor-intensive and time-consuming, needing long incubation periods and results being collected after 6 weeks of exposure and constant work, turning the reproduction tests more expensive. In contrast, the avoidance test presents very fast results of environmental effects and a high level of sensitivity. However, the avoidance test is not intended to replace the reproduction test but to provide faster screenings of environmental effects in different levels of sensitivity. For a complete and better understanding of the effects of a toxicant on the soil, it is possible to use the avoidance test as the first information screen and the reproduction test to verify the chronic sublethal effects.

4. Conclusions

The reproduction behavior of *Enchytraeus albidus* presented the highest numbers when worms were in soil moistened with the reconstituted freshwater as a control, followed by tap water and, in last, the demineralized water. The type of water used to leach the components influenced the reproduction behavior of *E. albidus*. The best performance in the number of juveniles is achieved when the standard soil is moistened with liquids containing physic-chemical conditions more similar to those found in the environment.

Evaluating the reproduction results, it can be concluded that the eluates from the waterproof material did not present high toxicity to the reproduction behavior of *E. albidus;* besides, a decrease in the number of juveniles was measured. The leachings from waterproof TW presented a higher number of juveniles in comparison to waterproof DW. The leachings from silicate TW and silicate DW were toxic for the reproduction behavior of *E. albidus.* Eluates from silicate DW had the lowest number of juveniles and inhibited more than 50% of the population of *E. albidus.*

The avoidance results for all the eluates presented a similar sensitivity for *E. albidus* when compared to the reproduction results. The organisms avoided soils containing high concentrations of chemicals. Eluates of the silicate resin product in soil presented a limited habitat function with avoidance results >80%.

Looking at these results, it can be concluded that it is important to track biomarkers for *Enchytraeids* in order to assess the possible hazard effects of construction products and understand the mechanism of action in the environment.

5. Final Comments

The time and effort required for the preparation of the leachings in accordance with DIN CEN/TS 16637-2 should also be emphasized: it is extensive and intensive work, and therefore this study recommends for screening the use of other methods of leaching as the EN 12457-4 [33], which lead to similar results from a biological point of view.

It is important to note that a difference in the final products and consequently in the eluates may occur due to different handling of the starting products. The air gun and mixers used in the laboratories are prepared to handle a much smaller sample volume than those used in the construction industry. For this reason, it is recommended here to collect samples from real applications in order to clarify whether the products have possible toxicity. Another problem is the solubility of these components in the environment. The laboratory tests try to consider as far as possible the aspects from a real scenario situation, but it is still likely that this will not give the same results as in the real environment and that a much higher dilution of these components would occur and, therefore, lower toxicity would be detected.

The organisms used in these tests may never come into contact with this type of component in a real scenario. Nevertheless, it is the responsibility of ecotoxicology to identify the possible causes and influences that could occur in case of such an event. This study does not condemn the use of construction products, nor does it attempt to restrict or prevent their use. The sole purpose is to clarify human activities and their consequences for the environment and to find sustainable solutions that would help to focus on satisfying the needs of the present without compromising the ability of future generations to meet their needs.

Author Contributions: Conceptualization, D.S.; methodology, M.A.v.W.; formal analysis, M.A.v.W.; investigation, M.A.v.W.; resources, D.S.; data curation, M.A.v.W. and D.S.; writing—original draft preparation, M.A.v.W.; writing—review and editing, M.A.v.W. and D.S.; supervision, D.S.; project administration, D.S.; funding acquisition, D.S. Both authors have read and agreed to the published version of the manuscript.

Funding: This research was funded by "Forschungsinitiative Zukunft Bau des Bundesinstitutes für Bau-, Stadt- und Raumforschung", grant number SWD-10.08.18.7-17.49.

Institutional Review Board Statement: Not Applicable.

Informed Consent Statement: Not Applicable.

Data Availability Statement: Data availability within this article.

Acknowledgments: The authors thank Daria Blizniukova for helping in the cultivation of *Enchytraeids* and the collection of leachings data.

Conflicts of Interest: The authors declare no conflict of interest.

References

1. Bandow, N.; Gartiser, S.; Ilvonen, O.; Schoknecht, U. Evaluation of the impact of construction products on the environment by leaching of possibly hazardous substances. *Environ. Sci. Eur.* **2018**, *30*, 14. [CrossRef] [PubMed]
2. van der Sloot, H.A.; Comans, R.N.J.; Hjelmar, O. Similarities in the leaching behaviour of trace contaminants from waste, stabilized waste, construction materials and soils. *Sci. Total Environ.* **1996**, *178*, 111–126. [CrossRef]
3. *DIN EN 15804: 2012-04—Sustainability of Construction Works—Environmental Product Declarations—Core Rules for the Product Category of Construction Products*; European Committee for Standardization: Brussels, Belgium, 2012; German version EN 15804:2012.
4. Kabirifar, K.; Mojtahedi, M.; Wang, C.; Tam, V.W.Y. Construction and demolition waste management contributing factors coupled with reduce, reuse, and recycle strategies for effective waste management: A review. *J. Clean. Prod.* **2020**, *263*, 121265. [CrossRef]
5. Galvín, A.P.; Ayuso, J.; García, I.; Jiménez, J.R.; Gutiérrez, F. The effect of compaction on the leaching and pollutant emission time of recycled aggregates from construction and demolition waste. *J. Clean. Prod.* **2014**, *83*, 294–304. [CrossRef]
6. *Regulation (EU) No 305/2011 of the European Parliament and of the Council of 9 March 2011*; Official Journal of the European Union: Berlin, Germany, 2011.
7. DIN CEN/TS Deutsches Institut für Normung e. V. *Construction Products—Assessment of Release of Dangerous Substances—Part 2: Horizontal dynamic Surface Leaching Test*; Beuth Verlag GmbH: Berlin, Germany, 2016; (16637-2).
8. Gartiser, S.; Heisterkamp, I.; Schoknecht, U.; Bandow, N.; Burkhardt, N.M.; Ratte, M.; Ilvonen, O. Recommendation for a test battery for the ecotoxicological evaluation of the environmental safety of construction products. *Chemosphere* **2017**, *171*, 580–587. [CrossRef] [PubMed]
9. Brameshuber, W.; Brockmann, J.; Rankers, R. *Emission von Umweltrelevanten Organischen Bestandteilen aus Betonen Mit Organischen Betonzusatzstoffen. Forschungsbericht F 626*. Bauforschung. 2000. Available online: https://www.baufachinformation.de/literatur/00019015585 (accessed on 15 October 2020).
10. Hartwich, P.; Vollpracht, A. Influence of leachate composition on the leaching behaviour of concrete. *Cem. Concr. Res.* **2017**, *100*, 423–434. [CrossRef]
11. Paulus, H.; Schick, J.; Poirier, J.-E. Assessment of dynamic surface leaching of monolithic surface road materials. *J. Environ. Manag.* **2016**, *176*, 79–85. [CrossRef] [PubMed]
12. Märkl, V.; Pflugmacher, S.; Reichert, A.; Stephan, D.A. Leaching of polyurethane systems for waterproofing purposes whilest curing. *Water Air Soil Pollut.* **2017**, *228*. [CrossRef]
13. Märkl, V.; Pflugmacher, S.; Stephan, D.A. Leaching of PCE-based superplasticiser from microfine cement: A chemical and ecotoxicological point of view. *Water Air Soil Pollut.* **2017**, *228*. [CrossRef]
14. Gartiser, S.; Heisterkamp, I.; Schoknecht, U.; Burkhardt, M.; Ratte, M.; Ilvonen, O.; Brauer, F.; Brückmann, J.; Dabrunz, A.; Egeler, P.; et al. Results from a round robin test for the ecotoxicological evaluation of construction products using two leaching tests and an aquatic test battery. *Chemosphere* **2017**, *175*, 138–146. [CrossRef] [PubMed]
15. Hybská, H.; Hroncová, E.; Ladomerský, J.; Balco, K.; Mitterpach, J. Ecotoxicity of concretes with granulated slag from gray iron pilot production as filler. *Materials* **2017**, *10*, 505. [CrossRef] [PubMed]
16. Didden, W.; Römbke, J. Enchytraeids as indicator organisms for chemical stress in terrestrial ecosystems. *Ecotoxicol. Environ. Saf.* **2001**, *50*, 25–43. [CrossRef] [PubMed]
17. Scanes, C.G.; Toukhsati, S. *Animals and Human Society. Invertebrates and Their Use by Humans*; Academic Press: Amsterdam, The Netherlands, 2017; ISBN 978-0-12-805247-1.
18. Edwards, C.A. Soil pollutants and soil animals. *Sci. Am.* **1969**, *220*, 88–99. [CrossRef] [PubMed]
19. ISO 17512:2008. *Soil Quality—Avoidance Test for Determining the Quality of Soils and Effects of Chemicals on Behaviour. Part 1: Test with Earthworms (Ei-senia Fetida and Eisenia Andrei)*; International Organization for Standardization: Geneva, Switzerland, 2008.
20. Kuperman, R.G.; Amorim, M.J.B.; Römbke, J.; Lanno, R.; Checkai, R.T.; Dodard, S.G.; Sunahara, G.I.; Scheffczyk, A. Adaptation of the enchytraeid toxicity test for use with natural soil types. *Eur. J. Soil Biol.* **2006**, *42*, S234–S243. [CrossRef]
21. ISO 16387 International Standard. *ISO 16387—Soil Quality—Effects of Contaminants on Enchytraeidae (Enchytraeus sp.)—Determination of Effects on Reproduction*, International Organization for Standardization: Geneva, Switzerland, 2014.
22. Henle, F.G.J. *Ueber Enchytraeus, Eine Neue Anneliden-Gattung*; Archiv für Anatomie, Physiologie und Wissenschaftliche Medicin: Berlin, Germany, 1837; pp. 74–90.
23. LUFA—Landwirtschaftliche Untersuchungs und Forschungsanstalt. *Analyses Data Sheet for Standard Soils according to GLP 2018, 1*; LUFA Speyer: Speyer, Germany, 2018.
24. Hu, R.; Xie, J.; Wu, S.; Yang, C.; Yang, D. Study of toxicity assessment of heavy metals from steel slag and its asphalt mixture. *Materials* **2020**, *13*, 768. [CrossRef] [PubMed]
25. Brameshuber, W.; Vollpracht, A. *Konzept zur Bewertung des Auslaugverhaltens mineralischer Baustoffe/-körper durch Sicker- und Kontaktgrundwasserprognose; Teil 2: Bestimmung der Quellstärke für Boden und Grundwasser in Kontakt mit Betonfundamenten und Bodeninjektionen: Abschlußbericht*; IBAC: Aachen, Germany, 2004.
26. *Grundsätze zur Bewertung der Auswirkungen von Bauprodukten auf Boden und Grundwasser: Teil II und Teil III*; Deutsches Institut für Bautechnik: Berlin, Germany, 2009. [CrossRef]
27. *Umwelteigenschaften Mineralischer Werkmörtel. Zugl.: Stuttgart, Univ., Diss., 2012*; Scherer, C. (Ed.) Fraunhofer Verl.: Stuttgart, Germany, 2009; ISBN 9783839604700.

28. Zhou, S.; Ding, Y.; Wang, Z.; Dong, J.; She, A.; Wei, Y.; Li, R. Weathering of roofing insulation materials under multi-field coupling conditions. *Materials* **2019**, *12*, 3348. [CrossRef] [PubMed]
29. Coz, A.; Andrés, A.; Soriano, S.; Viguri, J.R.; Ruiz, M.C.; Irabien, J.A. Influence of commercial and residual sorbents and silicates as additives on the stabilization/solidification of organic and inorganic industrial waste. *J. Hazard. Mater.* **2009**, *164*, 755–761. [CrossRef] [PubMed]
30. Römbke, J.; Moser, T. Validating the enchytraeid reproduction test: Organization and results of an international ringtest. *Chemosphere* **2002**, *46*, 1117–1140. [CrossRef]
31. Roembke, J.; Knacker, T. Aquatic toxicity test for enchytraeids. *Hydrobiologia* **1989**, *180*, 235–242. [CrossRef]
32. Amorim, M.J.B.; Sousa, J.P.; Nogueira, A.J.A.; Soares, A.M.V.M. Bioaccumulation and elimination of -lindane by Enchytraeus albidus in artificial (OECD) and a natural soil. *Chemosphere* **2002**, *49*, 323–329. [CrossRef]
33. DIN EN 12457-4 (2002): Charakterisierung von Abfällen: Auslaugung—Übereinstimmungsuntersuchung für die Auslaugung von körnigen Abfällen und Schlämmen Teil 4: Einstufiges Schüttelverfahren mit einem Flüssigkeits-/Feststoffverhältnis von 10 l/kg für Materialien mit einer Korngröße unter 10 mm (ohne oder mit Korngrößenreduzierung) Deutsche Fassung EN 12457-4:2002 Ausgabe 2003-01. 2002. Available online: https://www.din.de/de/mitwirken/normenausschuesse/naw/wdc-beuth:din21:52065766 (accessed on 8 January 2021).

Article

Conversion of Bivalve Shells to Monocalcium and Tricalcium Phosphates: An Approach to Recycle Seafood Wastes

Somkiat Seesanong [1], Banjong Boonchom [2,3,*], Kittichai Chaiseeda [4,*], Wimonmat Boonmee [5] and Nongnuch Laohavisuti [6]

1   Department of Plant Production Technology, Faculty of Agricultural Technology, King Mongkut's Institute of Technology Ladkrabang, Bangkok 10520, Thailand; ksesomki@yahoo.com
2   Advanced Functional Phosphate Material Research Unit, Department of Chemistry, Faculty of Science, King Mongkut's Institute of Technology Ladkrabang, Bangkok 10520, Thailand
3   Municipal Waste and Wastewater Management Learning Center, Faculty of Science, King Mongkut's Institute of Technology Ladkrabang, Bangkok 10520, Thailand
4   Organic Synthesis, Electrochemistry and Natural Product Research Unit (OSEN), Department of Chemistry, Faculty of Science, King Mongkut's University of Technology Thonburi, Bangkok 10140, Thailand
5   Department of Biology, Faculty of Science, King Mongkut's Institute of Technology Ladkrabang, Bangkok 10520, Thailand; bwimonmat@gmail.com
6   Department of Animal Production Technology and Fishery, Faculty of Agricultural Technology, King Mongkut's Institute of Technology Ladkrabang, Bangkok 10520, Thailand; nongnuch.la@kmitl.ac.th
*   Correspondence: kbbanjon@gmail.com (B.B.); kittichai@gmail.com (K.C.)

Citation: Seesanong, S.; Boonchom, B.; Chaiseeda, K.; Boonmee, W.; Laohavisuti, N. Conversion of Bivalve Shells to Monocalcium and Tricalcium Phosphates: An Approach to Recycle Seafood Wastes. *Materials* **2021**, *14*, 4395. https://doi.org/10.3390/ma14164395

Academic Editors: Franz-Georg Simon and Ute Kalbe

Received: 21 June 2021
Accepted: 3 August 2021
Published: 5 August 2021

Publisher's Note: MDPI stays neutral with regard to jurisdictional claims in published maps and institutional affiliations.

Abstract: The search for sustainable resources remains a subject of global interest and the conversion of the abundantly available bivalve shell wastes to advanced materials is an intriguing method. By grinding, calcium carbonate ($CaCO_3$) powder was obtained from each shell of bivalves (cockle, mussel, and oyster) as revealed by FTIR and XRD results. Each individual shell powder was reacted with H_3PO_4 and H_2O to prepare $Ca(H_2PO_4)_2 \cdot H_2O$ giving an anorthic crystal structure. The calcination of the mixture of each shell powder and its produced $Ca(H_2PO_4)_2 \cdot H_2O$, at 900 °C for 3 h, resulted in rhombohedral crystal β-$Ca_3(PO_4)_2$ powder. The FTIR and XRD data of the $CaCO_3$, $Ca(H_2PO_4)_2 \cdot H_2O$, and $Ca_3(PO_4)_2$ prepared from each shell powder are quite similar, showing no impurities. The thermal behaviors of $CaCO_3$ and $Ca(H_2PO_4)_2 \cdot H_2O$ produced from each shell were slightly different. However, particle sizes and morphologies of the same products obtained from different shells were slightly different—but those are significantly different for the kind of the obtained products. Overall, the products ($CaCO_3$, $Ca(H_2PO_4)_2 \cdot H_2O$, and $Ca_3(PO_4)_2$) were obtained from the bivalve shell wastes by a rapidly simple, environmentally benign, and low-cost approach, which shows huge potential in many industries providing both economic and ecological benefits.

Keywords: calcium phosphate; calcium carbonate; recycling; environmental problems; seashell

1. Introduction

Seafood productions contributed importantly to require for a source of protein worldwide, but it also creates huge waste quantities of solid and liquid in the processes [1]. For mollusk, shell wastes of over 13 million tons were manufactured yearly [2,3]. Three major types of mollusk informed by the Food and Agriculture Organization of the United Nations (FAO) Fisheries and Aquaculture Department are cockles, mussels, and oysters, which are consumed very largely around the world [1]. The main sources are generally from aquaculture upward than wild fisheries, amongst which oysters were dominant followed by mussels and cockles. In 2018, 5.8, 1.6, and 0.4 million tons of oyster, mussel, and cockle were produced, respectively [2]. Generally, bivalve shell wastes account for about 65–80% of live weight, which is expected to be over 5 million tons a year [4]. Large numbers of bivalve shells are dumped into public waters and/or landfills and create numerous environmental obstacles that contribute to pollution to coastal fisheries, public water surface, an unpleasant smell as a consequence of the decomposition of organics attached to the shells, and natural landscape affecting to health/sanitation problems [2–5]. Consequently, the disposal of bivalve shell wastes is getting an extremely fatal issue for the marine aquaculture industries and various consumer countries. The increasing knowledge of sustainable evolution and research attention in innovative technologies on the conversion of bivalve shell wastes into helpful and expensive chemicals and compounds have been starting in the 21st century [4]. So far, many researchers have studied on characterizations of bivalve (cockles, mussels, and oysters) shells and reported that chemical contents consist of primarily calcium carbonate (>95%) with various crystal phases [4]. Based on their environment, the different species of shells may comprise various quantities of cation contaminations such as silicon, magnesium, aluminum, strontium, phosphorus, sodium, or sulfur [6–8].

Calcium carbonate ($CaCO_3$) naturally occurs in rocks and shells of various organisms and is widely used in construction, papermaking, pharmaceuticals, agriculture, etc. This $CaCO_3$ compound occurs naturally in three polymorphs including calcite (β), aragonite (γ), and vaterite (μ) [6–8]. Nowadays, the principal calcium carbonate production is from mineral resources, which have the risk of heavy metal contamination and are non-renewable resources, unlike calcium carbonate from bio-derived shells which are generally abundant, renewable, inexpensive, and environmentally friendly [1,4]. Based on the above mentioned, recycling seashell wastes to $CaCO_3$ raw material offers many advantages and has potential application in various fields (Figure 1). Various worldwide research shows immense potential for applications of seashells. Recently, they have been used to produce hydroxyapatite [7], nano-hydroxyapatite (nHA) [8], apatite nanoparticles [9], calcite lime [10], CaO [11,12], bio-filler in polypropylene [13], matte glaze [14], cement clinker [15], cementitious construction materials [16], expansive additive in cement mortar [17], adsorbent for Pb(II) adsorption [18], adsorbent for sulfate and metals removal [19], covalently functionalized biogenic $CaCO_3$ [20], calcined mussel shell powder (CMSP) for antistatic oil-removal [21]. However, in Southeast Asia, especially Thailand, a country with the highest bivalve (cockles, mussels, and oysters) production, the waste shell recycling means is not created appropriately, and these wastes are mainly dumped in the near areas affecting an environmental issue [22,23]. Alarmed with the problems, the Ministry of Higher Education, Science, Research, and Innovation of Thailand planned to resolve according to the Bio-Circular-Green Economy (BCG) model and financed a program to set new strategies for recycling these wastes, including establishing factories for producing calcium compounds to increase the recycling quantity of bivalve shell wastes. However, only 30% of bivalve shell wastes are reused/recycled by these factories [24,25].

Figure 1. Potential applications of bivalve shell wastes.

Realizing this problem, our research focuses on the conversion of bivalve shells to calcium phosphates, which are used as nutritional supplements, catalysts for some chemical reactions, fertilizers, and animal feed minerals, the mineral basis of the tooth and bone tissues, as well as for creating materials with unique properties [1,4–9]. In Thailand, monocalcium phosphate (MCP) and tricalcium phosphate (TCP) are enormously used in many fields and both compounds are imported every year. MCP has been used in huge quantities in agriculture. It is called superphosphate fertilizer, classified by three levels of %P_2O_5 (single (9–20%), double (20–48%), and triple (48–58%) superphosphates), and P-21 for animal feed minerals [24,25]. Additionally, MCP is also used in large amounts in the food industry as a buffer, hardener, leavening reagent, yeast food, beverage, bakery, and nutrient [26]. TCP is widely used in huge amounts in the medical and pharmaceutical industries as medicine, tooth, bone, calcium supplement, in the animal feed industry as calcium additive and supplement, and in the food industry as various functions (acidity regulator, anticaking, emulsifying, firming, flouring, humectant, raising, stabilizer, and thickener) in many food substances under the number 341(iii) [27]. Both compounds have been synthesized to produce high-purity grades from various calcium compounds (chloride, carbonate, oxide, nitrate, acetate, etc.) and various phosphorus compounds (phosphoric acid, sodium, potassium, ammonium, etc.) by many methods including chemical precipitation [24], hydrothermal synthesis [28,29], microwave-assisted methods [30,31], precipitation of emulsions [32], sol-gel [33], crystallization of solutions [34], chemical deposition [35], electrodeposition [36], and mechanic-chemical synthesis [37]. The method performed in an individual event depends on the requisite kind of morphology, structure, and chemical content. However, the typical drawback of these synthesized methods is expensive raw materials resulting in the high cost of the obtained products which are not suitable to use for some applications such as fertilizer and animal feed industries.

Therefore, one of the solving keys for the mention-above points is to apply/recycle the bivalve shell wastes in such a route that it can be more valuable and resourcefully used to create these calcium phosphates which may solve some financial issues to purchase expensive compounds for many industries in Thailand. Although, MCP and TCP have been reportedly prepared from bivalve shell wastes such as oyster shells [38], and Mediterranean mussel shells [39], the methods used complex and high-cost processes which many parameters must be carefully controlled (concentration, pH, time, and temperature). The aim of this present work is to easily and quickly obtain calcium carbonate from bivalve shell wastes (cockles, mussels, oysters) and then subsequently use it to produce MCP and TCP by using an easy, cost-effective, and environmentally benign method. Moreover, this work also highlights some potential applications for shell wastes that can bring both economic and ecological benefits.

2. Materials and Methods

2.1. Starting Reagents

Raw reagents utilized for the current study were waste bivalve shells of cockle, mussel, and oyster collected from seafood restaurants of fishermen residing in areas of the Chonburi beaches in eastern Thailand. The individual kind of seashells derived in the primitive shape was carefully cleaned with triply distilled water and dried in an oven at 100 °C for 3 h. Each kind of dried seashell was pulverized to produce fine powder by using an agate mortar and pestle and then was sieved in 100 mesh (150 μm). All fine seashell powders were then characterized to identify the purity and solid phase of calcium carbonate before proceeding to the preparation of the calcium phosphates. Fine seashell powders of cockle, mussel, and oyster were $CaCO_3$ compounds and denoted with the sample codes CSP, MSP, and OSP, respectively.

2.2. Monocalcium Phosphate Monohydrate (Ca(H₂PO₄)₂·H₂O, MCPM) Preparation

A collection of monocalcium phosphate hydrate samples was synthesized by the reaction of individual seashell powder (cockle (CSP), mussel (MSP), and oyster (OSP) shells) with 85 wt% phosphoric acid and distilled water, using a constant-addition method modified from the generic reaction reported in previous work [10].

Briefly, 15 mL of 85 wt% H_3PO_4 was slowly added into a beaker (100 mL), which contains 9 g of CSP, and was constantly stirred with a Teflon stir bar. This mixing reaction exhibited an exothermic process noticed by increasing temperature (65 °C). Then, 12 mL of distilled water (H_2O) was immediately added into the resulting mixture with constant stirring until CO_2 gas bubbles are no evolved (about 30 min). The resulting reaction has been stayed in the open air for about 3 h to become dried powder of monocalcium phosphate hydrate without other processes such as filtration and drying with temperature control. The monocalcium phosphate hydrate obtained from CSP was labeled with the sample code MCP-C. For MSP and OSP, the processes were repeated in the same way as CSP, and the obtained products were labeled as MCP-M and MCP-O, respectively.

2.3. Tricalcium Phosphate Anhydrous ($Ca_3(PO_4)_2$, TCP) Preparation

A series of tricalcium phosphate samples was prepared by mixing powders of individual seashell powder (cockle (CSP), mussel (MSP), and oyster (OSP) shells) with their prepared monocalcium phosphate hydrate pair (MCP-C, MCP-M, and MCP-O). The generic reaction is:

$$2CaCO_3(s) + Ca(H_2PO_4)_2 \cdot H_2O(s) \rightarrow Ca_3(PO_3)_2(s) + H_2O(g) \tag{1}$$

In the typical way, 2.0 g of CSP and 2.52 g of MCP-C were weighed and then well mixing by grinding in a crucible. Then, the mixed powder was calcined at 900 °C for 3 h in a furnace. Its final product after heating is tricalcium phosphate, labeled as TCP-C. The tricalcium phosphate powders prepared from the mixed powders of MSP + MCP-M and OSP + MCP-O were prepared in the same way as CSP + MCP-C and the obtained products were labeled as TCP-M and TCP-O, respectively.

2.4. Sample Characterization

2.4.1. Thermogravimetric Analysis (TA)

The thermal behaviors of the dried fine seashell powders and monocalcium phosphate monohydrate were analyzed by a thermogravimetric/differential thermal analyzer (TG-DTA, Pyris Diamond, Perkin Elmer). The experiments were performed in the static air, at the heating rates of 10 °C min^{-1} over the temperature range from 30 to 900 °C and the O_2 flow rate of 100 mL min^{-1}. The sample mass of about 6.0–10.0 mg was filled into an alumina crucible without pressing. The thermogram of a sample was recorded in an open aluminum pan using Al_2O_3 as the reference material.

2.4.2. Fourier Transform Infrared (FTIR) Spectroscopy

The molecular structures were measured by a Fourier Transform Infrared Spectrophotometer (FTIR, Spectrum GX, Perkin Elmer), which were recorded in the range of 4000–400 cm^{-1} with eight scans and the resolution of 4 cm^{-1} using KBr pellets (spectroscopy grade, Merck).

2.4.3. Powder X-ray Diffraction (XRD)

The structure of the prepared samples was recorded by X-ray powder diffraction using an X-ray diffractometer (D8 Advance, Bruker AXS GmbH) with Cu K radiation ($\lambda = 0.1546$ nm) operating at the condition of 40 kV and 40 mA. The specimen was pulverized into a fine powder and used for the analysis. The diffraction angle was continuously scanned from 10° to 60° in 2θ at a scanning rate of 2°/min. A range of 10–60° is shown in the figures because no relevant peaks occurred in the excluded region.

2.4.4. Scanning Electron Microscopy (SEM)

The morphology of the selected resulting samples was determined by a scanning electron microscope using LEO SEM VP1450 after gold coating.

3. Results and Discussion

3.1. Characterization Results of Bivalve Shell Powders

Figure 2 displays TG/DTG curves of the CSP, MSP, and OSP samples, which are quite similar. TG lines of each sample show the mass loss in the region of 600–800 °C, which correspond to a strong single peak of DTG curves at 752, 772, 750 °C for the CSP, MSP, and OSP samples, respectively. Four DTG peaks observed at 515, 540, 569, and 625 °C for the MSP sample may have resulted from other cation contaminations that may be formed the mixing phase of metal carbonates, which can be not pointed out still clearly. The quantities of mass loss are found to be 43.1% for the CSP sample, 43.9% for the OSP sample, and 48.4% for the MSP sample. The thermal results were well consistent with those of the reference data of $CaCO_3$ and theoretical data [17,40,41]. The thermal behavior obtained indicates that the bivalve shell powders can be transformed to CaO by calcination at above 772 °C, which may be useful for the production of this compound to be used in specific applications.

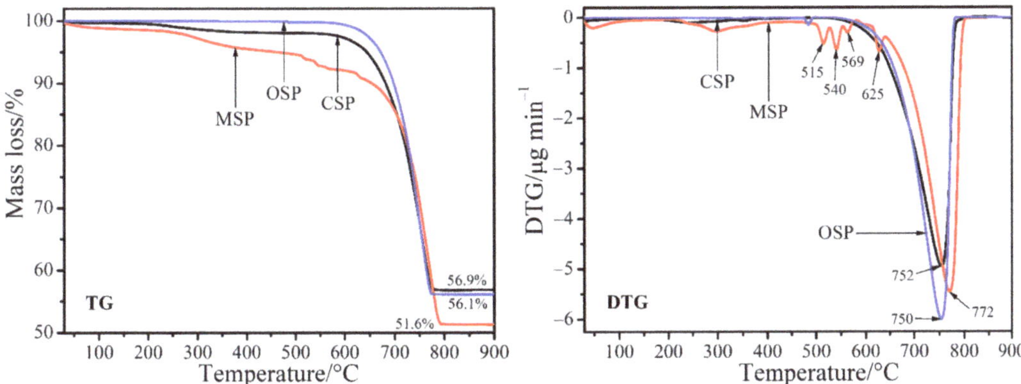

Figure 2. TG and DTG curves of the bivalve shell (CSP, MSP and OSP) powders.

Figure 3 illustrates FTIR spectra of the CSP, MSP, and OSP samples that are quite similar due to fundamental vibrational bands of CO_3^{2-} block unit in the $CaCO_3$ structure for each sample. The vibrational modes of the CO_3^{2-} anion are divided into three types [42]: (i) internal vibrational modes of (CO_3^{2-}) groups, (ii) hydroxyl vibrations (in the case of hydroxyl carbonates $\approx$ 900 cm^{-1}, 1500–1600 cm^{-1}, and 3400 cm^{-1}), and (iii) vibrational M-O modes from the interactions between the cation and oxygen of either (CO_3^{2-}) or OH$^-$ (external or lattice modes). The carbonate anion (CO_3^{2-}) is a nonlinear structure with four atoms resulting in six $(3 \times 4 - 6)$ normal modes of vibrations [43]. The six normal vibrational modes are a nondegenerate symmetric stretch (v_1; A$'_1$: Raman active), nondegenerate asymmetric (out of plane) bend (v_2; A$''_1$: IR active), doubly degenerate asymmetric stretch (v_3; E$'$: Raman and IR active), and doubly degenerate symmetric (in-plane) bend (v_4; E$'$: Raman and IR active). The FTIR spectra of the CSP, MSP, and OSP samples were analyzed according to this theory. Two strong intense bands at 696 cm^{-1} and 863 cm^{-1} are assigned to the v_4 and v_2 modes, respectively. A weak band at 1030 cm^{-1} is contributed to v_1 mode. A band at 1413 cm^{-1}, looking like a mountain, is related to v_3 mode. A weak band observed at 1782 cm^{-1} may be respected as the combination bands of $v_4 + v_1$ modes. A weak band at 2520 cm^{-1} and a broad band around 2875 cm^{-1} may be regarded as a combination of or/and overtone of v_4, v_3, and v_1 modes. A single band at 3453 cm^{-1} was assigned to the OH-stretching modes. For the FTIR results of all bivalve shells, the v_1 mode that appear normally active in Raman is observed and v_3 and v_4 modes are not shown doubly degenerate bands, which may be correlated with the atomic cation masses and the presence of molecules belonging to site symmetry of their structures [44]. The FTIR results obtained are very similar to that of the calcite phase of

CaCO$_3$ in literature [43,44], which indicates that the CSP, MSP, and OSP samples have the main content as this crystalline phase.

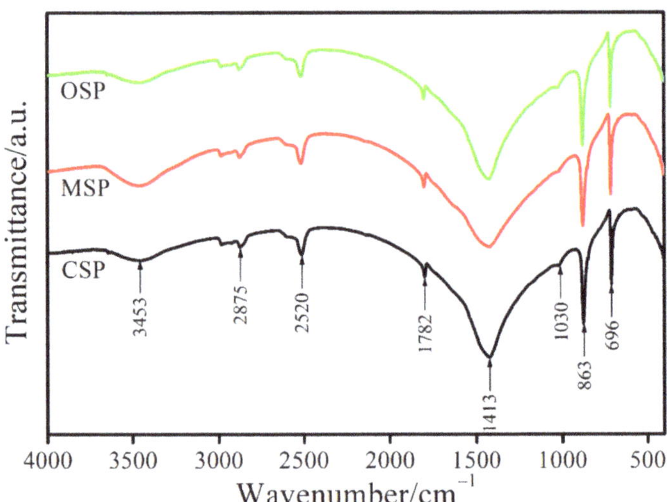

Figure 3. FTIR spectra of the bivalve shell (CSP, MSP and OSP) powders.

XRD patterns of the CSP, MSP, and OSP samples are very similar and are exhibited in Figure 4. Bivalve (cockle, mussel, and oyster) shells, biological wastes are primarily made up of calcium carbonate ($\geq$96wt%) and little contaminations of other chemical contents. It can clearly indicate that the main crystalline phases of the CSP, MSP, and OSP samples are calcite (β-CaCO$_3$) and minor aragonite are detected for OSP while little vaterite is detected for the other two samples, shown in Figure 4. The diffraction intensity is the calcite (002) $2\theta = 29.81$ and the next to strongest are the calcites at (111), (012), (202), (112) (200), and (202), respectively. The XRD analysis verified that the crystalline phase of CaCO$_3$ in the CSP, MSP, and OSP samples was calcitic polycrystals, which was found to match with the PDF data file of CaCO$_3$ (PDF no.72–1937) [41,45]. The XRD results are in well agreement with the FTIR data.

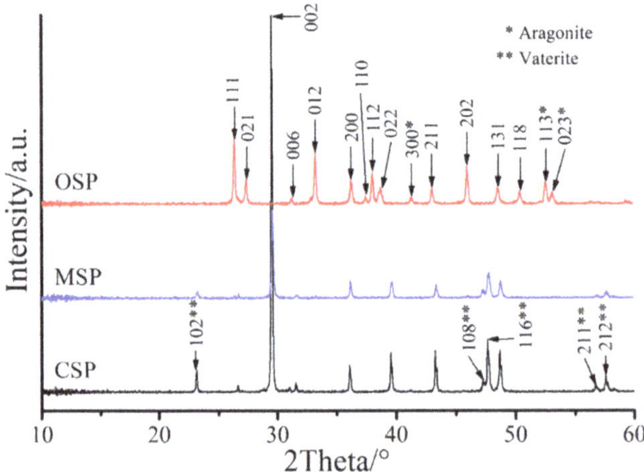

Figure 4. XRD patterns of the bivalve shell (CSP, MSP and OSP) powders.

SEM micrographs of the CSP, MSP, and OSP samples exhibit different morphological features and are shown in Figure 5. SEM image of CSP reveals an elongated, partly polyhedral morphology of rod-like crystals (up to 10 µm long) with different sizes and forms. SEM image of MSP reveals elongated, partly polyhedral morphology of sheet-like crystals with different sizes (5–10 µm). Finally, the SEM image of OSP reveals plate-like crystals of different sizes, which are agglomerated.

Figure 5. SEM micrographs of the bivalve shell (CSP, MSP and OSP) powders.

3.2. Characterization Results of Monocalcium Phosphates

TG/DTG curves of three monocalcium phosphate samples prepared from cockle (CSP), mussel (MSP), and oyster (OSP) shell powders, labeled as MCP-C, MCP-M, and MCP-O, respectively, are displayed in Figure 6. TG lines of all samples showing the mass loss in the range of 100–600 °C are similar. The mass losses found to be around 20% for each sample correspond to the loss of three water molecules in structure. Total mass losses found to be 18.9% for MCP-C, 25.0% for MCP-M and 17.3% for MCP-O are slightly different from that of theoretical value at 21.4% [41,46]. The obtained results indicate that the number of water molecules in $Ca(H_2PO_4)_2 \cdot nH_2O$ structure would be 0.7, 1.5, and 0.4 mole for the MCP-C, MCP-M, and MCP-O samples, respectively. The number of crystal water found in the range of 0-n < 2 is an impossible theory, which is consistent with that of the previous works [24,26,41,46]. The relative with TG data, DTG curves of the MCP-C, MCP-M, and MCP-O samples showing the numbers and peak positions of steps of thermal transformation are different. Four DTG peaks are observed at 127, 178, 224, and 330 °C for the MCP-C sample and 127, 185, 245, and 330 °C for the MCP-M sample while five DTG peaks occur at 127, 185, 224, 265, and 330 °C for the MCP-O sample. Two peaks that occurred below 200 °C correspond to the dehydration steps of about one molecule of water. Two/three peaks observed in the range of 200–330 °C relate to deprotonated steps of two dihydrogen phosphate ($H_2PO_4^-$) anions. General mechanism reactions of thermal transformation could be:

$$Ca(H_2PO_4)_2 \cdot H_2O(s) \xrightarrow{below\,200°C} Ca(H_2PO_4)_2(s) + H_2O(g) \qquad (2)$$

$$Ca(H_2PO_4)_2(s) \xrightarrow{200-330°C} Ca_3(PO_3)_2(s) + 2H_2O(g) \qquad (3)$$

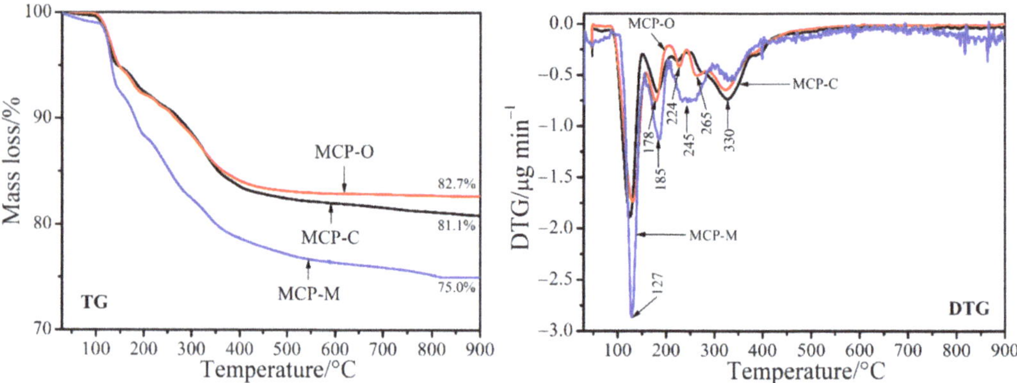

Figure 6. TG and DTG curves of MCP-C, MCP-M, and MCP-O samples.

The final decomposition of $Ca(H_2PO_4)_2 \cdot H_2O$ to calcium polyphosphate $Ca_3(PO_3)_2$, as revealed by reaction (3), occurred at above 400 °C. Many peaks in DTG curves observed for three $Ca(H_2PO_4)_2 \cdot H_2O$ samples may indicate the splitting of each step of the dehydration step (reaction 2) and deprotonated hydrogen phosphate reaction (reaction 3). This result could be regarded to be affected by different inter/intramolecular interactions due to the different surroundings of water and $H_2PO_4^-$ within the structure. Thermal properties of $Ca(H_2PO_4)_2 \cdot H_2O$ prepared from different bivalve shells giving the different results indicate clearly that this property is dependent on raw materials.

Figure 7 presents FTIR spectra of the MCP-C, MCP-M, and MCP-O samples that are very similar because of fundamental vibrational bands of $H_2PO_4^-$ and H_2O block units in the $Ca(H_2PO_4)_2 \cdot H_2O$ structure for each sample. The vibrational modes of the $H_2PO_4^-$ anion are characterized by two types [47,48]: (i) the PO_4^{3-} (T_d symmetry) internal vibrations, and (ii) the vibrations involving OH motions. The dihydrogen phosphate anion ($H_2PO_4^-$) is a nonlinear structure containing seven atoms, which must have 15 ($3 \times 7 - 6$) normal vibrational modes [43]. The nine vibrations coming from the PO_4^{3-} (T_d symmetry) contain well-known normal modes: symmetric stretching ($\nu_1(A_1)$), symmetric bending ($\nu_2(E)$), asymmetric stretching ($\nu_3(F_2)$), and asymmetric bending ($\nu_4(F_2)$) modes. The existence of two P-OH bonds results in a decreasing molecular symmetry of the $H_2PO_4^-$ anion from its highest possible symmetry of the C_{2v} point group. As a result, the degenerate modes of $\nu_2(E)$, $\nu_3(F_2)$, and $\nu_4(F_2)$ are fully lifted: $\nu_2(E)$ separates into two modes ($A_1 + A_2$) and $\nu_3(F_2)$ and $\nu_4(F_2)$ into three modes ($A_1 + B_1 + B_2$) each. These eight vibrations happen from the intra-ionic coupling interaction of two longer P- OH and two shorter P-O bonds for the PO_4 stretching vibrations, which may also be led to additional modes as $\nu_s(P(OH)_2)$, $\nu_{as}(P(OH)_2)$, $\nu_s(PO_2)$, and $\nu_{as}(PO_2)$ for each $H_2PO_4^-$ group. The six vibrations linking OH motions are characteristic for the $H_2PO_4^-$ anion consisting of three modes ($\nu(OH)$, $\delta(OH)$, and $\gamma(OH)$) for each POH group. For water molecules, fundamental vibrations contain three normal vibrations: symmetric stretching ($\nu_1(OH)$), symmetric bending($\nu_2(HOH)$), and asymmetric stretching ($\nu_3(OH)$) and three vibrations (wagging, rocking, and twisting). The bands observed in spectra of each sample are 493, 565, 690, 862, 963, 1091, 1164, 1237, 1388, 1679, 1700, 2311, 2440, 2960, 3263, 3470 cm^{-1}, which are assigned to $\nu_2(PO_4{}^{3-})$, $\nu_4(PO_4{}^{3-})$, $L_1(H_2O)$, $\gamma(OH)$, $\nu_{as}(P(OH)_2)$, $\nu_s(PO_2)$, $\nu_{as}(PO_2)$, $\delta(OH)$ (1), $\delta(OH)$ (2), $\nu_2(HOH)$, $\nu(OH)$ or C band, $\nu(OH)$ or B band, $\nu(OH)$ or B band, $\nu(OH)$ or A band, $(\nu_1(OH))$ of H_2O, and $(\nu_3(OH))$ of H_2O, respectively [49]. The FTIR results obtained are very similar to that of the $Ca(H_2PO_4)_2 \cdot H_2O$ in literature [49], which confirms that the MCP-C, MCP-M, and MCP-O samples have major content as this crystal phase.

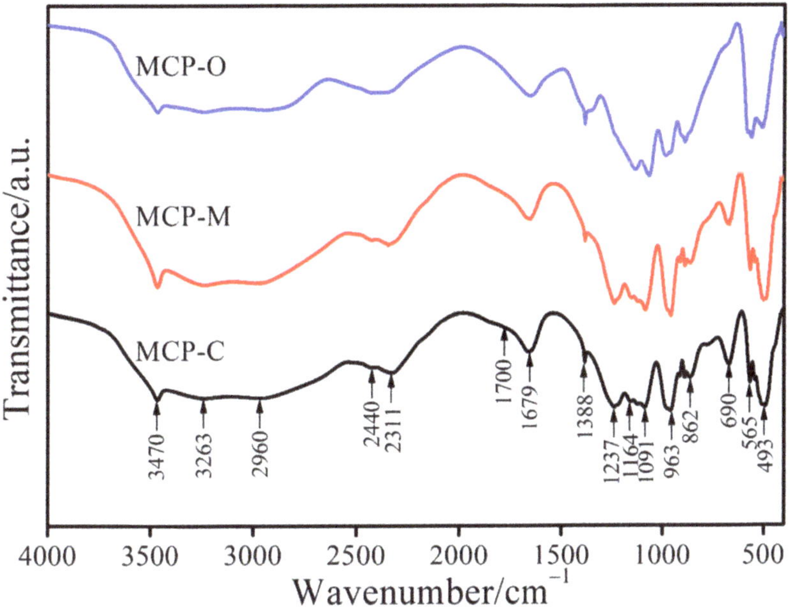

Figure 7. FTIR spectra of MCP-C, MCP-M, and MCP-O samples.

X-ray diffraction patterns of the MCP-C, MCP-M, and MCP-O samples (Figure 8) are the same 2θ positions but intense peaks are different. All detectable peaks of the obtained MCP-C, MCP-M, and MCP-O samples indexed as the $Ca(H_2PO_4)_2 \cdot H_2O$ structure match with the standard data of PDF no. 70–0090 [41,46]. The XRD patterns exhibit two sharp characteristic peaks at 2θ = 22.95 and 24.18° corresponding to (0–21) and (210) reflections for anorthic crystal structure of $Ca(H_2PO_4)_2 \cdot H_2O$. The labeled diffraction peaks can be indexed according to standard XRD data and XRD peaks of other phases were not observed, confirming the pure compounds obtained under study. The XRD results and the FTIR data are well coincident.

Figure 9 presents the typical micrographs of the three selected powder (MCP-C, MCP-M, and MCP-O) samples. As shown in Figure 9, the MCP-C particles resemble polyhedral morphologies of sheet shapes with smooth surfaces. The MCP-M particles seem to have inherited the polyhedral morphologies of the plate-like microstructures with smooth surfaces. The MCP-O particles show polyhedral morphologies of lamellar-like shapes with smooth surfaces. Morphologies of three $Ca(H_2PO_4)_2 \cdot H_2O$ samples prepared different bivalve (cockle, mussel, and oyster) shells are slightly different in shape and particle size but these morphologies are significantly from those of raw material powders.

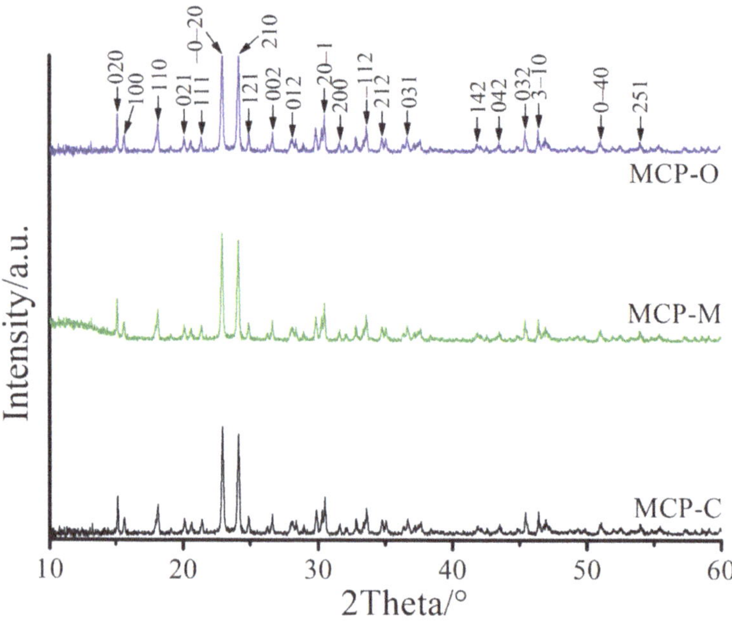

Figure 8. XRD patterns of MCP-C, MCP-M, and MCP-O samples.

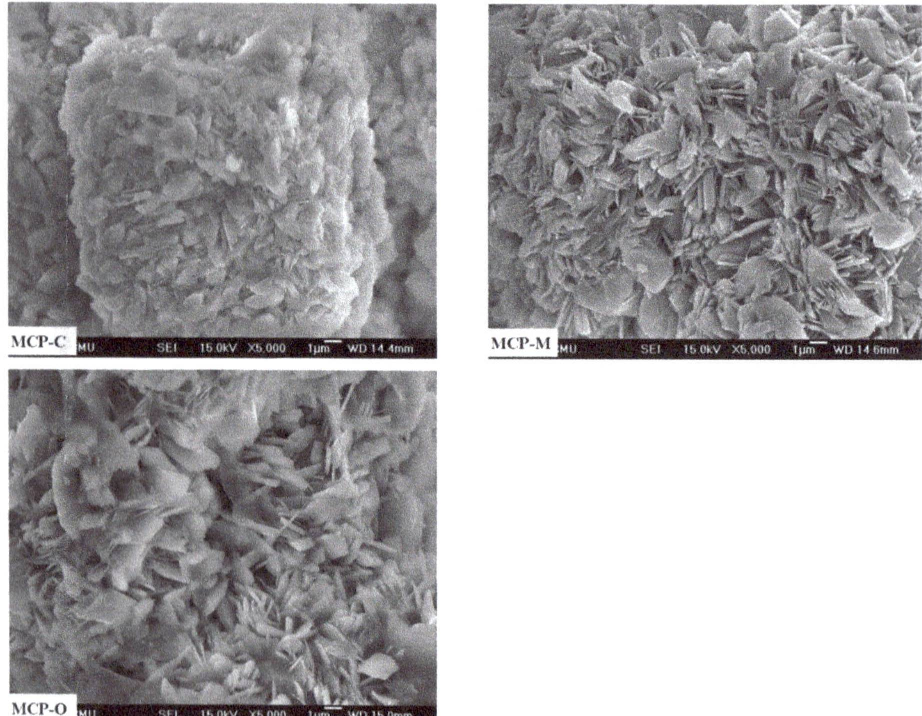

Figure 9. SEM micrographs of MCP-C, MCP-M and MCP-O samples.

3.3. Characterization Results of Tricalcium Phosphates

For FTIR spectra of tricalcium phosphates prepared from the calcination of the mixture of the produced monocalcium phosphates with its calcium raw materials (cockle, mussel, and oyster shells), labeled as the TCP-C, TCP-M, and TCP-O and shown in Figure 10. The FTIR spectra of each sample are very similar due to the fundamental vibrating unit, PO_4^{3-} anion containing within the structure. Vibrational modes are discussed similarly with the previous section and 9 ($3 \times 5 - 6$) normal vibrational modes of each phosphate group will be assigned [41]. In theory, the $\nu_3(F_2)$ and $\nu_4(F_2)$ modes are active in infrared while the $\nu_1(A_1)$ and $\nu_2(E)$ modes are active in Raman [42]. Vibrational bands of the PO_4^{3-} anion for all prepared products detected in the regions of 700–450 and 1250–900 cm^{-1} are defined to the $\nu_4(PO_4^{3-})$ and $\nu_3(PO_4^{3-})$ modes, respectively. Various vibrational peaks in these frequency regions insist on the presence of distinct nonequivalent phosphate block units in each structure and the loss of the degenerate modes of vibration resulting from correlation field splitting [42,43]. Additionally, a strong $\nu_s(POP)$ band (721 cm^{-1}) occurred is known to be the most striking characteristic of polyphosphate vibration.

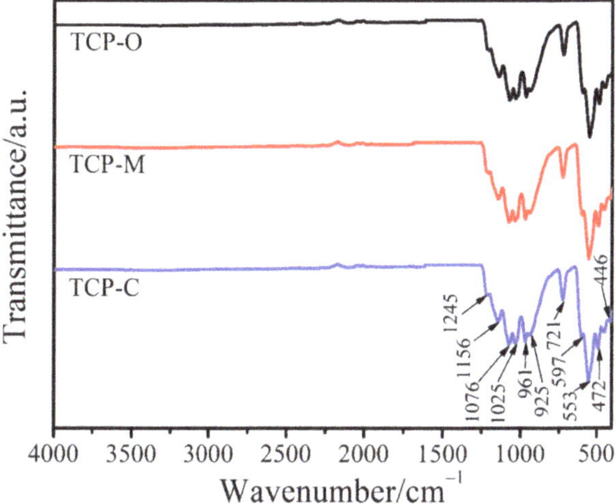

Figure 10. FTIR spectra of TCP-C, TCP-M and TCP-O samples.

The XRD patterns of TCP-C, TCP-M, and TCP-O samples were indexed as $Ca_3(PO_4)_2$ structures, which are classified using the standard data from the International Centre for Diffraction Data (ICDD), exhibited in Figure 11. The powders performed by different shells match with PDF no. 55–0898. In all the prepared samples, the maximum peak relating with the crystallinity of beta phase was detected at $2\theta = 31.5°$, corresponding to the (021) planes, and indicating rhombohedral crystal phase, which agrees with the previous reports [50,51]. From the XRD data of the prepared TCP samples, no other identified peaks related to impurities and any intermediate or remaining raw materials are noted, which further insists on the purity of the synthesized β-$Ca_3(PO_4)_2$ products. The obtained FTIR data of all the synthesized samples are consistent with the XRD results, which verify the classification of each compound.

The typical micrographs of the three selected powder (TCP-C, TCP-M, and TCP-O) samples are presented in Figure 12. The TCP-C particles look like the coarse surface on the grain-like shape and are highly agglomerated. The TCP-M particles were like polyhedral morphology of grain shape with uniform particles of 0.5–5.0 μm in size and agglomerations appear. The TCP-O particles were like polyhedral granular with identical particles of 0.2–5.0 μm in size and smooth surfaces. The difference of particle sizes and morphologies of three $Ca_3(PO_4)_2$ samples was caused by the different kinds of used bivalve shells as raw reagents. The difference of particle sizes and morphologies of the three $Ca_3(PO_4)_2$ samples was caused by the different kinds of used bivalve shells as raw reagents. The SEM micrograph data of the obtained products are significantly different from those of the raw material powders.

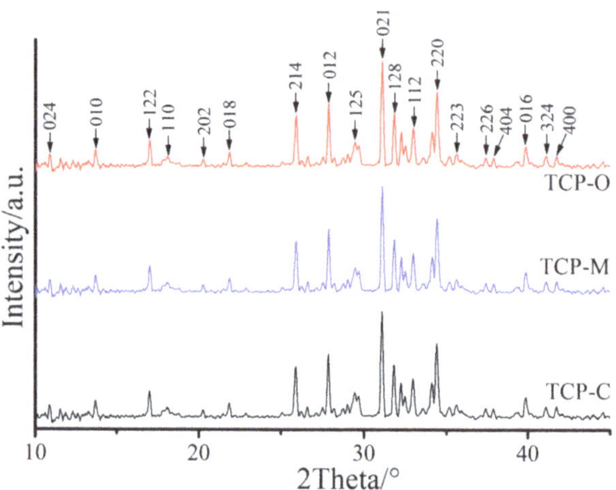

Figure 11. The XRD patterns of TCP-C, TCP-M, and TCP-O samples.

Figure 12. SEM micrographs of TCP-C, TCP-M and TCP-O samples.

4. Conclusions

With depleting natural resources, it is important to find new sustainable sources of materials and in this research, MCP and TCP were successfully produced from bivalve shells that include cockle, mussel, and oyster. In addition, by recycling these seafood wastes we also help remove large amounts of shell wastes that could pose health and environmental hazards. The method reported herein is simple, rapid, cost-effective, and environmentally friendly. MCP and TCP obtained from this method showed resemble characteristics to previous reports, indicating high purities. The slight differences in thermal properties of the $CaCO_3$ and MCP prepared depend on the starting shell powders. The morphologies of all the prepared samples were significantly different clearly indicating the raw material's effect on this property. Overall, MCP and TCP converted from these bivalve shell wastes, have a huge potential to be used in many industrials bringing about both environmental and economic benefits.

Author Contributions: S.S. conceived and designed the experiments; B.B. performed the experiments, analyzed the data and wrote the manuscript; K.C. contributed SEM observations and proof-read the manuscript; W.B. and N.L. discussed the experiments and the manuscript. All authors have read and agreed to the published version of the manuscript.

Funding: This research was funded by the Thailand Science Research and Innovation (TSRI), grant number RE-KRIS/008/64.

Institutional Review Board Statement: Not applicable.

Informed Consent Statement: Not applicable.

Data Availability Statement: The data presented in this study are available on request from the corresponding author.

Acknowledgments: The authors would like to thank the Scientific Instruments Center KMITL for supporting TGA, FTIR, XRD, and SEM techniques.

Conflicts of Interest: The authors declare no conflict of interest.

References

1. Yang, H.; Yan, N. Transformation of Seafood Wastes into Chemicals and Materials. *Encycl. Sustain. Sci. Technol.* **2018**. [CrossRef]
2. FAO. *The State of World Fisheries and Aquaculture 2020. Sustainability in Action*; FAO: Rome, Italy, 2020.
3. Yan, N.; Chen, X. Sustainability: Don't waste seafood waste. *Nature* **2015**, *524*, 155–157. [CrossRef] [PubMed]
4. Miculescu, F.; Mocanu, A.-C.; Maidaniuc, A.; Dascălu, C.-A.; Miculescu, M.; Voicu, Ş.I.; Ciocoiu, R.-C. Biomimetic Calcium Phosphates Derived from Marine and Land Bioresources. *Hydroxyapatite Adv. Compos. Nanomater. Biomed. Appl. Its Technol. Facet.* **2017**. [CrossRef]
5. Vecchio, K.S.; Zhang, X.; Massie, J.B.; Wang, M.; Kim, C.W. Conversion of bulk seashells to biocompatible hydroxyapatite for bone implants. *Acta Biomater.* **2007**, *3*, 910–918. [CrossRef]
6. Zhang, X.; Vecchio, K.S. Conversion of natural marine skeletons as scaffolds for bone tissue engineering. *Front. Mater. Sci.* **2013**, *7*, 103–117. [CrossRef]
7. Wu, S.-C.; Hsu, H.-C.; Wu, Y.-N.; Ho, W.-F. Hydroxyapatite synthesized from oyster shell powders by ball milling and heat treatment. *Mater. Charact.* **2011**, *62*, 1180–1187. [CrossRef]
8. Shavandi, A.; Bekhit, A.E.-D.A.; Ali, A.; Sun, Z. Synthesis of nano-hydroxyapatite (nHA) from waste mussel shells using a rapid microwave method. *Mater. Chem. Phys.* **2015**, *149–150*, 607–616. [CrossRef]
9. Bhattacharjee, B.N.; Mishra, V.K.; Rai, S.B.; Parkash, O.; Kumar, D. Structure of apatite nanoparticles derived from marine animal (crab) shells: An environment-friendly and cost-effective novel approach to recycle seafood waste. *ACS Omega* **2019**, *4*, 12753–12758. [CrossRef] [PubMed]
10. Ferraz, E.; Gamelas, J.A.F.; Coroado, J.; Monteiro, C.; Rocha, F. Recycling waste seashells to produce calcitic lime: Characterization and wet slaking reactivity. *Waste Biomass Valor.* **2019**, *10*, 2397–2414. [CrossRef]
11. Dampang, S.; Purwanti, E.; Destyorini, F.; Kurniawan, S.B.K.B.; Abdullah, S.R.S.; Imron, M.F. Analysis of optimum temperature and calcination time in the production of CaO using seashells waste as $CaCO_3$ source. *J. Ecol. Eng.* **2021**, *22*, 221–228. [CrossRef]
12. Janković, B.; Smičiklas, I.; Manić, N.; Mraković, A.; Mandić, M.; Veljović, Đ.; Jović, M. Thermo-oxidative evolution and physico-chemical characterization of seashell waste for application in commercial sectors. *Thermochim. Acta* **2020**, *686*, 178568. [CrossRef]
13. Li, Y.; Huang, P.; Guo, S.; Nie, M. A promising and green strategy for recycling waste oyster shell powder as bio-filler in polypropylene via mycelium-enlightened interfacial interlocking. *J. Clean. Prod.* **2020**, *272*, 122694. [CrossRef]

14. Peksen, C.; Koroglu, L.; Kartal, H. Utilization of seashells in matte glaze preparation. *Int. J. Appl. Ceram. Technol.* **2020**, *17*, 1940–1947. [CrossRef]
15. Her, S.; Park, T.; Zalnezhad, E.; Bae, S. Synthesis and characterization of cement clinker using recycled pulverized oyster and scallop shell as limestone substitutes. *J. Clean. Prod.* **2021**, *278*, 123987. [CrossRef]
16. Wang, J.; Liu, E. Upcycling waste seashells with cement: Rheology and early-age properties of Portland cement paste. *Resour. Conserv. Recycl.* **2020**, *155*, 104680. [CrossRef]
17. Seo, J.H.; Park, S.M.; Yang, B.J.; Jang, J.G. Calcined oyster shell powder as an expansive additive in cement mortar. *Materials* **2019**, *12*, 1322. [CrossRef]
18. Wang, Q.; Jiang, F.; Ouyang, X.-K.; Yang, L.-Y.; Wang, Y. Adsorption of Pb(II) from aqueous solution by mussel shell-based adsorbent: Preparation, characterization, and adsorption performance. *Materials* **2021**, *14*, 741. [CrossRef] [PubMed]
19. Larraguibel, A.; Navarrete-Calvo, A.; García, S.; Armijos, V.F.; Caraballo, M.A. Exploring sulfate and metals removal from Andean acid mine drainage using $CaCO_3$-rich residues from agri-food industries and witherite ($BaCO_3$). *J. Clean. Prod.* **2020**, *274*, 123450. [CrossRef]
20. Magnabosco, G.; Giuri, D.; Di Bisceglie, A.P.; Scarpino, F.; Fermani, S.; Tomasini, C.; Falini, G. New material perspective for waste seashells by covalent functionalization. *ACS Sustain. Chem. Eng.* **2021**, *9*, 6203–6208. [CrossRef]
21. Wei, D.; Zhang, H.; Cai, L.; Guo, J.; Wang, Y.; Ji, L.; Song, W. Calcined mussel shell powder (CMSP) via modification with surfactants: Application for antistatic oil-removal. *Materials* **2018**, *11*, 1410. [CrossRef]
22. FAO. *The State of World Fisheries and Aquaculture 2018-Meeting the Sustainable Development Goals*; FAO: Rome, Italy, 2018.
23. Jovic, M.; Mandic, M.; Sljivic-Ivanovic, M.; Smiciklas, I. Recent trends in application of shell waste from mariculture. *Stud. Mar.* **2019**, *32*, 47–62.
24. Seesanong, S.; Laosinwattana, C.; Chaiseeda, K.; Boonchom, B. A simple and rapid transformation of golden apple snail (*Pomacea canaliculata*) shells to calcium carbonate, monocalcium and tricalcium phosphates. *Asian J. Chem.* **2019**, *31*, 2522–2526. [CrossRef]
25. Seesanong, S.; Laosinwattana, C.; Boonchom, B. Microparticles of calcium carbonate $CaCO_3$, calcium hydrogen phosphate hydrate $CaHPO_4 \cdot 1.9H_2O$ and tricalcium phosphate $Ca_3(PO_4)_2$ prepared from golden apple snail shells (*Pomacea canaliculata*). *Res. J. Chem. Environ.* **2020**, *24*, 1–6.
26. Sánchez-Enríquez, J.; Reyes-Gasga, J. Obtaining $Ca(H_2PO_4)_2 \cdot H_2O$, monocalcium phosphate monohydrate, via monetite from brushite by using sonication. *Ultrason. Sonochem.* **2013**, *20*, 948–954. [CrossRef]
27. Dorozhkin, S.V. Biphasic, Triphasic, and Multiphasic Calcium Orthophosphates. In *Advanced Ceramic Materials*; Tiwari, A., Gerhardt, R.A., Szutkowska, M., Eds.; Scrivener Publishing: Beverly, MA, USA, 2016; pp. 33–95.
28. Hsu, Y.-S.; Chang, E.; Liu, H.-S. Hydrothermally-grown monetite ($CaHPO_4$) on hydroxyapatite. *Ceram. Int.* **1998**, *24*, 249–254. [CrossRef]
29. Gonzalez-McQuire, R.; Chane-Ching, J.-Y.; Vignaud, E.; Lebugle, A.; Mann, S. Synthesis and characterization of amino acid-functionalized hydroxyapatite nanorods. *J. Mater. Chem.* **2004**, *14*, 2277–2281. [CrossRef]
30. Liu, J.; Li, K.; Wang, H.; Zhu, M.; Yan, H. Rapid formation of hydroxyapatite nanostructures by microwave irradiation. *Chem. Phys. Lett.* **2004**, *396*, 429–432. [CrossRef]
31. Kong, X.-D.; Sun, X.-D.; Lu, J.-B.; Cui, F.-Z. Mineralization of calcium phosphate in reverse microemulsion. *Curr. Appl. Phys.* **2005**, *5*, 519–521. [CrossRef]
32. Boonchom, B.; Danvirutai, C. The morphology and thermal behavior of calcium dihydrogen phosphate monohydrate ($Ca(H_2PO_4)_2 \cdot H_2O$) obtained by a rapid precipitation route at ambient temperature in different media. *J. Optoelectron. Biomed. Mater.* **2009**, *1*, 115–123.
33. Liu, D.-M.; Yang, Q.; Troczynski, T.; Tseng, W.J. Structural evolution of sol–gel-derived hydroxyapatite. *Biomaterials* **2002**, *23*, 1679–1687. [CrossRef]
34. Sivakumar, G.R.; Girija, E.K.; Narayana Kalkura, S.; Subramanian, C. Crystallization and characterization of calcium phosphates: Brushite and monetite. *Cryst. Res. Technol.* **1998**, *33*, 197–205. [CrossRef]
35. Rohanizadeh, R.; LeGeros, R.Z.; Harsono, M.; Bendavid, A. Adherent apatite coating on titanium substrate using chemical deposition. *J. Biomed. Mater. Res. A* **2005**, *72*, 428–438. [CrossRef] [PubMed]
36. Djošić, M.S.; Mišković-Stanković, V.B.; Kačarević-Popović, Z.M.; Jokić, B.M.; Bibić, N.; Mitrić, M.; Milonjić, S.K.; Jančić-Heinemann, R.; Stojanović, J. Electrochemical synthesis of nanosized monetite powder and its electrophoretic deposition on titanium. *Colloids Surf. A Physicochem. Eng. Asp.* **2009**, *341*, 110–117. [CrossRef]
37. Suchanek, W.L.; Shuk, P.; Byrappa, K.; Riman, R.E.; TenHuisen, K.S.; Janas, V.F. Mechanochemical–hydrothermal synthesis of carbonated apatite powders at room temperature. *Biomaterials* **2002**, *23*, 699–710. [CrossRef]
38. Onoda, H.; Nakanishi, H. Preparation of calcium phosphate with oyster shells. *Nat. Resour.* **2012**, *3*, 71–74. [CrossRef]
39. Macha, I.J.; Ozyegin, L.S.; Chou, J.; Samur, R.; Oktar, F.N.; Ben-Nissan, B. An alternative synthesis method for di calcium phosphate (monetite) powders from Mediterranean mussel (*Mytilus galloprovincialis*) shells. *J. Aust. Ceram. Soc.* **2013**, *49*, 122–128.
40. Karunadasa, K.S.P.; Manoratne, C.H.; Pitawala, H.M.T.G.A.; Rajapakse, R.M.G. Thermal decomposition of calcium carbonate (calcite polymorph) as examined by in-situ high-temperature X-ray powder diffraction. *J. Phys. Chem. Solids* **2019**, *134*, 21–28. [CrossRef]
41. Seesanong, S.; Laosinwattana, C.; Boonchom, B. A simple rapid route to synthesize monocalcium phosphate monohydrate using calcium carbonate with different phases derived from green mussel shells. *J. Mater. Environ. Sci.* **2019**, *10*, 113–118.

42. Nakamoto, K. *Infrared and Raman Spectra of Inorganic and Coordination Compounds: Part A: Theory and Applications in Inorganic Chemistry*, 6th ed.; John Wiley & Sons: Hoboken, NJ, USA, 2008.
43. Cotton, F.A. *Chemical Applications of Group Theory*, 3rd ed.; Wiley-Interscience: New York, NY, USA, 1990.
44. Koura, N.; Kohara, S.; Takeuchi, K.; Takahashi, S.; Curtiss, L.A.; Grimsditch, M.; Saboungi, M.-L. Alkali carbonates: Raman spectroscopy, ab initio calculations, and structure. *J. Mol. Struct.* **1996**, *382*, 163–169. [CrossRef]
45. Rahman, M.A.; Oomori, T. Structure, crystallization and mineral composition of sclerites in the alcyonarian coral. *J. Cryst. Growth* **2008**, *310*, 3528–3534. [CrossRef]
46. Boonchom, B. Parallelogram-like microparticles of calcium dihydrogen phosphate monohydrate ($Ca(H_2PO_4)_2 \cdot H_2O$) obtained by a rapid precipitation route in aqueous and acetone media. *J. Alloys Compd.* **2009**, *482*, 199–202. [CrossRef]
47. Koleva, V.; Stefov, V.; Najdoski, M.; Cahil, A. Vibrational spectra of cobalt dihydrogen phosphate dihydrate, $Co(H_2PO_4)_2 \cdot 2H_2O$. *Vib. Spectrosc.* **2012**, *62*, 229–237. [CrossRef]
48. Koleva, V.; Stefov, V. Phosphate ion vibrations in dihydrogen phosphate salts of the type $M(H_2PO_4)_2 \cdot 2H_2O$ (M=Mg, Mn, Co, Ni, Zn, Cd): Spectra–structure correlations. *Vib. Spectrosc.* **2013**, *64*, 89–100. [CrossRef]
49. Xu, J.; Gilson, D.F.R.; Butler, I.S. FT-Raman and high-pressure FT-infrared spectroscopic investigation of monocalcium phosphate monohydrate, $Ca(H_2PO_4)_2 \cdot H_2O$. *Spectrochim. Acta A* **1998**, *54*, 1869–1878. [CrossRef]
50. Rangavittal, N.; Landa-Cánovas, A.R.; González-Calbet, J.M.; Vallet-Regí, M. Structural study and stability of hydroxyapatite and β-tricalcium phosphate: Two important bioceramics. *J. Biomed. Mater. Res.* **2000**, *51*, 660–668. [CrossRef]
51. Prevéy, P.S. X-ray diffraction characterization of crystallinity and phase composition in plasma-sprayed hydroxyapatite coatings. *J. Therm. Spray Technol.* **2000**, *9*, 369–376. [CrossRef]

Article

Evaluation of Hydrogenation Kinetics and Life Cycle Assessment on Mg$_2$NiH$_x$–CaO Composites

Hyo-Won Shin, June-Hyeon Hwang, Eun-A Kim and Tae-Whan Hong *

Department of Materials Science and Engineering, College of Engineering,
Korea National University of Transportation, Chungju 27469, Korea; whun2288@ut.ac.kr (H.-W.S.);
hj930403@ut.ac.kr (J.-H.H.); euna_0106@ut.ac.kr (E.-A.K.)
* Correspondence: twhong@ut.ac.kr; Tel.: +82-43-841-5388

Abstract: Magnesium-based alloys are attractive as hydrogen storage materials due to their lightweight and high absorption, but their high operating temperatures and very slow kinetics are obstacles to practical applications. Therefore, the effect of CaO has improved the hydrogenation kinetics and slowed down the degradation. The Mg$_2$NiH$_x$–CaO composites were prepared by hydrogen-induced mechanical alloying (HIMA). Hydrogenation kinetics was performed by using an Automatic PCT Measuring System and evaluated in the temperature range of 423, 523, and 623 K. As a result of calculating the hydrogen absorption amounts through the hydrogenation kinetics curve, they were calculated as about 0.52 wt%, 1.21 wt%, and 1.59 wt% (Mg$_2$NiH$_x$–10 wt% CaO). In this study, the material environmental aspects of Mg$_2$NiH$_x$–CaO composites were investigated through life cycle assessment (LCA). LCA was performed analyzing the environmental impact characteristics of the manufacturing process by using Gabi software and the Eco-Indicator 99' and Centrum voor Milieuweten schappen (CML 2001) methodology. As a result, the contents of global warming potential (GWP) and fossil fuels were found to have a higher impact than other impact categories.

Keywords: hydrogen storage; kinetics; material life cycle assessment; Eco-Indicator 99'; CML 2001

Citation: Shin, H.-W.; Hwang, J.-H.; Kim, E.-A.; Hong, T.-W. Evaluation of Hydrogenation Kinetics and Life Cycle Assessment on Mg$_2$NiH$_x$–CaO Composites. *Materials* **2021**, *14*, 2848. https://doi.org/10.3390/ma14112848

Academic Editors: Ute Kalbe and Franz-Georg Simon

Received: 27 April 2021
Accepted: 24 May 2021
Published: 26 May 2021

Publisher's Note: MDPI stays neutral with regard to jurisdictional claims in published maps and institutional affiliations.

1. Introduction

Recently, due to greenhouse gas emissions and global warming problems, the need to develop a new and renewable energy source that can replace fossil fuels has increased, and hydrogen, a clean energy media, has attracted attention [1]. Hydrogen is known as a clean energy media because, unlike fossil fuels, it cannot be used directly in nature, and it is produced from primary energy sources to produce energy through internal combustion engines or fuel cells, and only water is produced as a by-product [2,3]. Accordingly, in January 2019, the government announced the "Hydrogen Economy Revitalization Roadmap" to realize a zero-carbon society and lead the transformation of the energy paradigm by utilizing hydrogen with a high energy storage density (33.3 kWh/kg H$_2$). To realize these hydrogen economies, technological innovation in the production, storage, transportation, and utilization of hydrogen are necessary. However, in the case of hydrogen storage technology, which is essential for safely storing and transporting hydrogen, secure technology is urgently needed because investment is not made relatively [4].

Among the hydrogen storage methods, metal hydrides belonging to hydrogen storage alloys are produced by the reaction between metal and hydrogen, and the metal adsorbs hydrogen gas, and when heated again, it releases hydrogen. The reaction in which hydrogen reacts with a metal to form metal hydride (MH) is an exothermic reaction [5]. In particular, Mg-based hydride has the advantages of having high hydrogen storage, a low cost, and being lightweight. The surface is thermodynamically stable and has a very slow hydrogenation reaction rate [6–8].

To improve these obstacles, studies on the catalytic effect were conducted by adding a transition metal, and in the case of metal hydride (Mg$_2$NiH$_4$), in order to change the thermodynamic stability, there have been studies on improving the storage and release characteristics of hydrogen by substituting various elements such as Ca and rare earth metals for Mg and Ni in the Mg_2Ni alloy [9]. In this study, Mg_2NiH_x–CaO composites were prepared by adding CaO to Mg_2NiH_x using hydrogen-induced mechanical alloying (HIMA). The effect of catalyst and oxidation resistance was investigated by paying attention to the hydrogenation behavior according to the added alkaline earth metal oxide.

Life cycle assessment (LCA) is an input and output to assess the environmental impact of a product or service throughout the entire process (raw material collection, product production, use, disposal), that is, resource depletion due to inputs, and environmental impacts caused by discharges. It can be said that it is a process of reviewing alternatives to improve environmental performance by preparing a quantitative data list of the data, evaluating the environmental impact [10]. In addition, as it forms the technical basis of the ISO 14000 series, it can be said to be an internationally important technique [11]. These results enable a fair comparison of products or processes and can also contribute to product design that minimizes environmental impact. Problems related to toxic emissions or waste can be solved by replacing materials or processes, as well as effects related to raw materials and energy consumed [12]. While LCA is a valuable way to measure environmental loads such as "cradle-to-grave", it has limitations when it comes to obtaining and evaluating data about the entire production process. LCA uses an internationally standardized methodological framework for analyzing the environmental impacts associated with the life cycle phases of products, processes, or activities over their entire life, typically from cradle-to-grave [13]. All products are made from materials, and one material is made using different technologies or used in different products [14]. Material life cycle assessment literally provides an important tool for material research as an environmental evaluation method that focuses on materials rather than processes. There is a case in which the potential environmental impact of recycling of indium tin oxide (ITO) transparent electrodes separated from the display panel has been studied with this material-focused environmental evaluation method [15]. Therefore, in this study, LCA was carried out to confirm that the hydrogenation kinetics of the Mg_2NiH_x–CaO composites were improved and to evaluate the potential environmental impact of the material.

2. Experimental Procedure

2.1. Specimen Preparation and Characterization

Mg (Sigma-Aldrich, St. Louis, MO, USA, 98%) and Ni (Sigma-Aldrich, St. Louis, MO, USA, 99.7%) powder was charged into a 1/2-inch STS304 container. At this time, the weight ratio of Mg and Ni powder was designed as 45:55 with reference to the Mg–Ni binary phase diagram. After making a vacuum up to 5×10^{-2} Torr using a rotary pump, hydrogen of 99.9999% purity was applied to a pressure of 3.0 MPa and alloyed for 96 h at a rotational speed of 200 rpm using a planetary ball mill (Pulverisette-5, FRITSCH Co., Idar-Oberstein, Germany), which is a hydrogen-induced mechanical alloying method. At this time, the ball to chips weight ratio (BCR) of 1/2-inch chrome steel balls and magnesium chips was set to 66:1 with reference to the preceding paper [16]. Then, the prepared powder and 5, 10 wt% CaO (Sigma-Aldrich, 99.7%) in the form of powder were charged into a container and alloyed for 24 h at a rotation speed of 200 rpm under the same conditions.

For the metallurgical characterization of the sample prepared through the alloying process, the crystal structure and phase of the sample were characterized using X-ray diffraction (XRD) analysis (D8 Advance, Bruker, Billerica, MA, USA), which was performed using a Cu $K\alpha$ radiation of 1.5405 Å (scanning speed: 3 deg/min, scanning angle: 20–80°). In order to observe the surface shape and particle size of the sample according to the alloying time, it was observed using scanning electron microscopy (SEM) (FEI Quanta 400, FEI, Hillsboro, OR, USA), and Brunauer–Emmett–Teller (BET) surface analysis (Micromeritics-3-Flex, Heidelberg, Germany) was used to measure the particle specific surface area, which has a large influence on hydrogen diffusion. After that, the dehydrogenation activation energy was measured by drawing an Arrhenius plot through the dehydrogenation temperature.

In addition, Sivert's type automatic PCT (pressure–composition–temperature) method, an automated volumetric measurement method, was used to measure the hydrogenation kinetics, and the hydrogen absorption reaction rate was evaluated at a temperature range of 423, 523, and 623 K for 1 h by applying a constant hydrogen pressure of 3.0 MPa.

2.2. Life Cycle Assessment (LCA)

The LCA is used to analyze the environmental impact of a product and can provide information on the stages of a product's life cycle [17]. LCA has an ISO standardization method (ISO 14040 2006; ISO 14044 2006), an LCA consists of four categories: (1) goal and scope definition, (2) life cycle inventory analysis, (3) impact assessment and (4) interpretation of results [18,19]. The quality of the LCA depends on the exact description of the production process to be analyzed. To know where each phase of the life cycle begins and ends properly, you need to collect and interpret its process data. The study used the Centrum voor Milieuweten schappen (CML 2001), a combined lifecycle impact assessment method developed by the University of Leiden to determine the environmental performance of the process under study. The CML method defines several impact categories for emissions and resource consumption as problem-oriented (midpoint).

2.2.1. Goal and Scope

The goal and scope outline the material to be studied, assess the environmental impact categories, and analyze the resulting limitations or assumptions. First of all, it is very important to first establish the decisions to be presented by the evaluation for material selection [20]. The life cycle inventory (LCI) consists of identifications of all unit processes, product-related flows [13]. Additionally, the methodology adopted for LCA analysis in this study complies with the following standards [21]:

ISO 14040, 14044: 2006—Environmental Management—Life Cycle Assessment—Requirements and Guidelines [18,19].

In this study, a cradle-to-gate approach of LCA was applied to assess Mg_2NiH_x–5, 10 wt% CaO composites that were manufactured and characterized, and an environmental assessment was carried out throughout the disposal process. The goal is to identify whether hydrogenation kinetics is improved by adding CaO in the synthesis process of Mg_2NiH_x–CaO composites, and quantify the resulting environmental load, and analyze the environmental characteristics.

The function of the prepared Mg_2NiH_x–CaO composites are the hydrogenation kinetics, and the functional unit, which is a unit representing the function, is set as hydrogen content (wt%), and the reference flow that satisfies the functional unit is 10 g of powder. Figure 1 is composed of a manufacturing step, a characteristic evaluation step, and a disposal step of the Mg_2NiH_x–CaO composites in the LCA. The raw material category includes Mg_2Ni and CaO. The energy category includes electricity used in the manufacturing and characterization stages, and air emissions are carbon monoxide (CO), carbon dioxide (CO_2), sulfur oxides (SO_x), nitrogen oxides (NO_x), and dust generated throughout the process. Data quality requirements were established by dividing the technical, temporal and regional scope into manufacturing, characterization, and disposal stages.

2.2.2. Impact Assessment Categories and Environmental Impact Methodology

In this study, the end-point concept CML 2001 methodology and Eco-Indicato'99 (EI99) methodology developed by Nederland and pre-consulting organizations were used. In this study, the CML 2001 and Eco-Indicator 99' (EI99) methodology with an end-point concept developed by a Dutch pre-consulting institution were used, and these are shown in Tables 1 and 2. The software Gabi 6 (Sphera, Stuttgart, AR, USA) was used to perform an environmental impact assessment on the process of synthesizing the composites. The EI99 methodology considers three damage categories: human health, ecosystem health, and resources, of which, the following types of damage are sorted into: carcinogenic, respiratory effects, climate change, radioactivity, ozone layer, ecotoxicity, acidification, land use, resource, and fuel replenishment. As an indicator, in the human health category, the index of the endpoint level is derived using the disability-adjusted life years (DALY) as an indicator. It is expressed as the probability (PDF $\times$ m^2 $\times$ yr) of the potential disappearance of species per area (m^2), and in the resource depletion category, the surplus energy input to harvest 1 kg of resources is selected as an indicator [22].

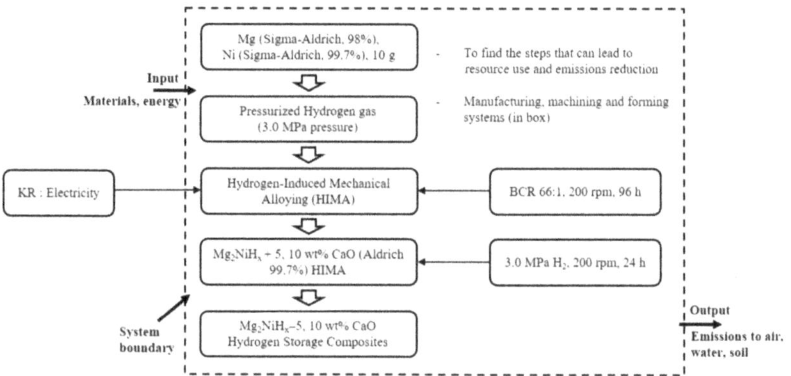

Figure 1. Process flow diagram for Mg$_2$NiH$_x$–5, 10 wt% CaO synthesis and analyses.

Table 1. Environmental impact categories applied using CML 2001.

Environmental Impact Categories	Unit	Life Cycle Environmental Impacts	
		Mg$_2$NiH$_x$–5 wt% CaO	Mg$_2$NiH$_x$–10 wt% CaO
Abiotic Resource Depletion (ARD)	Kg yr^{-1}	6.85×10^{-3}	8.76×10^{-3}
Global Warming Potential (GWP)	Kg CO$_2$ eq	3.86×10^{-2}	4.09×10^{-2}
Stratospheric Ozone Depletion Potential (ODP)	Kg CFC^{-11} eq	1.69×10^{-5}	2.57×10^{-5}
Photochemical Oxidation Potential (POCP)	Kg C$_2$H$_4$ eq	1.41×10^{-5}	3.64×10^{-5}
Acidification Potential (ACP)	Kg SO$_2$ eq	1.83×10^{-2}	1.45×10^{-2}
Eutrophication Potential (EUP)	Kg PO$_4$ eq	5.14×10^{-3}	5.69×10^{-3}
Fresh-water Aquatic Ecotoxicity Potential (FAETP)	Kg 1,4-DCB eq	1.57×10^{-8}	2.25×10^{-8}
Marine Aquatic Ecotoxicity Potential (MAETP)	Kg 1,4-DCB eq	1.89×10^{-8}	3.17×10^{-8}
Terrestrial Ecotoxicity Potential (TETP)	Kg 1,4-DCB eq	3.14×10^{-4}	5.33×10^{-4}
Human Toxicity Potential (HTP)	Kg 1,4-DCB eq	7.25×10^{-8}	7.74×10^{-8}

Table 2. Environmental impact categories Eco-Indicator 99' (EI99).

Damage Categories		Damage Unit	Life Cycle Environmental Impacts	
			Mg$_2$NiH$_x$–5 wt% CaO	Mg$_2$NiH$_x$–10 wt% CaO
Human health	Carcinogenic effect	DALY	1.12×10^{-8}	1.63×10^{-8}
	Respiratory (organic)	DALY	2.47×10^{-8}	2.68×10^{-8}
	Respiratory (inorganic)	DALY	2.98×10^{-2}	5.02×10^{-2}
	Climate change	DALY	1.16×10^{-1}	1.48×10^{-1}
	Ionizing radiation	DALY	2.03×10^{-2}	2.15×10^{-2}
	Ozone depletion	DALY	1.26×10^{-8}	1.29×10^{-8}
Ecosystem quality	Ecotoxicity	PDF $\times$ m^2 $\times$ yr	1.32×10^{-2}	2.11×10^{-2}
	Acidification/Eutrophication	PDF $\times$ m^2 $\times$ yr	1.84×10^{-2}	1.95×10^{-2}
	Land-use	PDF $\times$ m^2 $\times$ yr	1.16×10^{-2}	1.19×10^{-2}
Resources	Minerals	MJ	1.73×10^{-2}	1.82×10^{-2}
	Fossil	MJ	2.71×10^{-1}	3.18×10^{-1}

3. Results and Discussion

3.1. Evaluation of the Synthesized Composites

Figure 2 shows the XRD pattern of the Mg_2NiH_x–10 wt% CaO composites, which was ball-milled for 96 h in a hydrogen atmosphere to prepare Mg_2NiH_x, and then ball-milled for an additional 24 h by adding 5, 10 wt% CaO. As a result of the analysis, clear peaks of magnesium hydride and calcium oxide appeared, and Mg_2NiH, Mg_2NiH_4, and CaO peaks were identified through the International Centre for Diffraction Data (ICDD). Additionally, Mg_2NiH_4 has a monoclinic structure, and Mg_2NiH and CaO have a cubic structure.

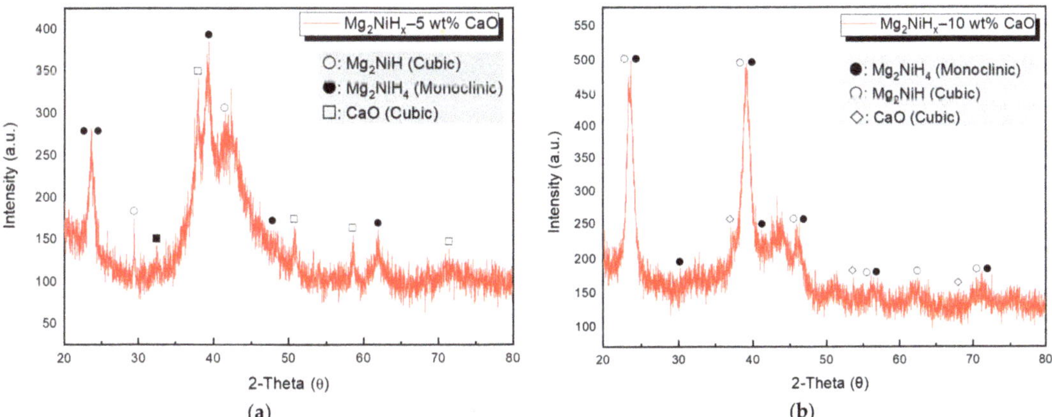

Figure 2. (**a**) Mg_2NiH_x–5 wt% CaO (reprinted with permission from. [23]. Copyright 2021. Copyright Shin, H.-W.), (**b**) Mg_2NiH_x–10 wt% CaO composites X-ray diffraction (XRD) image.

Figure 3 is a scanning electron microscope (SEM) surface shape observation photograph of Mg_2NiH_x–CaO composites. Particle sizes vary from 1 to 10 μm and a tendency to clumping has been observed due to the many nano-sized particles and the grinding process. It was pointed out that this cluster formation and irregular shape of the particle size distribution are evidence of the milling effect [24]. According to Huang et al., who studied the relationship between particle size and hydrogen diffusion, nano-sized particles and an increase in specific surface area promoted hydrogen absorption and desorption [25]. As a result of comparing the Mg_2NiH_x–5 wt% CaO and Mg_2NiH_x–10 wt% CaO composites, it can be seen that the particle size decreases as CaO is added. As the particle size decreases, the diffusion length of hydrogen decreases and the reaction surface area increases, so it is considered to be easier for hydrogen absorption and desorption.

Figure 4 is the result of specific surface area analysis (SSA) measuring the nitrogen absorption and desorption behavior of Mg_2NiH_x–CaO composites, and SSA is calculated as 2.955 m^2/g (Mg_2NiH_x–5 wt% CaO [23]), 3.773 m^2/g (Mg_2NiH_x–10 wt% CaO). Increasing the SSA of nanoparticles promote absorption and desorption of hydrogen but increases the alloying time and decreases the size of the particles, increasing the formation of nano and amorphous phases [26]. Comparing the Mg_2NiH_x–5, 10 wt% CaO composites, the specific surface area increases with the addition of CaO, and thus the absorption and desorption behavior of hydrogen is expected to be advantageous.

Figure 3. Mg$_2$NiH$_x$–5 wt% CaO (reprinted with permission from [23]. Copyright 2021. Copyright Shin, H.-W.). SEM morphologies: (**a**) ×2500, (**b**) ×5000, (**c**) ×10,000, Mg$_2$NiH$_x$–10 wt% CaO SEM morphologies: (**d**) ×2500, (**e**) ×5000, (**f**) ×10,000.

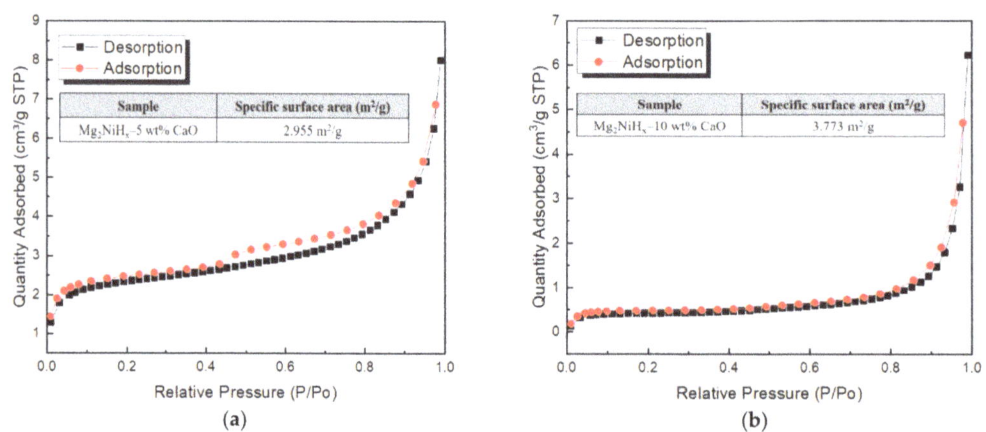

Figure 4. Bruner–Emmett–Teller surface area analysis results for (**a**) Mg$_2$NiH$_x$–5 wt% CaO (reprinted with permission from [23]. Copyright 2021 Copyright Shin, H.-W.), (**b**) Mg$_2$NiH$_x$–10 wt% CaO.

Figure 5a is the result of measuring the hydrogen absorption reaction kinetics of Mg$_2$NiH$_x$–10 wt% CaO composites under each temperature (423, 523, 623 K) condition. After the hydrogen pressure was kept constant at 3.0 MPa, the change in hydrogen absorption with time for 1 h was investigated. As a result, hydrogen absorption was highest at 623 K and lowest at 423 K. When the effective hydrogen storage amount was calculated through the hydrogenation kinetics curve, Mg_2NiH_x–5 wt% CaO at 423, 523, and 623 K temperatures were 0.51 wt%, 0.93 wt%, and 1.12 wt%. Mg_2NiH_x–10 wt% CaO showed 0.52 wt%, 1.21 wt%, 1.59 wt% of hydrogen absorption, through which it was confirmed that the hydrogenation reaction rate increased as CaO was added. Figure 5b is the result calculated through the van' t Hoff equation from the result of the hydrogen absorption reaction rate. The absorption enthalpy (ΔH) of the Mg_2NiH_x–5 wt% CaO was calculated as a value of 14.138 $\pm$ 0.67 kJ/mol [23], and the Mg_2NiH_x–10 wt% CaO showed a relatively high value of 20.617 $\pm$ 0.14 kJ/mol. Therefore, Mg_2NiH_x–10 wt% CaO composites exhibited superior kinetics characteristics compared to Mg_2NiH_x–5 wt% CaO due to the high heat of reaction.

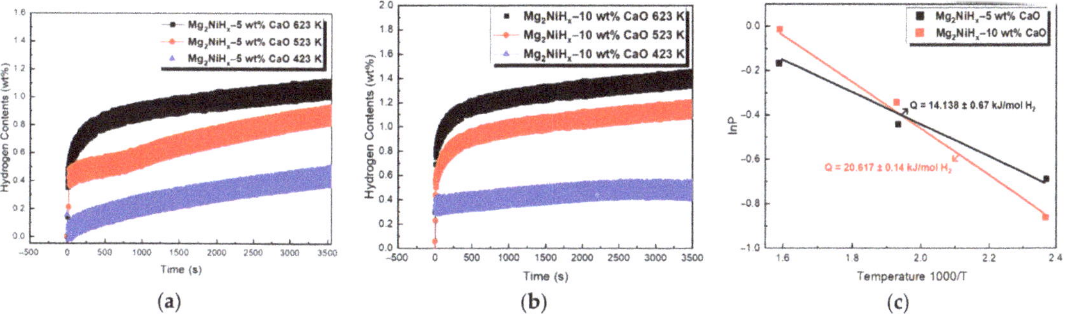

Figure 5. Hydrogen absorption kinetics of (**a**) Mg_2NiH_x–5 wt% CaO (reprinted with permission from [23]. Copyright 2021 Copyright Shin, H.-W.), (**b**) Mg_2NiH_x–10 wt% CaO, (**c**) calculation of van't Hoff plots on Mg_2NiH_x–5 wt% CaO (reprinted with permission from [23]. Copyright 2021 Copyright Shin, H.-W.) and Mg_2NiH_x–10 wt% CaO.

3.2. Life Cycle Assessment on Composites Prepared

The LCA process uses classification, characterization, and normalization. The environmental impact is then derived according to this order, and major issues are then identified, based on this. In our work, the classification involved 10 impact categories, which included abiotic resource depletion (ARD), global warming potential (GWP), stratospheric ozone depletion potential (ODP), acidification potential (ACP), and eutrophication potential (EUP), ecotoxicity potential (ETP), and human toxicity potential (HTP). Among these, ecological toxicity included fresh-water aquatic ecotoxicity potential (FAETP), marine aquatic ecotoxicity potential (MAETP), and terrestrial ecotoxicity potential (TETP). In addition, 11 impact categories are included in human health, ecosystem quality, and resources [27].

In Figure 6, the normalization result of applying CML 2001 to the Mg_2NiH_x–CaO composite material is plotted as one graph through the comparison target. As a result, it was confirmed that both Mg_2NiH_x–5, 10 wt% CaO composites showed the highest GWP value, followed by ACP and ARD. As can be seen from the graph, Mg_2NiH_x–10 wt% CaO overall showed a higher value than Mg_2NiH_x–5 wt% CaO, which is considered to be the effect of CaO addition. In addition, the highest GWP value appears to be the use of electricity through multiple mechanical alloying (MA) processes. Therefore, it is necessary to study a process that can be synthesized by performing a single MA process when manufacturing Mg_2NiH_x–CaO composite materials, and to find a way to lower the global warming index by reducing electricity consumption.

Figure 7 shows the Mg_2NiH_x–CaO composites as a graph using the Eco-Indicator 99′ (EI99) methodology. As a result of the graph, fossil fuels were the largest. In addition, the impact categories were measured in the order of climate change and respiratory. In particular, Mg_2NiH_x–10 wt% CaO shows the greatest difference between fossil fuels and climate change values when compared to Mg_2NiH_x–5 wt% CaO, which appears to be highly related to global warming, similar to the previous CML 2001 methodology. In view of the fact that the remaining environmental impact figures show almost similar values, it is judged that the additional amount of 5 wt% CaO does not have a significant effect. As mentioned above, efforts are required to minimize unnecessary energy consumption during the manufacturing process and to reduce the use of electricity as much as possible to reduce a large amount of environmental load.

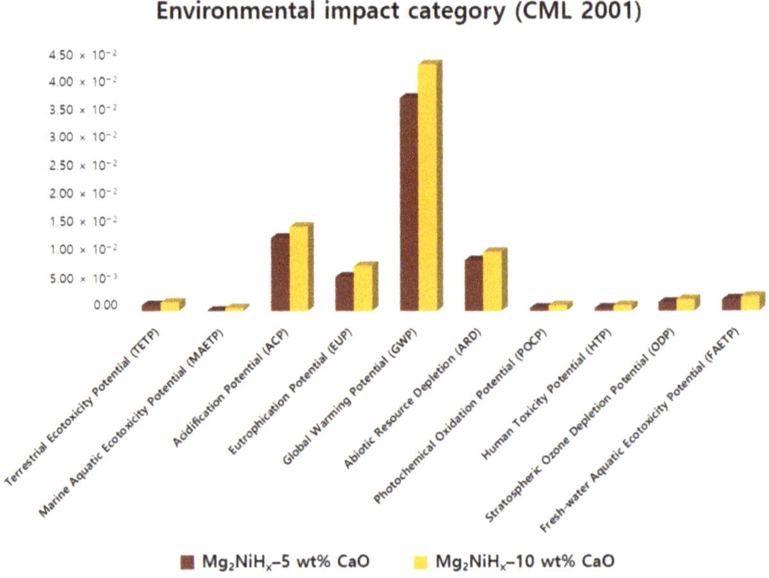

Figure 6. Normalization of Mg_2NiH_x–5 wt% CaO (reprinted with permission from [28]. Copyright 2021 Copyright Shin, H. W) and Mg_2NiH_x–10 wt% CaO composites by environmental impact category (CML 2001).

Figure 8 shows the CO_2 value of global warming impact by synthesized Mg_2NiH_x–CaO composites. The carbon dioxide emission of Mg_2NiH_x–5 wt% CaO [28] was 0.0378 kg, Mg_2NiH_x–10 wt% CaO was 0.0413 kg. Accordingly, it was found that Mg_2NiH_x–10 wt% CaO with a higher CaO content had higher CO_2 emissions than Mg_2NiH_x–5 wt% CaO. In order to decrease the occurrence of such global warming, various efforts for sustainable development have been reported in developed countries around the world to prevent environmental pollution by reducing CO_2 emissions [29]. Therefore, it is necessary to improve the manufacturing process of the Mg_2NiH_x–CaO composites and to find the optimum mass ratio that can reduce the amount of CaO and increase the efficiency.

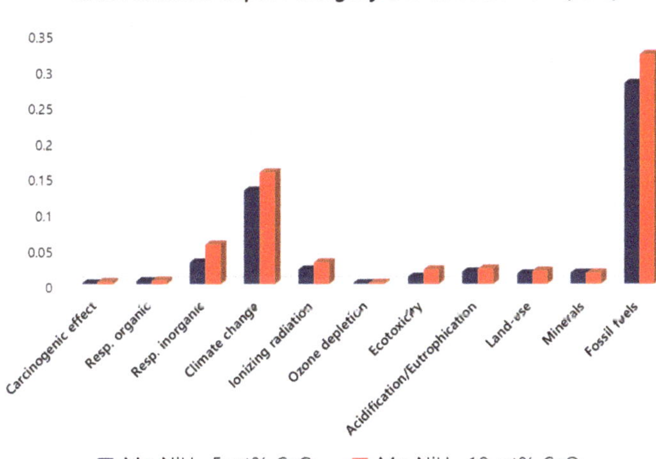

Figure 7. Normalization of Mg_2NiH_x–5 wt% CaO (reprinted with permission from [28]. Copyright 2021 Copyright Shin, H.-W.) and Mg_2NiH_x–10 wt% CaO compo-sites by environmental impact category Eco-Indicator 99' (EI99).

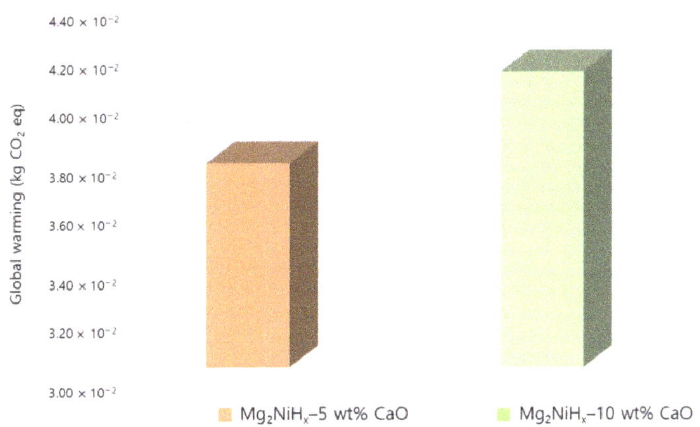

Figure 8. Comparing CO_2 value of global warming impact by Mg_2NiH_x–CaO composites.

4. Conclusions

In this study, environmental pollution caused by the amount of CaO added during the synthesis of Mg_2NiH_x–CaO composites was evaluated through the LCA process. We generated 11 impact categories assessed using the EI99 methodology and 10 impact categories assessed using the CML 2001 methodology. As a result of the CML 2001 methodology, Mg_2NiH_x–5, 10 wt% CaO had the highest GWP value, and the resulting carbon dioxide generation was 0.0378 kg (Mg_2NiH_x–5 wt% CaO), 0.0413 kg (Mg_2NiH_x–10 wt% CaO). In addition, it can be seen that all environmental load values were overall higher depending on the amount of CaO added. Accordingly, in order to lower the global warming potential, it is necessary to save electricity or to find a more environmentally friendly material than CaO. The EI99 methodology showed the highest levels of fossil fuels, followed by climate change, followed by respiratory. This seems to be closely related to the aforementioned global warming, and the remaining environmental load values are very low and show similar results, so it is judged that the amount of CaO added did not have a significant effect. Therefore, efforts to reduce electricity usage as much as possible are expected to minimize unnecessary energy consumption in the synthesis process and to reduce large amounts of environmental impact figures. Ultimately, when comparing Mg_2NiH_x–5, 10 wt% CaO composites, Mg_2NiH_x–10 wt% CaO was better in terms of hydrogenation kinetics, but Mg_2NiH_x–5 wt% CaO showed better results in terms of environment. Therefore, it can be seen that the hydrogenation kinetics and the environmental load value are inversely proportional depending on the amount of CaO added. Ultimately, it is necessary to explore materials that exhibit excellent performance while using eco-friendly materials, and research should be conducted considering environmental factors, not performance-oriented, when making out an alloy.

Author Contributions: H.-W.S., J.-H.H., E.-A.K. and T.-W.H. conceived and designed the experiments and hydrogenation kinetics and life cycle assessment; H.-W.S. performed the experiments and wrote the paper; J.-H.H. and E.-A.K. assisted the experiments; T.-W.H. advised H.-W.S. on experimental results and edited the paper. All authors have read and agreed to the published version of the manuscript.

Funding: This study was supported by the National Research Foundation of Korea (NRF) grant funded by the Korea government (MSIT) (No. 2019R1F1A1041405); the Ministry of SMEs and Startups, Republic of Korea (S3045542); and the Basic Science Research Capacity Enhancement Project (National Research Facilities and Equipment Center) through the Korea Basic Science Institute funded by the Ministry of Education (Grant No. 2019R1A6C1010047).

Institutional Review Board Statement: Not applicable.

Informed Consent Statement: Not applicable.

Data Availability Statement: The data presented in this study are available on request from the corresponding author.

Conflicts of Interest: The authors declare no conflict of interest.

Nomenclature

Eco-Indicator 99′	As a lifecycle impact assessment tool developed by Consultants B.V., designers can perform environmental assessments of products by calculating environmental mark scores for used materials and processes.
ISO 14000 series	Environmental management life cycle assessment principles and framework, Korean Agency for Technology and Standard, 2007.
Global warming potential (GWP)	In a relative sense, it is a type of index based on a simplified radiative characteristic that can be used to measure the possible future impacts of the evolution of various gases on the climate system in the future.
ODP (ozone layer depletion potential)	It is a numerical expression of the degree of destruction of compounds that destroy ozone. Based on the ozone depletion capacity of CFC-11 as 1, the destructive power of the remaining chemicals was assumed.
AP (acidification potential)	Acidification occurs primarily when nitrogen oxides (NO_x) and sulfur dioxide gas (SO_2) interact with other atmospheric components.
EP (eutrophication potential)	It refers to a phenomenon in which nutrients are excessively supplied to water due to the inflow of chemical fertilizers or sewage, causing rapid growth or extinction of plants and killing organisms by taking away oxygen from the water.
ETP (ecotoxicity potential)	Ecotoxicity refers to the ecological impacts of chemicals, pesticides, and pharmaceuticals on freshwater organisms and the possible risks to the aquatic ecosystem.
Human toxicity potential (HTP)	Total emissions are assessed in terms of benzene and toluene equivalents, but potential doses include multiple routes of exposure, including inhalation, ingestion, and dermal contact of fish and meat.

References

1. Ryi, S.K.; Han, J.Y.; Kim, C.H.; Lim, H.; Jung, H.Y. Technical Trends of Hydrogen Production. *Clean Technol.* **2017**, *23*, 121–132.
2. Balat, M. Potential Importance of Hydrogen as a Future Solution to Environmental and Transportation Problems. *Int. J. Hydrogen Energy* **2008**, *33*, 4013–4029. [CrossRef]
3. Demirbas, A.; Dincer, K. Sustainable Green Diesel: A Futuristic View. *Energy Sources Part A* **2008**, *30*, 1233–1241. [CrossRef]
4. Yun, C.W. Liquid Organic Hydrogen Carrier (LOHC) technology for realization of a hydrogen society. *News Inf. Chem. Eng.* **2019**, *37*, 471–485.
5. Lee, P.J.; Kim, J.W.; Bae, K.K.; Jeong, S.U.; Kang, K.S.; Jung, K.J.; Park, C.S.; Kim, Y.H. Heat Transfer Characteristics and Hydrogen Storage Kinetics of Metal Hydride-Expended Graphite Composite. *Trans. Korean Hydrog. New Energy Soc.* **2020**, *31*, 564–570. [CrossRef]
6. Zhi, C.; Chao, T.; Hui, P.; Huabin, Y. Rehydrogenation performance of an MgH_2–Nb_2O_5 system modified by heptane and acetone. *Int. J. Hydrogen Energy* **2010**, *35*, 8289–8294. [CrossRef]
7. Yang, W.N.; Shang, C.X.; Guo, Z.X. Site density effect of Ni particles on hydrogen desorption of MgH_2. *Int. J. Hydrogen Energy* **2010**, *35*, 4534–4542. [CrossRef]
8. Park, H.R.; Song, M.Y. Reaction Rate with Hydrogen and Hydrogen-storage Capacity of an 80Mg+14Ni+6TaF5 Alloy Prepared by High-energy Ball Milling in Hydrogen. *Trans. Korean Hydrog. New Energy Soc.* **2017**, *28*, 137–143.
9. Jang, S.Y.; Kang, K.M.; Hayato, O.; Shigeharu, K. Mg-based hydrogen storage alloy. *Gas Ind. Technol.* **2003**, *6*, 40–47.
10. Berlin, J. Environmental life cycle assessment (LCA) of Swedish semi-hard cheese. *Int. Dairy J.* **2002**, *12*, 939–953. [CrossRef]
11. Lee, S.H.; Jo, Y.M. Review of National Policies on the Utilization of Waste Metal Resources. *KIC News* **2010**, *13*, 2–9.
12. Lee, S.S.; Hong, T.W. Life Cycle Assessment for Proton Conducting Ceramics Synthesized by the Sol-Gel Process. *Materials* **2014**, *7*, 6677–6685. [CrossRef] [PubMed]
13. Rhys, J.T.; Marco, L.L.; Andrew, N.; Kevin, D.P.; Ian, H. An evaluation of life cycle assessment and its application to the closed-loop recycling of carbon fibre reinforced polymers. *Compos. Part B Eng.* **2020**, *184*, 107665.
14. Lee, N.R.; Lee, S.S.; Kim, K.I.; Hong, T.W. Environmental Assessment of Chemically Strengthened Glass for Touch Screen Panel by Material Life Cycle Assessment. *Clean Technol.* **2012**, *18*, 301–306. [CrossRef]
15. Lee, S.S.; Lee, N.R.; Kim, K.I.; Hong, S.J.; Hong, T.W. MLCA (Material Life Cycle Assessment) for ITO Recycling. *Mater. Sci. Forum.* **2012**, *724*, 12–16. [CrossRef]
16. Hong, T.W.; Lim, J.W.; Kim, S.K.; Kim, Y.J.; Park, H.S. Formation of Mg_2NiH_x hydrogen absorbing materials by hydrogen induced mechanical alloying. *J. Korean Inst. Met. Mater.* **1999**, *37*, 369–376.
17. Yasin, S.; Behary, N.; Perwuelz, A.; Guan, J. Life cycle assessment of flame retardant cotton textiles with optimized end-of-life phase. *J. Clean. Prod.* **2018**, *172*, 1080–1088. [CrossRef]
18. International Organisation for Standardisation. *ISO 14040: 2006—Environmental Management—Life Cycle Assessment—Principles and Framework*; International Organisation for Standardisation: Geneva, Switzerland, 2006; Volume 1, pp. 1–32.
19. International Organisation for Standardisation. *ISO 14044: 2006—Environmental Management—Life Cycle Assessment—Requirements and Guidelines*; International Organisation for Standardisation: Geneva, Switzerland, 2006; Volume 1, pp. 1–46.
20. Meng, F.; McKechnie, J.; Turner, T.A.; Pickering, S.J. Energy and environmental assessment and reuse of fluidised bed recycled carbon fibres. *Compos. Part A Appl. Sci. Manuf.* **2017**, *100*, 206–214. [CrossRef]
21. Demertzi, M.; Silvestre, J.D.; Durão, V. Life cycle assessment of the production of composite sandwich panels for structural floor's rehabilitation. *Eng. Struct.* **2020**, *221*, 111060. [CrossRef]
22. Jeong, S.J.; Lee, J.Y.; Shon, J.S.; Hur, T. Life Cycle Assessments of Long-term and Short-term Environmental Impacts for the Incineration of Spent Li-ion Batteries (LIBs). *J. Korean Ind. Eng. Chem.* **2006**, *17*, 163–169.
23. Shin, H.W.; Hwang, J.H.; Kim, E.A.; Hong, T.W. Hydriding Kinetics on Mg_2NiH_x-5wt% CaO Composites. *Trans. Korean Hydrog. New Energy Soc.*, submitting.
24. Lee, J.K.; Kim, S.K. Effect of CaO addition on the ignition resistance of Mg-Al alloys. *Mater. Trans.* **2011**, *52*, 1483–1488. [CrossRef]
25. Huang, Z.G.; Guo, Z.P.; Calka, A.; Wexler, D.; Lukey, C.; Liu, H.K. Effects of iron oxide (Fe_2O_3, Fe_3O_4) on hydrogen storage properties of Mg-based composites. *J. Alloys Compd.* **2006**, *422*, 299–304. [CrossRef]
26. Sahoo, S.K.; Parveen, S.; Panda, J.J. The present and future of nanotechnology in human health care. *Nanomed. Nano Technol. Biol. Med.* **2007**, *3*, 20–31. [CrossRef] [PubMed]
27. Rashidi, A.M.; Nouralishahi, A.; Khodadadi, A.A.; Mortazzavi, Y.; Karimi, A.; Kashefi, K. Modification of single wall carbon nanotubes (SWNT) for hydrogen storage. *Int. J. Hydrogen Energy* **2010**, *35*, 9489–9495. [CrossRef]
28. Shin, H.W.; Hwang, J.H.; Kim, E.A.; Hong, T.W. Material Life Cycle Assessment on Mg_2NiH_x 5 wt% CaO Hydrogen Storage Composites. *Clean. Technol.*, submitting.
29. Kim, M.G.; Son, J.T.; Hong, T.W. Evaluation of TiN-Zr Hydrogen Permeation Membrane by MLCA (Material Life Cycle Assessment). *Clean. Technol.* **2018**, *24*, 9–14.

Article

Effect of Low Zeolite Doses on Plants and Soil Physicochemical Properties

Alicja Szatanik-Kloc [1], Justyna Szerement [2], Agnieszka Adamczuk [1] and Grzegorz Józefaciuk [1,*]

1   Institute of Agrophysics of Polish Academy of Sciences, Doswiadczalna 4 Str., 20-290 Lublin, Poland; a.kloc@ipan.lublin.pl (A.S.-K.); a.adamczuk@ipan.lublin.pl (A.A.)

2   Faculty of Geology, Geophysics and Environmental Protection, AGH University of Science and Technology, 30-059 Krakow, Poland; jsze@agh.edu.pl

*   Correspondence: jozefaci@ipan.lublin.pl; Tel.: +48-81-744-50-61

Abstract: Thousands of tons of zeolitic materials are used yearly as soil conditioners and components of slow-release fertilizers. A positive influence of application of zeolites on plant growth has been frequently observed. Because zeolites have extremely large cation exchange capacity, surface area, porosity and water holding capacity, a paradigm has aroused that increasing plant growth is caused by a long-lasting improvement of soil physicochemical properties by zeolites. In the first year of our field experiment performed on a poor soil with zeolite rates from 1 to 8 t/ha and N fertilization, an increase in spring wheat yield was observed. Any effect on soil cation exchange capacity (CEC), surface area (S), pH-dependent surface charge (Qv), mesoporosity, water holding capacity and plant available water (PAW) was noted. This positive effect of zeolite on plants could be due to extra nutrients supplied by the mineral (primarily potassium—1 ton of the studied zeolite contained around 15 kg of exchangeable potassium). In the second year of the experiment (NPK treatment on previously zeolitized soil), the zeolite presence did not impact plant yield. No long-term effect of the zeolite on plants was observed in the third year after soil zeolitization, when, as in the first year, only N fertilization was applied. That there were no significant changes in the above-mentioned physicochemical properties of the field soil after the addition of zeolite was most likely due to high dilution of the mineral in the soil (8 t/ha zeolite is only ~0.35% of the soil mass in the root zone). To determine how much zeolite is needed to improve soil physicochemical properties, much higher zeolite rates than those applied in the field were studied in the laboratory. The latter studies showed that CEC and S increased proportionally to the zeolite percentage in the soil. The Qv of the zeolite was lower than that of the soil, so a decrease in soil variable charge was observed due to zeolite addition. Surprisingly, a slight increase in PAW, even at the largest zeolite dose (from 9.5% for the control soil to 13% for a mixture of 40 g zeolite and 100 g soil), was observed. It resulted from small alterations of the soil macrostructure: although the input of small zeolite pores was seen in pore size distributions, the larger pores responsible for the storage of PAW were almost not affected by the zeolite addition.

Keywords: clinoptilolite; soil; water; CEC; specific surface; zeolitization

Citation: Szatanik-Kloc, A.; Szerement, J.; Adamczuk, A.; Józefaciuk, G. Effect of Low Zeolite Doses on Plants and Soil Physicochemical Properties. *Materials* **2021**, *14*, 2617. https://doi.org/10.3390/ma14102617

Academic Editors: Franz-Georg Simon and Ute Kalbe

Received: 22 April 2021
Accepted: 14 May 2021
Published: 17 May 2021

1. Introduction

Zeolites are natural or artificial crystalline aluminosilicates exhibiting an open, highly porous structure containing cations balancing high electrostatic charge of the framework of silica and alumina tetrahedral units. The internal surface area of the zeolite framework can reach as many as several hundred square meters per gram [1]. These unique physicochemical properties of zeolites make them extremely effective cation exchangers, water sorbents and adsorbents of uncharged molecules, which, coupled with the abundance of zeolites in sedimentary deposits and in rocks derived from volcanic parent materials, have made them useful in many industrial, environmental and agricultural applications [2–5].

An analysis of the www-sites concerning world zeolite market [6–9] indicates that the annual production of zeolite is about 3 million tons. The main contributors are China, South Korea, Japan, Jordan, Turkey, Slovakia and the US. The global zeolite market size, of which natural zeolites account for around 30%, was valued at USD 2.9 billion in 2016 and USD 4.3 billion in 2019. It is expected to grow at a compound annual growth rate of 2.5–4.7% (depending on the forecasting institution). Agriculture is the main end-user of the total world production of natural zeolites (25%), and together with water treatment and air purification covers around 70% total demand. Within the agricultural sector, major applications concern animal husbandry (litter and fodder additives). Around 30% of the mineral is used as a soil conditioner with the belief that zeolite addition improves the physical and chemical properties of soil for a long time [2]. Zeolites are considered to improve many soil physicochemical properties. They increase soil infiltration rate, saturated hydraulic conductivity, water holding capacity, aeration and many others [3–5,10,11]. Various researchers reported that zeolite increases soil cation exchange capacity and water retention in the root zone [12,13], decreases mineral components leaching [5,14] and traps significant amounts of heavy metals and organic pollutants in contaminated soils [15,16]. Zeolite is not acidic but slightly alkaline and its use with fertilizers can help buffer soil pH levels, thus reducing the need for lime application [17,18]. Zeolites' action as slow-release fertilizers are reported as well [3]. Unlike other soil amendments (e.g., lime), zeolite does not break down over time but remains in the soil to improve nutrient retention. Therefore, its addition to soil significantly reduces water and fertilizer costs by retaining beneficial nutrients in the root zone [4]. The above-mentioned effects of zeolites on soil properties are frequently used for explanations of their positive effects on plant yield, growth and survival of abiotic stresses (drought, metals toxicity, etc.) that have been reported in many papers for sunflower, soybean, tomatoes, radish, beans, potatoes, clover, maize, winter wheat, sugar cane and other plants [5,17–26]. These effects are achieved at different doses of the mineral: from less than 10 t/ha [27–31] up to 120 t/ha [32].

We believe that with low zeolite doses, at which the mineral dilution in soil is very high, changes in soil physicochemical properties should be insignificant and therefore their effect on plant growth and yield should be negligible. Assuming that the soil bulk density is 1.5 g/cm^3 and the 15 cm soil layer weight is 2250 tons/ha, one ton of the zeolite dose per hectare is equivalent to an addition of 0.044 g of the zeolite to 100 g of the soil.

To study this problem in more detail, we performed a field experiment to observe the reactions of a plant and a soil on low zeolite doses. Particular attention was placed on these physicochemical properties which are the most important for soil–plant interactions: CEC, specific surface area, water sorption energy, water retention and mesoporosity. According to the authors' knowledge, similar literature reports are lacking. Because wheat is the most common cereal crop in Europe, a popular variety of spring wheat was selected as the testing plant, for which yield quantitative and qualitative parameters were measured. Wheat is very sensitive to water and nutrient deficits [33], and therefore a large impact of the zeolite addition was expected. From about 40 known types of natural zeolites, clinoptilolite is the most common, the cheapest and the most frequently applied [34]; thus, this mineral was used in the experiment.

2. Materials and Methods

A 3-year experiment (2014, 2015 and 2016) was conducted in Rogozno (51°38' N, 22°56' E, 168 m a.s.l.) on acidic brown soil of pH (KCl) = 4.9 and loamy texture containing 51% sand, 42% silt and 7% clay determined by laser diffractometry using Mastersizer 2000 Malvern UK apparatus according to the method described in detail by Ryzak and Bieganowski [35]. Soil organic matter level was 1.2% as determined according to Walkley and Black [36] and Nelson and Sommers [37]. The content of phosphorus was 9.8 mg P_2O_5/100 g soil that lies on a boundary between low and average phosphorus level. The low content of potassium was 7.7 mg K_2O/100 g soil. This rather poor soil was selected with the belief that more pronounced effects of the further zeolite addition will be observed.

Ground, 1 mm sieved zeolite material was prepared from a clinoptilolitic tuff deposited in Socirnica (Ukraine). The zeolite was purchased from Andalusia Ltd., Warsaw, Poland. The exchangeable cations content of the mineral were: 396 mmol/kg calcium, 377 mmol/kg potassium, 91 mmol/kg sodium and 58 mmol/kg magnesium, and its pH was 7.4. The exchangeable cations were measured by AAS in 1 M ammonium acetate extract, according to Chapman [38].

Scanning electron microphotographs of the studied zeolite using Phenom ProX desktop SEM (Thermo Fisher Scientific, Waltham, MA, USA) are presented in Figure 1.

Figure 1. SEM picture of the studied clinoptilolite.

The X-ray diffraction spectrum of the studied zeolite registered using Panalytical Philips X'pert PRO APD MPD XRD (PANalytical, Kassel, Germany) spectrometer, thanks to the courtesy of Professor Wojciech Franus from Lublin Polytechnical University, is shown in Figure 2, wherein characteristic clinoptilolite peaks are abbreviated by the letter C.

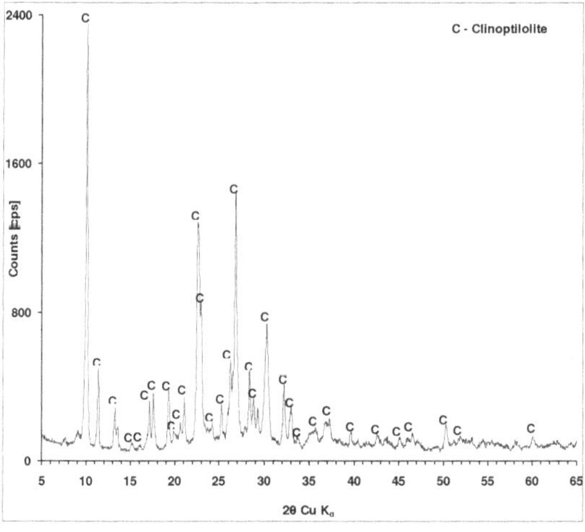

Figure 2. XRD spectrum of the studied zeolite.

Nitrogen adsorption/desorption isotherms of the studied zeolite measured using 3Flex 3500 Surface Characterization Analyzer (Micromeritics Instrument Corporation, Norcross, GA, USA) are presented in Figure 3.

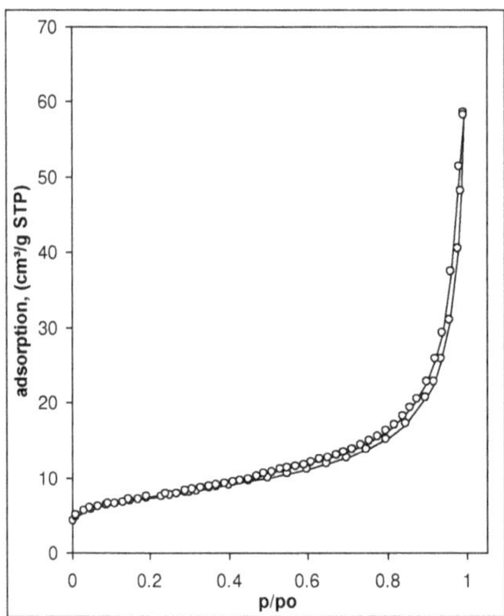

Figure 3. Nitrogen adsorption/desorption isotherms of the studied zeolite.

The content of clinoptilolite in the zeolite used is about 75%, and it was recognized from characteristic distances d_{hkl} = 8.95; 7.94; 3.96 and 3.90 Å. The other mineral components are opal, quartz and feldspars. Clinoptilolite occurs in forms of thin plates (around 10 µm), sometimes of distinct hexagonal shapes. The BET surface area derived from nitrogen adsorption was 26.1 m^2 g^{-1} and the micropore volume (from t-plot) was 0.001 cm^3 g^{-1}. The average pore radius (from desorption isotherm) was 6.78 nm. The above data certify that the zeolite used was a high grade clinoptilolite.

The zeolite was added to the soil in amounts of 1, 2, 4 and 8 t/ha. The rates of zeolite were chosen taking into account the financial possibilities of Polish farmers. The minimum price of one ton of a zeolite is around EUR 100. One ton of wheat costs around EUR 200. An average yield of wheat is around 4 tons per hectare. Even excluding the other costs of cropping, to apply 8 tons of a zeolite per hectare, a farmer must invest his whole income.

Three replicates of a set of six 2 m × 10 m plots separated from each other by 2 m distance from each side were arranged on the experimental field. From each set, one plot was leaved as a control (abbreviated further as FZ0), four were fertilized with the above-mentioned doses of the zeolite (FZ1, FZ2, FZ4 and FZ8, respectively) and one was fertilized with nitrogen, potassium and phosphorus (FNPK). The plots within each replicate had random locations of different fertilization variants, as illustrated in Scheme 1.

1st replicate	FZ0	FNPK	FZ1	FZ2	FZ4	FZ8
2nd replicate	FZ4	FZ8	FZ0	FZ1	FNPK	FZ4
3rd replicate	FZ8	FZ2	FNPK	FZ4	FZ0	FZ1

Scheme 1. Design of the field experiment—scheme of the location of plots.

The upper soil layer of the whole experimental field was mixed by 15 cm deep disking. The 3-year field experiment was designed to:

- Compare NPK and zeolite action on soil and plants (first year of the experiment);
- Look for eventual advantages of zeolite plus NPK over standard NPK fertilization (second year of the experiment). In this year we decided to apply NPK on all but FZ0 plots to prevent exhaustion of soil nutrients;
- Look for the eventual presence of the long-term zeolitization effects (third year of the experiment).

In the first year of the experiment before sowing, all plots (including the control ones) were supplied with 150 kg/ha of slow-release nitrogen fertilizer Saletrosan®26macro (26% N and 13% S) to avoid plant growth differentiation by this macroelement. The FNPK plots were additionally enriched with 100 kg/ha of Agrafoska PK 20–30 (33% P_2O_5 and 30% K_2O). This is worth noting that one ton of the zeolite contains 14.7 kg of exchangeable potassium (17.7 kg K_2O); thus, 2 tons of the zeolite supplies a similar amount of potassium as 100 kg of Agorafoska. Although the above amounts are similar, the plant availability of exchangeable potassium (zeolite) and soluble potassium (Agorafoska) may be different.

In the second year of the experiment, all but the control plots (only N fertilization, as in the previous year) were fertilized with NPK (100 kg/ha of Agrafoska PK 20-30 + 150 kg/ha of Saletrosan® 26 macro). The NPK was applied to avoid exhaustion of the nutrients from the poor soil on which the experiment was performed and to observe eventual effects of combined zeolite/NPK treatment. It was also suspected that zeolite will accumulate some nutrients added as NPK and then act as the long-term fertilizer, as this is stated in the literature [3–5].

In the third year of the experiment, the soil fertilization mode was the same as in the first year. We expected that in this year of the experiment, the long-term effect of the zeolite already present in the soil will be exhibited. The general view of the fertilization treatments in particular years of the field experiment is summarized in Scheme 2.

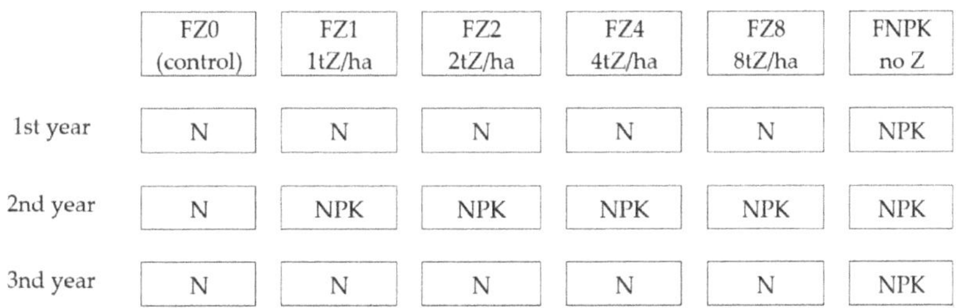

	FZ0 (control)	FZ1 1tZ/ha	FZ2 2tZ/ha	FZ4 4tZ/ha	FZ8 8tZ/ha	FNPK no Z
1st year	N	N	N	N	N	NPK
2nd year	N	NPK	NPK	NPK	NPK	NPK
3nd year	N	N	N	N	N	NPK

Scheme 2. Fertilization treatments in particular years of the field experiment on particular plots. The first row shows the name of the plot and the zeolite (Z) dose. The next rows, attributed to particular years, show the fertilizers applied (N—nitrogen, P—phosphorus, K—potassium).

In all 3 years, at the beginning of April, 190 kg/ha of spring wheat (*Triticum aestivum* L) cv. Izera were sown on the same plots in 10 cm rows by a seeder. At the wheat BBCH 1 (12) phase [39], 300 L/ha aqueous suspension containing 0.8 L of Puma Universal 069 EW were applied, and in the BBCH3 (31) phase, 400 L/ha 0.5% suspension of AMINOPIELIK SUPER 464 SL herbicides were applied. Whole plants were harvested in the first days of August from three randomly selected 0.5 m² areas located on each experimental plot at least 30 cm from its boundary. The number of ears per square meter (NE); number of florets in the ears (NFE); average length of straw (LS) and ears (LE); weight of plant (WHP), straw (WHS) and ear (WHE); number of grains per ear (NG); weight of 1000 grains (WTG); grain moisture (MG); volumetric grain density (D); total protein content in grain (TPC) (according to Polish standard PN-EN ISO 20483: 2007); wet gluten percentage (GW) and gluten index (GI), PN-A-74042/02: 1993); falling number (F) (PN-EN ISO 3093: 2010); and grain yield (YG) were measured.

Daily rainfall and average temperatures (average from the data at 8.00 and 18.00) were recorded during the growing seasons.

Soil properties were studied in two ways. In the first way (called later field soil experiment), soil samples taken directly from the experimental field, control plots (FZ0) and maximally zeolitized (FZ8) plots were examined. Roughly 1.5 kg of the soil samples were taken from 0–15 cm depth layer, from the centers of all 0.5 m^2 areas from which the plants were harvested, 2 mm sieved and homogenized. Here, the soil parameters were measured in one replicate for each soil sample (for each treatment it was 1 sample from each of 3 areas from a given plot $\times$3 replicates of each plot). In the second way, (called later laboratory soil experiment) an average soil sample (soil taken from all control plots mixed together) with much higher doses of zeolite—0, 1, 4, 8, 20 and 40 g of the zeolite per 100 g of the soil—was examined. These soil–zeolite mixtures (abbreviated further as Z0, Z1, Z4, etc.) were carefully homogenized by hand and subjected to five drying/wetting cycles. Here, each parameter was measured in three replicates.

The following parameters were determined in both field soil and laboratory soil–zeolite mixtures, as well as for the zeolite itself:

- pH in KCl (1:5 soil:solution ratio);
- Cation exchange capacity, CEC, at pH = 8.2 using Ba^{+2} as index cations (Mehlich method);
- Soil variable surface charge (called also pH-dependent charge), Qv (μmol g^{-1}), measured from potentiometric titration curves registered under nitrogen atmosphere using an auto-titrator SM Titrino 702 (Metrohm, AG-Switzerland). The suspensions of the studied material in 1 mol·dm^{-3} NaCl solution were adjusted to pH = 2.95 (not changing within 5 min) and slowly titrated to pH = 10 with 0.1 mol·dm^{-3} NaOH. The amount (Mole) of the base consumed by the whole suspension, Nsusp, was used for neutralization of acidic groups of the solid surface, NS, and the acids present in the supernatant, Nsol. The NS value (NS = Nsusp-Nsol) measured between any two pH values is equivalent to variable surface charge developed by a soil in the given pH range. It is responsible for changes of soil CEC with changes in soil reaction and for a part of soil buffering capacity. More details on the method and calculations are given in Józefaciuk et al. [40];
- Amount of gravitational, GW, and plant available, PAW, water from selected points of water moisture versus water potential curves (called water retention or pF curves) were measured according to the procedure described in Richards [41] and Mullins et al. [42]. For water potential (pF) measurements, tensiometers placed in the soil were used and the soil moisture was measured with TDR hygrometer [43]. The pF is defined as a logarithm of a pressure necessary to remove water from soil pores (macropores). The pressure, understood as a water suction (F), is expressed as water height (in cm). The water retention curve provides the best characteristic of soil water storage at high moistures. Based on the above curve, one can distinguish different kinds of water stored in the studied medium. Gravitational water (GW) stored between pF = 0 and pF = 2.2 can easily flow down the soil profile under gravitation force and it is generally not used by plants. Water available for plants (PAW) occurs between soil water potential corresponding to pF = 2.2 (field water capacity, FWC) and pF = 4.2 (permanent wilting point);
- Mesopore volume, V (cm^3 g^{-1}), and average mesopore radius, R (μm), measured by mercury intrusion porosimetry using Micromeritics Autopore IV 9500 (Norcross, GA, USA) porosimeter. Pores detected by mercury intrusion porosimetry (MIP) belong to the range between 3 nm and 200 μm, roughly. The MIP measurements were carried out for 8 mm diameter and 8 mm height cylindrical aggregates prepared from the homogenized soil material and subjected to five wetting–drying cycles to stabilize the structure. The volume of mercury intruded into the aggregate at the maximum pressure was assumed to be equal to the mesopore volume. The average mesopore radius was calculated from pore size distribution functions obtained as the porosimeter reports. Details on porosimetric studies are described in Sridharan and Venkatappa Rao [44]. Mesopores play a crucial role in the formation of soil structure. They govern soil water, air and solute transport, soil compaction, aeration, root growth and many others;

- Surface area, S (m^2 g^{-1}), and water vapor adsorption energy, E, estimated from water vapor adsorption/desorption isotherms for the soil aggregates (as in MIP). The isotherms were measured by weighing the samples after stepwise equilibration at different relative water vapor pressures, p/p$_0$, at 20 °C. The surface area was calculated from the linear form of the standard BET equation [45]. The average adsorption energy, E, was calculated from the energy distribution function, f(E), derived from adsorption isotherms plotted in energy coordinates, assuming that adsorption energy at a given p/p$_0$ equals to ln(p$_0$/p). More details on the calculations are given in Józefaciuk et al. [46]. The water vapor adsorption isotherm provides the best characteristic of soil water content at low moistures that are most often met in upper soil layers at normal weather conditions. The surface area expresses the summary surface of all soil particles. Particularly high input to the surface area are clay minerals, amorphous phases and organic matter. Except for water binding, surface area is responsible for sorption of humic acids, pesticides and herbicides, immobilization of contaminants and soil catalytic properties. It frequently correlates with soil CEC. The water adsorption energy reflects water binding forces. Systems with higher adsorption energy may grasp water from systems with lower adsorption energy.

3. Results and Discussion

3.1. Weather Conditions

The climatic data on the experimental site during the experiment are shown in Table 1 in comparison to long-term data. It is worth noting that the rainfall in April and particularly in May 2014 markedly exceeded the long-term data.

Table 1. Rainfall (R) and temperature (T) on the experimental site.

Year	Climatic Data	April	May	June	July
2014	R, mm	55	150	31	74
	T, °C	9.0	12.6	14.9	17.8
2015	R, mm	38	55	71.0	75
	T, °C	9.1	14.0	19.0	21.0
2016	R, mm	41	56	74	73
	T, °C	9.5	15.0	19.0	20.0
Long-term data	R, mm	39.8	59.9	66.5	80.6
	T, °C	8.0	13.5	16.1	18.2

3.2. Effect of Zeolite on Plants

The yield structure of the studied wheat is characterized in Table 2.

Table 2. Yield structure parameters of the studied wheat in the applied fertilization variants. The data presented are averages from three replicates ±SD. The same superscript letters in each row indicate no significant differences by Tukey's HSD test for α = 0.05.

	First Year of the Experiment (2014)					
Parameter *	FZ0	FZ1	FZ2	FZ4	FZ8	FNPK
NE	472.7 ± 12 [ab]	474.7 ± 7 [abc]	475 ± 12 [abc]	483.7 ± 6 [abcde]	495.3 ± 5 [defg]	505.7 ± 9.2 [efg]
WHP	74 ± 7 [a]	92 ± 11 [a]	96 ± 12 [a]	101 ± 11 [ab]	113 ± 17 [b]	144 ± 21 [c]
WHS	34 ± 3.3 [a]	41 ± 7.8 [a]	45 ± 5.4 [a]	46 ± 5.0 [a]	52 ± 8.7 [ab]	64 ± 1.5 [a]
WHE	41 ± 3.1 [a]	51 ± 9.7 [a]	51 ± 7.0 [a]	55 ± 7.1 [a]	60 ± 8.4 [ab]	79 ± 14.9 [b]
LS	58 ± 3.7 [a]	61 ± 6.1 [a]	65 ± 1.9 [ab]	68 ± 2.5 [ab]	68 ± 4.0 [ab]	73 ± 7.2 [b]
LE	4.3 ± 0.1 [a]	4.9 ± 0.1 [a]	5.1 ± 0.2 [ab]	5.3 ± 0.4 [bc]	5.3 ± 0.2 [bc]	6.1 ± 0.5 [c]
NFE	10 ± 0.9	13 ± 0.8	12 ± 0.8	12 ± 1.7	13 ± 1.3	16 ± 4.9
NG	17 ± 4.5 [a]	17 ± 6.5 [a]	23 ± 1.8 [ab]	24 ± 1.5 [ab]	26 ± 3.3 [b]	30 ± 3.9 [b]
YG	2.8 ± 0.1 [a]	3.4 ± 0.1 [b]	3.4 ± 0.1 [b]	3.5 ± 0.1 [b]	3.8 ± 0.1 [c]	4.2 ± 0.2 [d]

Table 2. *Cont.*

Second year of the experiment (2015)					
Parameter *	FZ0	FZ1	FZ2	FZ4	FZ8	FNPK
NE	475 ± 9 [a]	505.3 ± 12 [cefg]	501.8 ± 4 [defg]	502 ± 7 [efg]	501 ± 8 [efg]	511 ± 10 [g]
WHP	76 ± 5.5 [ab]	132 ± 23 [bc]	136 ± 35 [bc]	130 ± 29 [bc]	138 ± 35 [bc]	141 ± 26 [c]
WHS	37 ± 3.9 [a]	43 ± 7.[ab1]	42 ± 8.3 [ab]	48 ± 2.8 [ab]	48 ± 10.1 [ab]	70 ± 21 [b]
WHE	44 ± 6.5 [a]	51 ± 3.7 [ab]	64 ± 10.7 [abc]	57 ± 8.5 [abc]	54 ± 14 [abc]	84 ± 19 [c]
LS	59 ± 6.4 [ab]	69 ± 2.9 [abc]	64 ± 11 [abc]	71 ± 4.6 [abc]	67 ± 7.9 [abc]	77 ± 2.7 [c]
LE	4.5 ± 0.3 [a]	5.3 ± 0.3 [ab]	5.2 ± 0.6 [abcd]	5.5 ± 0.5 [bcde]	5.8 ± 0.5 [bcde]	6.5 ± 0.2 [e]
NFE	14 ± 3.1	13 ± 1.5	14 ± 2.5	15 ± 1.5	13 ± 5.7	14 ± 1.5
NG	19 ± 3.6 [abc]	29 ± 4.9 [d]	27 ± 2.0 [d]	31 ± 1.0 [d]	32 ± 2.1 [d]	32 ± 2.5 [d]
YG	2.86 ± 0.4 [a]	4.3 ± 1.2 [bcd]	4.3 ± 1.2 [bcd]	4.1 ± 0.3 [bcd]	4.5 ± 0.64 [cd]	4.71 ± 0.37 [d]

Third year of the experiment (2016)					
Parameter *	FZ0	FZ1	FZ2	FZ4	FZ8	FNPK
NE	471 ± 10 [abc]	473 ± 4 [abcd]	471 ± 9 [abc]	494 ± 4 [bcd]	481 ± 13 [abcde]	508 ± 10 [fg]
WHP	74 ± 6.2 [a]	91 ± 20.4 [abc]	110 ± 18 [bc]	96 ± 15 [abc]	97 ± 11 [bc]	139 ± 28 [bc]
WHS	35 ± 5.1 [a]	41 ± 1.4 [ab]	45 ± 5.1 [ab]	41 ± 2.4 [ab]	48 ± 10.8 [ab]	63 ± 18 [b]
WHE	41 ± 8.1 [a]	47 ± 6.8 [a]	56 ± 14 [abc]	62 ± 15.1 [abc]	53 ± 7.3 [abc]	81 ± 15.1 [bc]
LS	57 ± 3.9 [a]	64 ± 8.1 [abc]	62 ± 3.8 [abc]	61 ± 9 [abc]	67 ± 8.4 [abc]	75 ± 4.8 [abc]
LE	4.3 ± 0.3 [a]	4.8 ± 0.4 [ab]	5.1 ± 0.3 [abc]	4.4 ± 0.4 [a]	5.3 ± 0.9 [bcde]	6.3 ± 0.4 [cde]
NFE	11 ± 1.5 [a]	11 ± 1.5 [a]	14 ± 2.5 [a]	12 ± 1.0 [a]	12 ± 2.5 [a]	14 ± 2.0 [a]
NG	17 ± 5.6 [a]	28 ± 3.6 [abcd]	23 ± 1.5 [abcd]	24 ± 2.1 [abcd]	18 ± 7.4 [ab]	30 ± 3.1 [cd]
YG	2.8 ± 0.3 [a]	3.0 ± 0.2 [ab]	3.1 ± 0.9 [abc]	3.1 ± 0.6 [abc]	2.9 ± 0.8 [abc]	4.5 ± 0.2 [d]

* NE number of ears per 1 m^2; WHP weight of 100 plants (g); WHS weight of 100 straws (g); WHE weight of 100 ears (g); LS length of straws (cm); LE length of ears (cm); NFE number of floors in the ears; NG grain number in ears; YG grain yield (t/ha). The same superscript letters in each row indicate no significant differences by Tukey's HSD test for $\alpha = 0.05$.

In the first year of the field experiment, zeolite addition to the soil (along with nitrogen fertilizer) significantly increased the spring wheat grain yield per hectare in respect to the control plots. The yield at the highest zeolite dose (8 t/ha) was almost as high as that of NPK fertilization. The largest input to this increase had the number of ears per 1 m^2 and the number of grains in ears. The latter parameter increased by 58% for the maximum dose of zeolite and by 78% for the NPK fertilization in relation to the control variant. Similar results regarding the positive effects of zeolite on maize, wheat, sugar cane and/or oats were recognized by Kisic et al. [22], Cairo et al. [28] and Szatanik-Kloc et al. [31]. Similarly to our studies, Fotyma and Fotyma [47] observed that the weight of 1000 grains did not affect the yield of winter wheat grain. However, Nouri-Ganbalani et al. [48] observed a close relationship between the yield and the weight of 1000 grains and the number of stems with ears. The cereal grain yield shows generally a high variability due to diverse influences of fore crop, thermal conditions, precipitation, soil conditions and agrotechnical measures, as well as genetic properties of the species, so various authors may present different opinions on the components influencing grain yield of cereals [49,50]. Generally, the best quantitative and qualitative parameters of wheat yield were observed in NPK-fertilized objects, similar to that of 8 t/ha zeolite fertilization, which is consistent with Aaina et al.'s [26] report for maize. In the second year of the field experiment, all plant parameters were the worst on FZ0 plots, which were fertilized only with nitrogen. On all other plots, the plant parameters were similar, regardless of the presence and the amount of the zeolite. In the third year of the experiment, the plant parameters were the best on NPK-fertilized plots and the worst on FZ0 plots. On all other plots, regardless the dose of the zeolite, the plant parameters appeared to be similar.

Effects of fertilization on grain quality parameters are summarized in Table 3.

Table 3. Grain quality parameters in applied fertilization variants. The data presented are averages from three replicates ±SD. The same superscript letters in each row indicate no significant differences by Tukey's HSD test for α = 0.05.

First Year of the Experiment (2014)					
Parameter *	FZ0	FZ1	FZ2	FZ4	FZ8	FNPK
WTG	32.8 ± 1.9 [a]	33.0 ± 2.2 [a]	32.0 ± 1.3 [a]	31.6 ± 1.1 [a]	32.1 ± 2.1 [a]	35.6 ± 3.3 [a]
MG	12 ± 0.6 [a]	12 ± 0.4 [a]	11 ± 0.1 [a]	12 ± 0.1 [a]	12 ± 0.4 [a]	12 ± 0.7 [a]
DV	76.9 ± 1.5 [a]	74.4 ± 1.4 [a]	74.5 ± 0.9 [a]	73.0 ± 2.2 [a]	73.0 ± 1.5 [a]	73.4 ± 0.9 [a]
FN	227 ± 8.2 [a]	262 ± 44 [ab]	274 ± 7 [ab]	288 ± 8.1 [ab]	307 ± 15 [b]	309 ± 6.8 [b]
TPC	11.0 ± 0.5 [abcd]	11.2 ± 0.2 [abcd]	11.3 ± 0.8 [abcd]	11.4 ± 0.1 [abcd]	12.2 ± 0.5 [abcd]	12.4 ± 0.7 [bcd]
GW	23 ± 1 [a]	24.4 ± 1 [ab]	25.1 ± 0.7 [abc]	26.5 ± 1.4 [bcd]	28.2 ± 0.9 [cde]	29.7 ± 1 [de]
GI	73 ± 1 [a]	78 ± 1 [ab]	82 ± 1 [bc]	87 ± 1 [cd]	91 ± 1.5 [d]	85 ± 0.8 [cd]

Second year of the experiment (2015)					
Parameter *	FZ0	FZ1	FZ2	FZ4	FZ8	FNPK
WTG	30.6 ± 1.3 [a]	34.9 ± 2.3 [ab]	36.2 ± 4.5 [ab]	36.3 ± 3.7 [b]	35.9 ± 3.9 [ab]	37.6 ± 2.8 [b]
MG	12 ± 0.2 [a]	12 ± 0.5 [a]	12 ± 0.3 [a]	12 ± 0.7 [a]	12 ± 1.1 [a]	12 ± 0.3 [a]
DV	76.7 ± 2.3 [a]	74.0 ± 3.4 [a]	74.6 ± 3.2 [a]	73.0 ± 1.7 [a]	72.8 ± 1.5 [a]	71.8 ± 1.6 [a]
FN	228 ± 3.2 [a]	306 ± 34 [b]	301 ± 8.8 [b]	271 ± 31 [ab]	277 ± 18 [ab]	310 ± 9.6 [b]
TPC	12.0 ± 0.8 [abcd]	12.2 ± 0.3 [abcd]	12.3 ± 1.2 [bcd]	12.1 ± 0.6 [abcd]	12.5 ± 1 [cd]	12.5 ± 0.5 [d]
GW	22.8 ± 0.8 [a]	30.1 ± 1.5 [de]	29.2 ± 1.3 [de]	29.3 ± 0.5 [de]	30.4 ± 1.9 [de]	30.1 ± 0.2 [d]
GI	73 ± 1.7 [a]	85 ± 2.4 [c]	84.0 ± 2.4 [c]	86 ± 1.6 [cd]	84 ± 3.5 [cd]	85 ± 0.8 [cd]

Third year of the experiment (2016)					
Parameter *	FZ0	FZ1	FZ2	FZ4	FZ8	FNPK
WTG	30.7 ± 1.9 [a]	30.8 ± 1.2 [a]	31.9 ± 1.6 [a]	30.9 ± 2.9 [a]	31.8 ± 2 [a]	36.6 ± 2.7 [a]
MG	12.5 ± 0.02 [a]	12.5 ± 0.5 [a]	12.0 ± 0.3 [a]	12.3 ± 0.7 [a]	12.4 ± 1.1 [a]	12.0 ± 0.3 [a]
DV	76.4 ± 2.3 [a]	74.5 ± 4.8 [a]	73.3 ± 2.9 [a]	72.6 ± 2.7 [a]	72.6 ± 2.6 [a]	73.1 ± 0.6 [a]
FN	228 ± 2.1 [a]	261 ± 26 [ab]	288 ± 17 [b]	272 ± 41 [ab]	278 ± 37 [ab]	309 ± 4.1 [b]
TPC	10.2 ± 0.5 [a]	10.3 ± 0.6 [ab]	10.4 ± 1 [abcd]	10.5 ± 0.9 [abcd]	10.4 ± 0.5 [abc]	12.4 ± 1 [bcd]
GW	23.0 ± 0.6 [a]	23.2 ± 1.2 [a]	23.5 ± 0.5 [ab]	23.2 ± 1.1 [a]	23.3 ± 1.1 [ab]	29.7 ± 1 [de]
GI	73.4 ± 2.9 [a]	78 ± 4.4 [bc]	72 ± 1 [a]	77 ± 1 [b]	73 ± 1.4 [a]	86 ± 1 [cd]

* WTG weight of 1000 grains (g); MG grain moisture (%); D volumetric density (kg/hl); F falling number (s); TPC total protein content (%); GW wet gluten (%); GI gluten index. The same superscript letters in each row indicate no significant differences by Tukey's HSD test for α = 0.05.

In the first year of the experiment, grain volumetric density, falling number and the amount and quality of gluten significantly differed in various fertilization variants. The weight of 1000 grains was the highest in NPK variant and it practically did not differ in the other plots. However, Jelic et al. [29] reported that NPK fertilization did not affect the weight of 1000 grains. Slightly higher grain volumetric density was noted on the control than on the other plots. A decrease in volumetric density of the grains with increasing zeolite doses was noted. It indicated that the seed coat was weaning from the endosperm and worse filling of the caryopses occurred; thus, smaller grains of worse sowing value were produced. Zeolite had a positive effect on the falling number, which was the highest in traditionally NPK-fertilized plots and plots with the maximum dose of zeolite. There was no effect of zeolite fertilization on the total protein content, which is consistent with our previous findings for oats [28]. Similarly, Ozbahce et al. [32] observed no effect of zeolite on protein content in beans. Jelic et al. [29] reported that NPK fertilization did not affect the protein content in oat grains. In contrast, Sepaskhah and Barzegar [30] reported that the use of nitrogen and zeolite increased protein content in rice. The grain protein content and quality are determined by the supply of plants with nitrogen and phosphorus—the main component of nucleic acids, as well as potassium and magnesium, which activate enzymes involved in protein synthesis [51,52]. The qualitative and quantitative parameters of gluten improved after the application of zeolite and NPK. The lowest amount of wet gluten was observed in the control variant. No significant differences in the above parameter were observed in the other plots; however, a tendential increase in wet gluten with the zeolite rate can be seen. The gluten index reached a maximum at the highest dose of zeolite. The positive effect of zeolite on the amount and quality of gluten, a protein that determines flour quality parameters, observed in our experiment could result from better distribution of nitrogen and potassium in soil in the presence of zeolite, as postulated by Abdi et al. [27] and Ozbahce et al. [32]. The improvement of gluten properties increases the commercial value of the grains due to better bakery parameters of the flour. The decisive factor for the flour quality is also the falling number. The increase in this parameter in the fertilized objects may result from the limitation of the activity of amolytic enzymes, which is associated with smaller sprouting of grains.

It was expected that that the presence of the zeolite will influence the plant yield and grain quality parameters also in the second and the third years of the experiment due to a long-term effect of soil zeolitization, which has been mentioned in many previously cited papers. Some differences were observed between the variants with zeolite and the control soil (FZ0) in terms of WHP, WHS, WHE, LE and FN in the third year of the experiment, which can be caused by an exhaustion of the control field from cationic nutrients or some important microelements that were taken by plants and removed with the harvest. However, the most important information is that the general effect of zeolite on plant growth and yield parameters was principally not observed in the second and the third years (regardless of the amount of zeolite present in the soil, the plant and grain parameters were similar). We think that the latter finding could be due to high amounts of rainfall in the first year of the experiment: if the soil was fully saturated with water, rapid ion exchange reactions [53] between the zeolite and the soil occurred, and the native cations present on the zeolite surface could be replaced mainly by multivalent soil aluminum and/or calcium cations, and the ionic composition of both exchange complexes became equalized. The temporary excess of nutrients replaced from the zeolite into the solution is either taken by plants or leached out the rhizosphere. Therefore, a favorable effect of any nutrients brought by the zeolite could be observed only in the first year of the experiment. In the next years, the excess of each cationic nutrient on the zeolite disappeared and it could act no more as a long-term fertilizer: zeolite supplied the same ions as the soil did. One might also suspect that the zeolite could accumulate some potassium added in the second year of the experiment and release it in the third year (to check it, the experiment with full NPK fertilization in the second year was designed); however, the above effect was also absent.

3.3. Effect of Zeolite on Soil

3.3.1. Field Soil

Physicochemical properties of the field soil sampled from the control and the maximum zeolite dose plots are shown in Table 4, along with the properties of the zeolite itself.

Table 4. Physicochemical properties of the soil sampled from the control field (FZ0) and from the field with the maximum dose of the zeolite (FZ8). The data presented are averages from three replicates $\pm$SD. The same superscript letters in each row indicate no significant differences by Tukey's HSD test for $\alpha = 0.05$.

Year of the Experiment		2014		2015		2016	
Parameter *	Zeolite	FZ0	FZ8	FZ0	FZ8	FZ0	FZ8
pH	7.4 ± 0.11 [a]	4.7 ± 0.35 [b]	4.9 ± 0.42 [b]	4.8 ± 0.25 [b]	5.0 ± 0.33 [b]	4.8 ± 0.36 [b]	$5.0 + 0.42$ [b]
S, $m^2 g^{-1}$	125 ± 6.3 [a]	51 ± 3.7 [b]	52 ± 3.8 [b]	52 ± 5.0 [b]	53 ± 1.5 [b]	51 ± 3.6 [b]	53 ± 4.3 [b]
E, RT	4.9 ± 0.33 [a]	5.0 ± 0.42 [b]	5.1 ± 0.27 [b]	5.0 ± 0.44 [b]	5.1 ± 0.38 [b]	5.0 ± 0.33 [b]	5.1 ± 0.36 [b]
Q_v. cMol kg^{-1}	1.5 ± 0.11 [a]	1.7 ± 038 [b]	1.7 ± 0.42 [b]	1.7 ± 0.21 [b]	1.6 ± 031 [b]	1.8 ± 0.44 [b]	1.7 ± 047 [b]
CEC, cMol kg^{-1}	127 ± 5.3 [a]	6.6 ± 0.49 [b]	6.8 ± 0.43 [b]	6.7 ± 0.37 [b]	6.8 ± 0.36 [b]	6.6 ± 0.41 [b]	6.9 ± 0.28 [b]
V_{por}, $cm^3 g^{-1}$	0.5 ± 0.17 [a]	0.4 ± 0.2 [b]	0.4 ± 0.12 [b]	0.4 ± 0.16 [b]	0.4 ± 0.09 [b]	0.4 ± 0.13 [b]	0.5 ± 0.17 [b]

Table 4. *Cont.*

Year of the Experiment		2014		2015		2016	
Parameter *	Zeolite	FZ0	FZ8	FZ0	FZ8	FZ0	FZ8
R_{av}, μm	1.4 ± 0.19 [a]	3.0 ± 0.24 [b]	2.9 ± 0.33 [b]	3.0 ± 0.14 [b]	2.9 ± 0.10 [b]	3.1 ± 0.21 [b]	3.0 ± 0.11 [b]
GW%	10 ± 0.4 [a]	11 ± 1.6 [a]	11 ± 0.8 [a]	10 ± 0.9 [a]	12 ± 1.1 [a]	11 ± 1.8 [a]	12 ± 1.3 [a]
PAW%	21 ± 0.2 [a]	10 ± 1.4 [b]	10 ± 1.3 [b]	9 ± 1.2 [b]	10 ± 1.7 [b]	9 ± 1.1 [b]	10 ± 1.7 [b]

S—surface area; E—water adsorption energy; Qv—variable surface charge; CEC—cation exchange capacity; V_{por}—pore volume; R_{av}—average pore radius; GW—amount of gravitational water; PAW—amount of plant available water. The same superscript letters in each row indicate no significant differences by Tukey's HSD test for $\alpha = 0.05$. Rather high standard deviations calculated for the above values indicate high inhomogeneity of the studied soil. In all studied plots, regardless on the zeolite addition, the physicochemical parameters practically did not differ between each other, in contrast to the majority of the literature reports on the effect of zeolite on improving soil physicochemical properties.

Although the studied soil had very low cation exchange capacity, in our field experiment no effect of the added zeolite (up to 8 t/ha) on soil CEC was observed. However, an increase in CEC after zeolite addition has been frequently reported [7,47]. Abdi et al. [27] showed that the amount of exchange K+, Ca^{2+}, Mg^{2+} in the soil (that can be roughly considered as the CEC) had increased significantly due to the use of 10 t/ha zeolite. Ravali et al. [54] observed around a 30% increase in CEC of a sandy loam after addition of 7.5 t/ha zeolite. Fudlel et al. [55] reported the positive effect of 5 t/ha zeolite on CEC of an Alfisol. Chomczynska et al. [56] observed around a 20% increase in CEC after 1%w/w zeolite addition to two degraded soils (edge of sand mine excavation and the fallow land).

The small effect of the zeolite on soil pH observed in the field studies is in line with Litaor et al.'s [24] results for peat soil after several months of a field experiment. Tallai et al. [25] also reported that in a 2-year experiment, the pH of zeolitized soil increased only by 0.1 unit. The above observations are, however, in contrast to the results presented by most researchers [5,16,17,57], who generally observed an increase in soil pH after zeolite addition. The small effect of the zeolite on soil pH may be connected with low buffering properties of the zeolite itself which can be concluded from the small variable charge of the zeolite, even smaller than that of the studied sandy soil. Since variable charge is responsible for pH buffering, soil amended with zeolite buffers less pH than the soil itself. In our field experiment, no effect of the added zeolite on the specific surface area and water holding capacity of the soil was observed.

3.3.2. Laboratory Soil Studies

Soil laboratory studies revealed a small influence on pH of the examined mixtures. At the maximum zeolite dose (Z40), the soil pH increased to 5.13. The cation exchange capacity of the studied zeolite was 127 cMol kg^{-1}, around 20 times higher than that of the soil. The increase in CEC was proportional (R2 = 0.99) to the zeolite fraction, x, in the soil–zeolite mixture: CEC = 112.5x + 6.65 cMol kg^{-1}. In the above equation, the proportionality coefficient reflects the CEC of the zeolite and the intercept reflects the CEC of the soil. Both above values coincide well with the measured values of both parameters measured for the zeolite and for the control soil.

Dependencies of variable surface charge on pH calculated from potentiometric titration curves for the zeolite and the soil with different mineral doses are shown in Figure 4.

The total amount of the variable charge measured between pH 3 and pH 10 for the studied zeolite was 4.56 cMol kg^{-1}, which is around four times smaller than the variable charge of the soil. High variable charge of the soil itself is most likely connected with high pH-dependent charge of soil organic matter created mostly by surface carboxylic groups of rather high acidic strength. Inorganic surface hydroxyls present on the zeolite surface are weaker and their dissociation at any pH is smaller, and thus less variable charge is created. The addition of zeolite decreases soil variable surface charge; however, contrary to the surface area, the dependence of CEC on the fraction of zeolite is not linear, i.e., the variable charge is not additive.

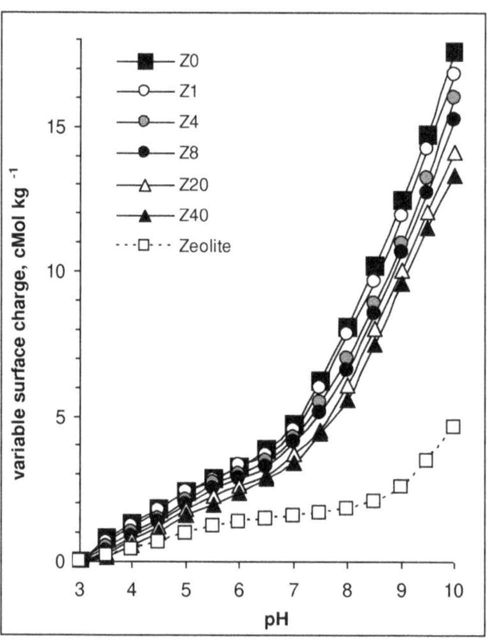

Figure 4. Dependence of variable charge of the studied zeolite and laboratory studied soil-zeolite mixtures on pH. Z0 abbreviates the control soil (no zeolite added), Z1, Z4 ... etc. denote mixtures of 100 g soil with 2 g, 4 g ... etc. of the zeolite. The data points cover the error bars.

Water vapor adsorption isotherms on zeolite and the soil with different zeolite doses are shown in Figure 5.

In the whole relative pressure range, the adsorption is the highest for zeolite and the lowest for the control soil. The addition of zeolite increases water vapor adsorption on the soil. The studied zeolite has very large surface area, 125.5 m^2 g^{-1}. The addition of zeolite to the soil results in an increase in surface area proportional (R2 = 0.99) to the zeolite fraction, x, in the soil–zeolite mixture: S = 128.9x + 52.1 m^2 g^{-1}. The slope of the above equation reflects the surface area of the zeolite and the intercept reflects the surface area of the soil. Both above values coincide well with the measured values of both parameters measured for the zeolite and for the control soil. Low percentage of the zeolite added to the soil, even at its highest rate (8 t/ha), was a reason that no effects of soil zeolitization were noted in the field. It means that only very high amendments of zeolite can significantly increase soil specific surface. An increase in the specific surface area resulting from the addition of zeolite can be beneficial for an increase in organic compounds' sorption capacity of the soil, as postulated by Cairo et al. [28]. Binding natural soil humic compounds limits their mineralization and loss of soil organic carbon [58]. From this point of view, even a small addition of zeolite to the soil can have a positive effect on the environment quality. Large surface area is also a crucial feature for high water sorption of the mineral; however, the adsorbed water is not used by plants. High water sorption by the zeolite may be dangerous for plants: dry zeolite can confiscate water from a rainfall coming after dry periods.

Water vapor adsorption energy of the zeolite, 4.98 RT (R is universal gas constant and T = 293 K is the temperature of the measurements) is practically equal to that of the soil, meaning that the zeolite and the soil bind water with similar forces. No difference in water binding between the zeolite and the soil indicates that at low moistures, where adsorption energy governs water binding, the zeolite of large specific surface can grasp huge amounts of water from the soil.

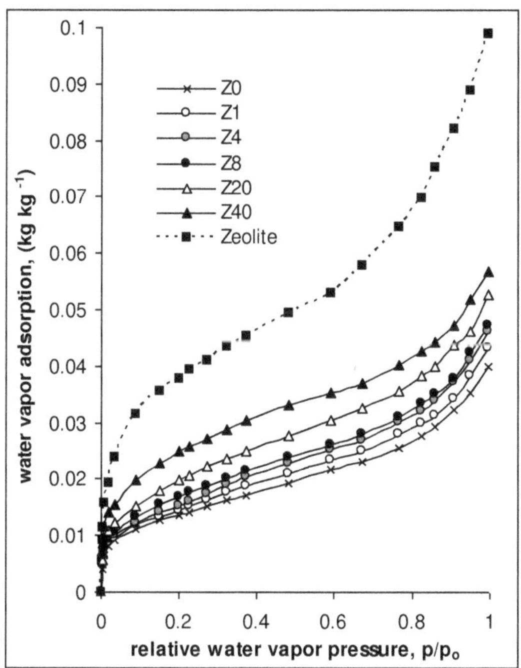

Figure 5. Water vapor adsorption isotherms of the studied zeolite and laboratory studied soil-zeolite mixtures. Z0 abbreviates the control soil (no zeolite added), Z1, Z4 . . . etc. denote mixtures of 100 g soil with 2 g, 4 g . . . etc. of the zeolite. The data points cover the error bars.

Water retention curves for the studied soil–zeolite mixtures are shown in Figure 6.

While zeolite can hold water up to 50% of its volume, the amount of plant available water is much less (20.9%) compared to soils of heavier textures. The amount of gravitational water (8.7%) is less than in the studied soil. The highest dose of zeolite decreases the GW of the soil from 11.27 ± 0.33% (Z0) to 9.33 ± 0,56 (Z40) and increases the amount of PAW from 9.48 ± 0.29 (Z0) to 13.1 ± 0.61% (Z40). As it is seen, even in laboratory experiments, high zeolite doses had rather small effects on gravitational and plant available water. Practically all of the literature reports are in contrast to these results. The increase in water in the soil in the available humidity zone due to the high porosity of the crystalline structure of zeolites was reported by Yasuda et al. [59] and Bernardi et al. [60]. According to Kralova et al. [7], soils enriched with natural zeolite were able to increase the water retention capacity by 18–19%, and for sandy soils even up to 50% [61]. More examples of the positive effect of zeolites on soil water retention are presented by Nakhli et al. [62] in their excellent review of the subject. Both gravitational and plant available soil water depend on soil macrostructure (large pores range, gravitational water) and soil mesostructure (medium pore range, plant available water) [63]. The slight effect of zeolite on plant available water is most likely connected with its small impact on soil macro- and mesostructure, which can be seen from the porosimetric measurements presented below. It is worth mentioning that around 20% of water bound by the zeolite (the part of moisture located above pF = 4.2 line in Figure 5) cannot be used by plants. This water occupies the finest pores present in the zeolite mineral network.

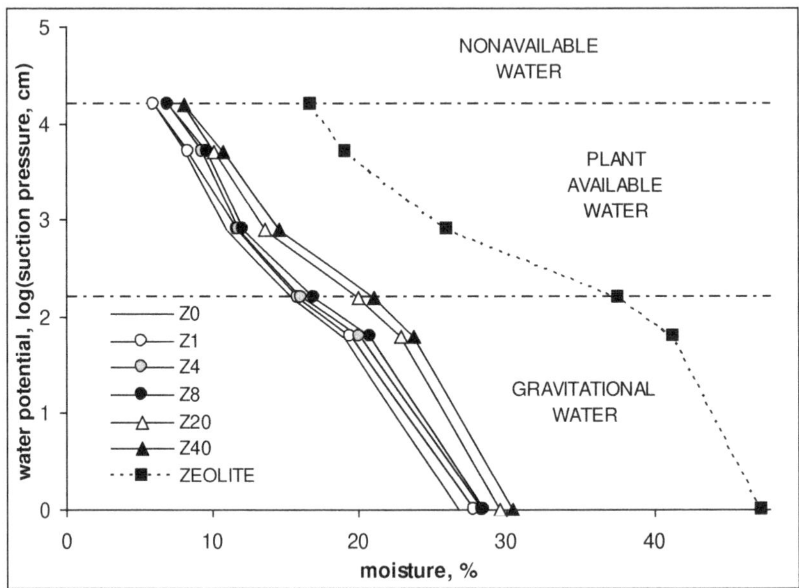

Figure 6. Water retention curves of the studied zeolite and laboratory studied soil–zeolite mixtures. The water potential is a logarithm of water suction pressure expressed in cm of water height. Z0 abbreviates the control soil (no zeolite added), Z1, Z4 . . . etc. denote mixtures of 100 g soil with 2 g, 4 g . . . etc. of the zeolite. The data points cover the error bars.

Pore size distribution functions derived from mercury intrusion porosimetry are shown in Figure 7. Not to shadow the lines, the results are presented only for the control soil, the zeolite and the soil–zeolite mixture with the maximum zeolite dose.

In the studied soil, practically only large pores are present, located around 3.2 μm (maximum frequency at $\log(R) = 0.5$). The zeolite poses dominant pores around 6 μm (maximum frequency at $\log(R) = 0.8$), a significant amount of medium-sized pores, around 0.032 μm (at $\log(R) = -1.5$), and some very fine pores, around 0.002 μm (at $\log(R) = -2.7$). Even the highest amendment of the zeolite affects the soil pore size distribution to a small extent: in the Z40 sample, the large pores peak is conserved and the frequency of finer pores occurrence increases slightly. The above results indicate that both the macro- and mesostructure of the soil-zeolite system were not significantly altered by the added mineral. We believe that this is due to the fact that the added zeolite had rather large grains. Its addition is equivalent to the enrichment of the soil with sand-size material that does not affect larger pores. Possibly, the large-grained zeolite addition to clayey soils may improve soil aeration due to soil texture coarsening. In our opinion, these results suggest that for an improvement in the retention of plant available water, zeolites of very fine granulation should be applied; however, high costs of grinding and difficulties in management may limit their application.

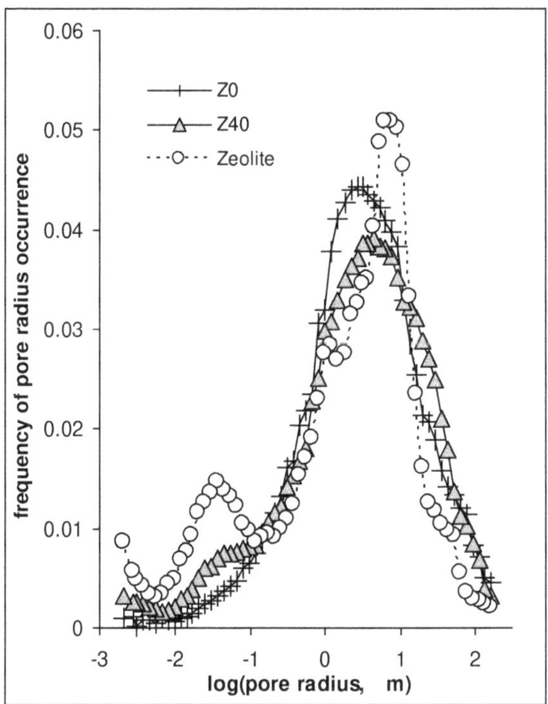

Figure 7. Pore size distributions for the studied zeolite, the control soil and the laboratory studied soil–zeolite mixture containing 100 g soil and 40 g zeolite (Z40). The data points cover the error bars.

General Remarks

The significant effect of zeolite application on plant yields in the first year of the experiment does not seem to be caused by the improvement of soil physicochemical characteristics. As mentioned before, 1 t/ha zeolite dose is equivalent to the presence of about 0.04% of the zeolite in the soil (0.044 g zeolite per 100 g of the soil). Therefore, it is unlikely that this amount of zeolite can affect the physicochemical parameters of a soil because of very high dilution of the added zeolite in soil material. Very large rates of this mineral are required. For example, Ozbahce et al. [32] reported that soil CEC increased only after the addition of zeolite in a dose of 90 t/ha. The observed improvement of spring wheat yield in the first year of the experiment may result from the release of potassium, calcium, magnesium and some microelements to the root zone by the zeolite exchange complex. Because all field plots were fertilized with the same amounts of nitrogen, it should not differentiate the yield; however, the combined effects of zeolite and nitrogen might have been important in this respect, as stated by Hung and Petrovic [64], who observed that the use of zeolite improved the effectiveness of nitrogen application in the soil by about 16–22% due to better nitrogen distribution in the soil. Cairo et al. [28] also reported that zeolite plus nitrogen increased the yield of sugar cane by almost 80%. Sepaskhah and Barzegar [30] attributed the higher yield of rice grain to the improved N retention in the soil by the added zeolite. Zeolites can also increase the availability of phosphorus [26] and limit the mobility of toxic heavy metals [15].

It seems that the paradigm of the improvement of water storage after soil zeolitization is wishful thinking rather than a real fact—this was proven based on water retention curves for the laboratory studied soil amended with different zeolite doses and supported by porosity measurements.

Author Contributions: Conceptualization, G.J and A.S.-K.; methodology, G.J., A.S.-K. and J.S.; formal analysis, G.J. and A.A.; investigation, J.S. and A.A.; data curation, G.J. and A.S.-K.; writing—original draft preparation, G.J., A.S.-K. and J.S.; writing—review and editing, A.A.; visualization, G.J. and A.A.; supervision, G.J.; funding acquisition, G.J. All authors have read and agreed to the published version of the manuscript.

Funding: This research was partly financed by Project No IPBU.01.01.00-06-570/11-00 "Developing an innovative model of the cross-border use of zeolitic tuff" performed within the Cross-border Cooperation Programme Poland-Belarus-Ukraine 2007–2013, co-financed by European Union and Ministry of Science and Higher Education Poland FN12/ILT/2019 and partly by the statutory activity of the Institute of Agrophysics PAS.

Institutional Review Board Statement: Not applicable.

Informed Consent Statement: Not applicable.

Data Availability Statement: All data are presented in the paper.

Conflicts of Interest: The authors declare no conflict of interest.

References

1. Yates, D.J.C. Studies on the surface area of zeolites, as determined by physical adsorption and X-ray crystallography. *Can. J. Chem.* **1968**, *64*, 1695–1701. [CrossRef]
2. Torii, K. Utilization of natural zeolites in Japan. In *Natural Zeolites: Occurrence, Properties, Use*; Sand, L.B., Mumpton, F.A., Eds.; Pergamon Press: Elmsford, NY, USA, 1978; pp. 441–450.
3. Mumpton, F.A. La Roca Magica: Uses of natural zeolites in agriculture and industry. *Proc. Natl. Acad. Sci. USA* **1999**, *96*, 3463–3470. [CrossRef] [PubMed]
4. Polat, E.; Karaca, M.; Demir, H.; Onus, A.N. Use of natural zeolite (clinoptilolite) in agriculture. *J. Fruit Ornam. Plant. Res. Spec.* **2004**, *12*, 183–189.
5. Ramesh, K.; Reddy, D.D. Zeolites and their potential uses in agriculture. *Adv. Agron.* **2011**, *113*, 219–240.
6. Available online: https://www.mordorintelligence.com/industry-reports/zeolites-market (accessed on 30 December 2020).
7. Available online: https://www.grandviewresearch.com/industry-analysis/zeolites-market (accessed on 30 December 2020).
8. Available online: https://www.greatmining.com/zeolite-mining.html (accessed on 30 December 2020).
9. Available online: https://www.fortunebusinessinsights.com/industry-reports/zeolite-market-101921 (accessed on 30 December 2020).
10. Kralova, M.; Hrozinkova, A.; Ruzek, P.; Kovanda, F.; Kolousek, D. Synthetic and natural zeolites affecting the physical-chemical soil properties. *Rostl. Vyrob.* **1994**, *40*, 131–138.
11. Jakkula, V.S.; Wani, S.P. Zeolites: Potential soil amendments for improving nutrient and water use efficiency and agriculture productivity. *Sci. Rev. Chem. Commun.* **2018**, *8*, 119–126.
12. Rehakova, M.; Cuvanova, S.; Dzivak, M.; Rimar, J.; Gavalova, Z. Agricultural and agrochemical uses of natural zeolite of the clinoptilolite type. *Curr. Opin. Solid State Mater. Sci.* **2004**, *8*, 397–404. [CrossRef]
13. Xiubin, H.; Zhanbin, H. Zeolite applications for enhancing water infiltration and retention in loess soil. *Resour. Conserv. Recycl.* **2001**, *34*, 45–52. [CrossRef]
14. Azough, A.; Marashi, S.K.; Babaeinejad, T. Growth characteristics and response of wheat to cadmium, nickel and magnesium sorption affected by zeolite in soil polluted with armaments. *J. Adv. Environ. Health Res.* **2017**, *5*, 163–171.
15. Tahervand, S.; Jalali, M. Sorption and desorption of potentially toxic metals (Cd, Cu, Ni and Zn) by soil amended with bentonite, calcite and zeolite as a function of pH. *J. Geochem. Explor.* **2017**, *181*, 148–159. [CrossRef]
16. Mohd, H.B.; Asima, J.; Arifin, A.; Hazandy, A.H.; Mohd, A.K. Elevation and variability of acidic sandy soil pH: Amended with conditioner, activator, organic and inorganic fertilizers. *Afr. J. Agric. Res.* **2013**, *8*, 4020–4024.
17. Ahmed, O.H.; Sumalatha, G.; Majid, N.M.A. Use of zeolite in maize Zea mays. cultivation on nitrogen, potassium, and phosphorus uptake and use efficiency. *Int. J. Phys. Sci.* **2010**, *5*, 2393–2401.
18. Gholamhoseini, M.; Ghalavand, A.; Khodaei-Joghan, A.; Dolatabadian, A.; Zakikhani, H.; Farmanbar, E. Zeolite-amended cattle manure effects on sunflower yield, seed quality, water use efficiency and nutrient leaching. *Soil Tillage Res.* **2013**, *126*, 193–202. [CrossRef]
19. Khan, H.; Khan, A.Z.; Khan, R.; Matsue, N.; Henmi, T. Soybean Leaf Area, Plant Height and Reproductive Development as Influenced by Zeolite Application and Allophanic Soil. *J. Plant. Sci.* **2008**, *3*, 277–286. [CrossRef]
20. Moore, A.D.; Olsen, N.L.; Carey, A.M.; Leytem, A.B. Residual effects of fresh and composted dairy manure applications on potato production. *Am. J. Potato Res.* **2011**, *88*, 324–332. [CrossRef]
21. Baninasab, B. Effects of the application of natural zeolite on the growth and nutrient status of radish *Raphanus sativus* L. *J. Hortic. Sci. Biotechnol.* **2009**, *84*, 13–16. [CrossRef]
22. Kisic, I.; Mesic, M.; Basic, F.; Butorac, A.; Vadic, Z. The Effect of Liming and Fertilization on Yields of Maize and Winter Wheat. *Agric. Conspec. Sci.* **2004**, *69*, 51–57.

23. Junrungreang, S.; Limtong, O.; Wattanaprapat, K.; Patsarayeangyong, T. Effect of zeolite and chemical fertilizer on the change of physical and chemical properties on Lat Ya soil series for sugar cane. In Proceedings of the 17th WCSS Symposium, Bangkok, Thailand, 12–14 August 2002.

24. Litaor, M.I.; Katz, L.; Shenker, M. The influence of compost and zeolite co-addition on the nutrients status and plant growth in intensively cultivated Mediterranean Soils. *Soil Use Manag.* **2017**, *33*, 72–80. [CrossRef]

25. Tallai, M.; Olah Zsuposne, A.; Sandor, Z.; Katai, J. The effect of using zeolite on some characteristics of sandy soil and on the Amount of the test plant biomass. *Ann. Ser. Agric. Silvic. Vet.* **2017**, *6*, 115–120.

26. Aaina, H.N.; Ahmed, O.H.; Majid, N.M.A. Effects of clinoptilolite zeolite on phosphorus dynamics and yield of Zea Mays L cultivated on an acid soil. *PLoS ONE* **2018**, *13*, e0204401.

27. Abdi, G.; Khosh-Khui, M.; Eshghi, S. Effects of Natural Zeolite on Growth and Flowering of Strawberry Fragariaxananassa Duch. *Int. J. Agric. Res.* **2006**, *1*, 384–389. [CrossRef]

28. Cairo, P.C.; Machado de Armas, J.; Artiles, P.T.; Martin, B.D.; Carrazana, R.J.; Lopez, O.R. Effects of zeolite and organic fertilizers on soil quality and yield of sugarcane. *Aust. J. Crop. Sci.* **2017**, *11*, 733–738. [CrossRef]

29. Jelic, M.; Dugalic, G.; Millvojevic, J.; Djekic, V. Effect of liming and fertilzation on yield and quality of oat Avena Sativa l on an acid luvisol soil. *Rom. Agric. Res.* **2013**, *30*, 249–258.

30. Sepaskhah, A.R.; Barzegar, M. Yield, water and nitrogen-use response of rice to zeolite and nitrogen fertilization in a semi-arid environment. *Agric. Water Manag.* **2010**, *98*, 38–44. [CrossRef]

31. Szatanik-Kloc, A.; Ambrożewicz-Nita, A.; Franus, W.; Jozefaciuk, G. Early effect of clinoptilolite on yield and quality of oat *Avena sativa* L. *Int. Agrophysics* **2019**, *33*, 107–112. [CrossRef]

32. Ozbahce, A.; Tari, A.F.; Gonulal, E.; Simsekli, N.; Padem, H. The effect of zeolite applications on yield components and nutrient uptake of common bean under water stress. *Arch. Agron. Soil Sci.* **2015**, *61*, 615–626. [CrossRef]

33. Kapusta, F. Cereals as a component of agriculture and Polish economy. *Econ. XXI Century* **2016**, *2*, 121–137.

34. Eyde, T.H. Zeolites. *Min. Eng.* **2005**, *57*, 54–55.

35. Ryzak, M.; Bieganowski, A. Determination of particle size distribution of soil using laser diffraction–Comparison with areometric method. *Int. Agrophysics* **2010**, *24*, 177–181.

36. Walkley, A.; Black, I.A. An examination of the Degtjareff method for determining soil organic matter, and a proposed modification of the chromic acid titration method. *Soil Sci.* **1934**, *37*, 29–38. [CrossRef]

37. Nelson, D.W.; Sommers, L. Total carbon, organic carbon, and organic matter. *Methods Soil Anal. Part 2 Chem. Microbiol. Prop.* **1983**, *9*, 539–579.

38. Chapman, H.D. Cation-exchange capacity. *Methods Soil Anal. Part 2 Chem. Microbiol. Prop.* **1965**, *9*, 891–901.

39. Meier, U. *Growth Stages of Mono-and Dicotyledonous Plants: BBCH Monograph*, 2nd ed.; Federal Biological Research Centre for Agriculture and Forestry: Bonn, Germany, 2001.

40. Józefaciuk, G.; Szatanik-Kloc, A.; Lukowska, M.; Szerement, J. Pitfalls and Uncertainties of Using Potentiometric Titration for Estimation of Plant Roots Surface Charge and Acid-Base Properties. *Am. J. Plant. Sci.* **2014**, *5*, 1862–1876. [CrossRef]

41. Richards, L.A. Methods for measuring soil moisture tension. *Soil Sci.* **1949**, *68*, 95–112. [CrossRef]

42. Mullins, C.E.; Mandiringana, O.T.; Nisbet, T.R.; Aitken, M.N. The design, limitations, and use of a portable tensiometer. *J. Soil Sci.* **1986**, *37*, 691–700. [CrossRef]

43. Skierucha, W. Design and performance of psychrometric soil water potential meter. *Sens. Actuators A* **2005**, *118*, 86–91. [CrossRef]

44. Sridharan, A.; Venkatappa Rao, G. Pore size distribution of soils from mercury intrusion porosimetry data. *Soil Sci. Soc. Am. Proc.* **1972**, *36*, 980–981. [CrossRef]

45. Brunauer, S.; Emmett, P.H.; Teller, E. Adsorption of Gases in Multimolecular Layers. *J. Am. Chem. Soc.* **1938**, *60*, 309–319. [CrossRef]

46. Józefaciuk, G.; Lukowska, M.; Szerement, J. Determination of Energetic and Geometric Properties of Plant Roots Specific Surface from Adsorption/Desorption Isotherm. *Am. J. Plant. Sci.* **2013**, *4*, 1554–1561. [CrossRef]

47 Fotyma, M.; Fotyma, E. The structure of winter cereal yield depending on nitrogen fertilization. in Polish. *Fragm. Agron.* **1993**, *10*, 101–102.

48. Nouri-Ganbalani, A.; Hassanpanah, D.; Nouri-Ganbalani, G. Effects of drought stress condition on the yield and yield components of advanced wheat genotypes in Ardabil. *J. Food Agric. Environ.* **2009**, *7*, 228–234.

49. Brzozowska, I.; Brzozowski, J.; Cymes, I. Effect of weather conditions on spring triticale yield and content of macroelements in grain. *J. Elem.* **2018**, *234*, 1387–1397. [CrossRef]

50. Hruskova, M.; Svec, I.; Kocourkova, Z. Interaction between wheat variety and harvest year analysed by statistical methods. *Cereal Technol.* **2011**, *4*, 152–159.

51. Muste, S.; Modoran, C.; Man, S.; Muresan, V.; Birou, A. The influence of wheat genotype on its quality. *J. Agroaliment. Process. Technol.* **2010**, *16*, 99–103.

52. Allen, E.; Ming, D.; Hossner, L.; Henninger, D.; Galindo, C. Growth and nutrient uptake of wheat in a clinoptilolite-phosphate rock substrate. *Agron. J.* **1995**, *87*, 1052–1059. [CrossRef]

53. Helfferich, F.G.; Hwang, Y.L. Ion Exchange Kinetics. In *Ion. Exchangers. Part. I*; Dorfner, K., Ed.; De Gruyter: Basel, Switzerland, 1991; pp. 1277–1310.

54. Ravali, C.; Rao, K.J.; Anjaiah, T.; Suresh, K. Effect of zeolite on soil physical and physico-chemical properties. *Multilogic Sci.* **2020**, *XXXIII*, 776–781.

55. Fudlel, A.Y.; Minardi, S.; Hartati, S.; Syamsiyah, J. Studying the residual effect of zeolite and manure on alfisols cation exchange capacity and green bean yield. *J. Soil Sci. Agroclimatol.* **2019**, *16*, 181–190. [CrossRef]

56. Chomczynska, M.; Wasag, H.; Kujawska, J. Application of Spent Ion Exchange Sorbents for the Reclamation of Degraded Soils. *J. Ecol. Eng.* **2019**, *20*, 239–244.

57. Filcheva, E.G.; Tsadilas, C.D. Influence of clinoptilolite and compost on soil properties. *Commun. Soil Sci. Plant. Anal.* **2002**, *33*, 595–607. [CrossRef]

58. Lutzow, M.V.; Kogel-Knabner, I.; Ekschmitt, K.; Matzner, E.; Guggenberger, G.; Marschner, B.; Flessa, H. Stabilization of organic matter in temperate soils: Mechanisms and their relevance under different soil conditions—A review. *Eur. J. Soil Sci.* **2006**, *57*, 426–445. [CrossRef]

59. Yasuda, H.; Tazuma, K.; Mizuta, N.; Nishide, H. Water retention variety of dune sand due to zeolite addition. *Bull. Fac. Agric. Tottori Univ.* **1995**, *48*, 27–34.

60. Bernardi, A.C.C.; Oliveira, P.P.A.; Barros, F. Brazilian sedimentary zeolite use in agriculture. *Microporous Mesoporous Mater.* **2013**, *17*, 16–21. [CrossRef]

61. Lancellotti, I.; Toschi, T.; Passaglia, E.; Barbieri, L. Release of agronomical nutrient from zeolite substrate containing phosphatic waste. *Environ. Sci. Pollut. Res.* **2014**, *21*, 13237–13245. [CrossRef] [PubMed]

62. Nakhli, S.A.A.; Delkash, M.; Bakhshayesh, B.E.; Kazemian, H. Application of Zeolites for Sustainable Agriculture: A Review on Water and Nutrient Retention. *Water Air Soil Pollut.* **2017**, *228*, 464–473. [CrossRef]

63. Liu, C.; Tong, F.; Yan, L.; Zhou, H.; Hao, S. Effect of Porosity on Soil-Water Retention Curves: Theoretical and Experimental Aspects. *Geofluids* **2020**. [CrossRef]

64. Hung, Z.T.; Petrovic, A.M. Clinoptilolite zeolite effect on evaporation rate and shoot growth rate of bentgrass on sand base grass. *J. Turfgrass Manag.* **1995**, *25*, 35–39.

materials

Article

Leaching of Titanium Dioxide Nanomaterials from Agricultural Soil Amended with Sewage Sludge Incineration Ash: Comparison of a Pilot Scale Simulation with Standard Laboratory Column Elution Experiments

Boris Meisterjahn [1], Nicola Schröder [1], Jürgen Oischinger [2], Dieter Hennecke [1,*], Karlheinz Weinfurtner [1] and Kerstin Hund-Rinke [1]

[1] Fraunhofer Institute for Molecular Biology and Applied Ecology IME, Auf dem Aberg 1, 57392 Schmallenberg, Germany; boris.meisterjahn@ime.fraunhofer.de (B.M.); nicola.schroeder@ime.fraunhofer.de (N.S.); karlheinz.weinfurtner@ime.fraunhofer.de (K.W.); kerstin.hund-rinke@ime.fraunhofer.de (K.H.-R.)

[2] Fraunhofer Institute for Environmental, Safety and Energy Technology UMSICHT, An der Maxhütte 1, 92237 Sulzbach-Rosenberg, Germany; juergen.oischinger@umsicht.fraunhofer.de

* Correspondence: dieter.hennecke@ime.fraunhofer.de; Tel.: +49-2972-302-209

Citation: Meisterjahn, B.; Schröder, N.; Oischinger, J.; Hennecke, D.; Weinfurtner, K.; Hund-Rinke, K. Leaching of Titanium Dioxide Nanomaterials from Agricultural Soil Amended with Sewage Sludge Incineration Ash: Comparison of a Pilot Scale Simulation with Standard Laboratory Column Elution Experiments. *Materials* **2022**, *15*, 1853. https://doi.org/10.3390/ma15051853

Academic Editor: Stefano Lettieri

Received: 16 December 2021
Accepted: 25 February 2022
Published: 1 March 2022

Publisher's Note: MDPI stays neutral with regard to jurisdictional claims in published maps and institutional affiliations.

Abstract: Nanoscale titanium dioxide ($nTiO_2$ (Hombikat UV 100 WP)) was applied to sewage sludge that was incinerated in a large-scale waste treatment plant. The incineration ash produced was applied to soil as fertilizer at a realistic rate of 5% and investigated in pilot plant simulations regarding its leaching behavior for $nTiO_2$. In parallel, the applied soil material was subject to standard column leaching (DIN 19528) in order to test the suitability of the standard to predict the leaching of nanoscale contaminants from treated soil material. Relative to the reference material (similar composition but without $nTiO_2$ application before incineration) the test material had a total TiO_2 concentration, increased by a factor of two or 3.8 g/kg, respectively. In contrast, the TiO_2 concentration in the respective leachates of the simulation experiment differed by a factor of around 25 (maximum 91.24 mg), indicating that the added $nTiO_2$ might be significantly mobilisable. Nanoparticle specific analysis of the leachates (spICP-MS) confirmed this finding. In the standard column elution experiment the released amount of TiO_2 in the percolates between test and reference material differed by a factor of 4 to 6. This was also confirmed for the $nTiO_2$ concentrations in the percolates. Results demonstrate that the standard column leaching, developed and validated for leaching prediction of dissolved contaminants, might be also capable to indicate increased mobility of $nTiO_2$ in soil materials. However, experiments with further soils are needed to verify those findings.

Keywords: nano titanium dioxide ($nTiO_2$); engineered nanomaterial (ENM); sewage sludge incineration (SSI); ENM containing sewage sludge ash (SSA); leaching; column elution; agricultural use

1. Introduction

Engineered nanomaterials (ENMs) applied, e.g., in consumer products, can be released to the environment during their use (e.g., release of silver nanoparticles (AgNPs) from facade painting [1] or TiO_2-nanoparticles (TiO_2-NPs) from sunscreens [2]), while, after use, a major fraction of the ENMs is supposed to be released to wastewater streams [3] and, furthermore, becomes attached to sewage sludge during wastewater treatment [4,5]. The majority of sewage sludge is incinerated and ends up in landfills [6].

Sewage sludge incineration ash (SSA) can be used for the production of phosphorous fertilizer (e.g., according German fertilizer ordinance [7]). This recycling route will be increased in the future because the recovery of phosphorus from sewage sludge and sewage sludge ash is required by law in Germany and other countries of the EU.

The use of materials in soils always requires a specific assessment to prevent soil contamination by re-use of waste materials. Integral part of the assessment is the determination of the source strength from those materials for leaching of contaminants. This is performed, e.g., by a standard column elution according to DIN 19528 for dissolved heavy metals and a set of organic substances.

So far, very little has been published regarding the release of nanoparticles from SSA. In a very recent study, Wielinski et al. [8] investigated the release of different nanomaterials from SSA in column experiments, but the column experiments described therein are very different to the standard column elution used for evaluation of source strength in the scope of regulation. Further, laboratory column experiments need references from the "real world" to assess the robustness of the laboratory results. However, we could not find any study regarding the release of nanomaterials from SSA after use in agricultural soils as fertilizer. Respective studies mostly focus on heavy metal contaminations (not in the form of NPs) or availability of phosphorous from SSA. Thus, there were two knowledge gaps identified for our study: (i) the question of leaching of NPs that might enter agricultural soils by fertilization with NP-contaminated SSA under realistic and environmentally relevant conditions and (ii) the predictive power of a column elution procedure actually used in the German soil protection ordinance [9] for NP leaching.

These gaps were addressed by the present study. In order to investigate the potential leaching of a representative ENM nano titanium dioxide ($nTiO_2$) from SSA applied to soils, combined elution experiments in pilot and laboratory scale were performed. $nTiO_2$ was selected as test material, as it is among the most produced nanomaterials [10,11] and was used, e.g., in suncreens [2], as photocatalyst [12] or food additives [13]. Ash material was produced in a large-scale waste incinerator from incineration of sewage sludge amended with $nTiO_2$. The resulting SSA was mixed with a reference soil at a rate of about 5%. This is about a factor of 10 greater than the maximum allowed application rate [14] and was considered to be a worst case scenario. The SSA amended soil was used for a pilot scale bioreactor trial in order to investigate $nTiO_2$ leaching from the material under controlled realistic conditions. In a timelapse experiment, three annual summer/winter cycles were simulated within 250 days to verify the influence of seasons on the leaching behavior. Those pilot scale simulations were accompanied by standard laboratory column elution experiments. Based on the results obtained in both tests, the suitability of the standard column elution with regard to a prediction of $nTiO_2$ leaching from soil treated with SSA should be determined. In case of a significant correlation, the standard laboratory column elution could be extended for risk assessment of nanomaterial leaching from SSA amended soils. So far, this has not been considered in soil protection, since transport of nanomaterials in soil cannot be described by common leaching models, as the sorption theory used for dissolved chemicals does not apply for nanomaterials [15].

2. Materials and Methods

2.1. Preparation of ENM Containing Sewage Sludge Incineration Ash (SSA)

For the production of the ENM-containing sewage sludge ash (SSA), the product Hombikat UV 100 WP from Co. VENATOR was used. The aqueous Hombikat UV 100 WP dispersion consists of 42% (w/w) $nTiO_2$ and about 6.5% (w/w) of poly acrylate acting as stabilizer. As a basis for the formulation of Hombikat UV 100 WP, $nTiO_2$, in the form of the product Hombikat UV 100 with a primary particles size of <10 nm, was employed. Further information can be found in Börner et al. [16] and Oischinger et al. [17]. The application of Hombikat UV 100 WP lies mainly in the photocatalytic area, as the $nTiO_2$ is present in the anatase modification.

The sewage sludge incineration was performed at the Sewage Sludge Incineration Plant (SSIP) of the waste water treatment plant ZVK Steinhäule at Neu-Ulm. Annually a wastewater amount of 440,000 population equivalents is cleaned in the plant, resulting in 10,000 t of sewage sludge (dry matter (DM)) and 2500 t SSA [18]. The SSIP consists of a centrifuge, a dryer, a fluidised-bed incinerator with selective non-catalytic NO_x reduction, an electrostatic precipitator, a 2-stage scrubber and an activated-carbon reactor with fabric filter. A schematic diagram of the plant is depicted in [16].

For the production of the reference ash material on the 1st day, the average amount of sewage sludge mounted up to 2171 kg h^{-1} (DM) and, for the day with nTiO$_2$ injection, on the 2nd day, up to 2120 kg h^{-1} (DM). During the measurement with nTiO$_2$ injection, 630 kg of nTiO$_2$ dispersion was added to the sewage sludge with a peristaltic pump over 6 h. On both days, the measured background concentration of titanium was <0.1 wt%. Hence, an augmentation of titanium in the sewage sludge of about 1.25 wt% was achieved by the injection of the dispersion. The SSA was sampled on the basis of the recommendations of LAGA PN98 [19] for moving waste, as far as they were applicable to the large scale plant [20].

2.2. Simulation Experiments

2.2.1. Leaching in Pilot Scale Simulation Reactors

For investigation of the release of nanoparticles from SSA applied in soil-related applications, experiments in pilot-scale simulation reactors were conducted. For this purpose, the incineration residue (with and without treatment with nTiO$_2$) was mixed with an agricultural soil (refesol 04-A, details of the soil characteristics see Table 1).

Table 1. Refesol 04-A soil characteristics (loamy sand).

Soil Type	Soil Texture (DIN)			Soil Texture	C$_{org}$ (%)	N$_{total}$ (g/kg)	pH$_{CaCl_2}$	CEC$_{eff}$ (mol/kg)	WHCmax (g/kg)
	Sand (%)	Silt (%)	Clay (%)						
Refesol 04-A	79.7	14.9	5.4	loamy sand	3.04	1.76	5.11	0.0412	346

The sandy soil with higher organic carbon content was selected in order to simulate worst case conditions regarding potential nanoparticle release and mobility [21,22]. Per reactor, 700 kg of soil material (dry weight basis) were mixed with an amount of SSA, representing 5% with regard to the dry weight of the soil material, corresponding to approximately 190 t sludge ash per ha, taking into account an assumed plough share depth of 25 cm. At mixing, the soil was adjusted to a water content representing 50% of its water holding capacity (WHC), being optimal for microbial activity. The mixture was then subjected to an experiment simulating at least three annual seasonal cycles during a total incubation time of 250 days. Thus, the soil/ash mixture was incubated at 20–22 °C simulating a summer phase, followed by freezing of the mixture to −10 to −15 °C final temperature. The soil in the reactor was kept at this temperature for 14 days (simulated winter period). After each winter period and thawing of the soil/ash mixture, the soil was dug up and samples were taken for the column elution experiments (seasonal cycles and sampling dates see Figure 1). The soil was irrigated over a period of 12 days with 10 L portions until approximately 5 L of seepage/leachate water was collected. The leachate was analysed for total titanium concentration by ICP-OES after microwave assisted digestion and, in addition, was subjected to nanoparticle specific analysis by single particle (sp)ICP-MS. For the next cycle, the soil was again dried to a water content of 50% WHC by digging the soil up several times and leaving the reactor surface open while incubating at 20 ± 2 °C. After reaching the desired water content, the reactor was closed again and the next cycle started.

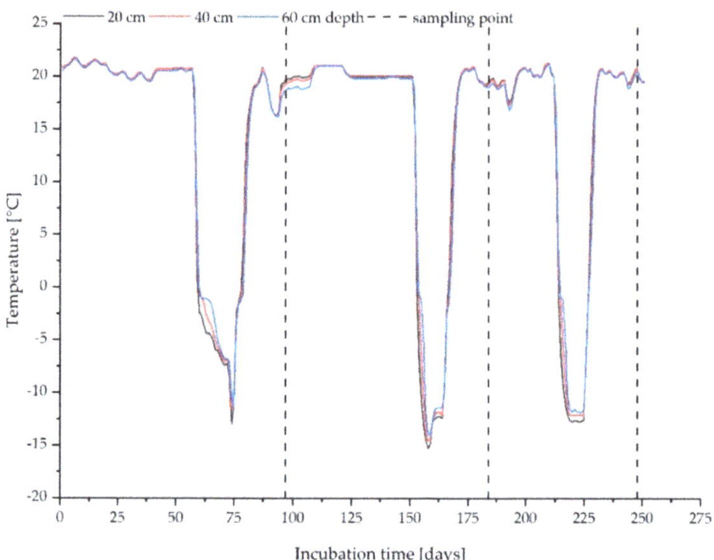

Figure 1. Temperature curve over the entire test period with three annual cycles, measured at 20, 40 and 60 cm distance from the edge of the reactor with treated soil ash/mixture. Dotted line marks the three sampling points after the winter phases (23 January, 24 April and 27 June).

2.2.2. Column Elution Tests

At three time points, samples of the soil/ash mixture in the pilot-scale reactors were sampled and filled into glass columns (diameter 5.5 cm) according to the instructions given by the standard DIN 19528 [23] for soil column leaching. At both ends of the glass columns, quartz sand was used as a filtration layer. The column was then subjected to an elution according to DIN 19528 [23]. The column was saturated with deionized water from the bottom to the top layer. After saturation, the column was eluted with deionized water and eluate samples taken at five different solid/water ratios (0.3, 1, 2, 4 and 10 L/kg). The titanium concentration in the eluate was determined as described in Section 2.3.1 without any previous eluate filtration or centrifugation. The amount of soil per soil column and the percolation rates for saturation and percolation are presented in Table 2. The percolation rates were calculated according to DIN 19528:

$$q = \frac{l \times \pi \times r^2 \times n}{t \times 60} \tag{1}$$

with q = percolation rate (mL min^{-1}); l = length of soil column; r = inner radius of column; n = porosity (0.43); t = time (2 h for saturation, 5 h for percolation).

Table 2. Mass of soil per soil column and percolation rates for saturation and percolation.

Soil	Mass per Column	Percolation Rate (mL min^{-1})	
	(g)	Saturation	Percolation
RefeSol 04-A	1100	2.64	1.06

2.3. Chemical Analysis

2.3.1. TiO$_2$ Concentrations in Reactor Leachates and Column Eluates

For determination of the total concentration of titanium in the leachates of the reactors, as well as for the column eluates, 10 mL of the respective sample was filled into a

Teflon^{TM} digestion vessel (MLS, Leutkirch, Germany) and evaporated at 105 °C to dryness. The residue was then mixed with 4.8 mL of HNO_3 (69%, supra pure, Roth) and 0.2 mL hydrofluoric acid (HF, 40%, supra pure, Roth) and digested in a microwave (Ultraclave II, MLS, Leutkirch, Germany; digestion parameters: 220 °C, 30 min, 100 bar). After digestion, 1 mL boric acid was added in order to complex the remaining HF; the mixture filled up to a final volume of 15 mL with ultrapure water. The concentration of Ti in the solution was then determined by ICP-OES (wavelength 334.941 nm, Instrument: Agilent 5110, Agilent technologies). The obtained results were converted into TiO_2 concentrations under the assumption that no other titanium-containing phase was present in the samples.

2.3.2. TiO_2 Concentrations in Soil/Ash-Mixed Samples

For determination of total titanium concentrations in mixtures of soil and sewage, sludge ash was applied for simulation in the pilot-scale reactors; in addition, column elution experiments, 5 aliquots of 5 g (fresh weight) each, were taken and mixed to obtain a representative sample. The samples were dried at 105 °C until constant weight. Aliquots of approximately 200 mg of the dried residues were then weighted in Teflon^{TM} digestion vessels and 1 mL HF (40% supra pure, Roth) and 4 mL HNO_3 (69%, supra pure, Roth) added. The samples were subjected to microwave-assisted digestion (Ultraclave II, MLS, digestion parameters: 220 °C, 30 min, 100 bar). After digestion, 5 mL of boric acid was added to complex remaining HF and the digestate filled up to a final volume of 15 mL with ultrapure water. The solution was analysed for titanium by ICP-OES at a wavelength of 334.941 nm.

2.3.3. Single Particle (sp)ICP-MS Analysis of Reactor Leachates and Column Eluates

In addition to the determination of the total titanium concentration, number-based particle size distributions were determined by single particle inductively coupled plasma mass spectrometry (spICP-MS) [24,25]. The aqueous samples were directly measured without any further sample preparation, except for dilution with ultrapure water. The analyses were performed using a triple-quadrupole ICP-MS instrument (ICP-QQQ-MS, Agilent 8900, Agilent Technologies, Waldbronn, Germany). The dwell time in the single particle measurement mode of the ICP-MS was set to 100 µs and time-resolved signals were recorded on the selected m/z for 60 s. Peak detection and integration was conducted automatically by the Agilent MassHuntersoftware. Conversion of signal heights of particle spikes into particle sizes were performed by application of a calibration with a dissolved Ti standard. To apply a dissolved calibration for size calculation of nanoparticles from their signal spikes, the nebulization or transport efficiency in the interface was determined by analyzing a gold nanoparticle standard of known concentration and size [24]. Dispersions of 60 nm gold nanoparticles (AuNPs 60 nm, BBI solutions, Kent, UK) were used for the determination of the nebulization efficiency and were prepared freshly on the day of measurement. The samples were diluted in ultrapure water by a factor of 10^2–10^5 for measurement in order to reach a particle concentration of 200–2000 particle events per minute.

Due to possible interferences for the most abundant titanium isotope ^{48}Ti caused by the calcium isotope ^{48}Ca present in the leachate and eluate matrices, titanium was measured in the MS/MS mode with ammonia as reaction gas (10% NH_3 + 1 mL/min He). Thus, titanium was measured as $[^{48}TiNH]^+$ with a m/z of 63. The threshold between background and particle signals was defined based on visual inspection of the measured signal distributions. The conversion of signal distributions into number-based size distributions, as well as particle number concentrations, was performed by the Agilent MassHunter software.

3. Results

3.1. Characterisation of the Produced SSA

The two SSA produced differed in their titanium content as expected. For the SSA, from the reference day, an average titanium concentration of about 0.3 wt% (dry mass

(DM)) with a standard deviation (SD) of 0.04 was determined, whereas the SSA treated with nTiO$_2$ had a mean value of 2.9 wt% (DM) with an SD of 0.19 [26]. Hence, the titanium content in the nTiO$_2$-SSA was increased compared to the reference measurement by a factor of about 9.7 [26]. Further information on characterisation of the used SSA can be found elsewhere [26].

3.2. Leaching from Pilot Scale Simulation Reactors

3.2.1. Determination of Total Titanium Content in Mixtures of Soil/SSA and in Leachates

Figure 2 shows the determined total titanium concentrations in both mixtures of soil with SSA with additional nTiO$_2$ or without nTiO$_2$ treatment (reference samples) which were filled into the pilot scale reactors. The SSA/soil mixture containing nTiO$_2$-treated ash shows a titanium concentration elevated by a factor of approximately two compared to the reference sample. Thus, at least this ratio should also be found in the leachates.

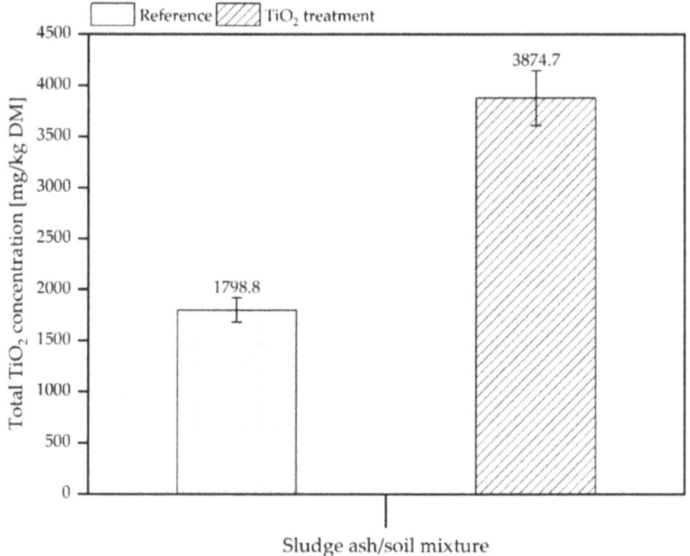

Figure 2. Total TiO$_2$ concentrations (converted Ti-concentrations) in sludge ash/soil mixtures. left column: Concentration in samples using sludge ash without treatment with nTiO$_2$, right column: Concentration in samples using SSA treated with nTiO$_2$. Error bars refer to standard deviation of four subsamples that were digested and analysed.

However, as shown in Figure 3, the titanium/converted TiO$_2$ content in the leachates of treatment and reference reactors collected at the end of the three simulated annual seasonal cycles was found to be higher for the treated samples compared to the references by a factor of ~25. Titanium-containing particles are released from the treatment reactor in a significantly higher amount than expected from the total titanium content of the respective mixtures, thus, indicating that the added nTiO$_2$ (or titanium containing particles) might be more mobilisable than the titanium-bearing particles (most likely also being TiO$_2$) already present in the soil and sludge ash.

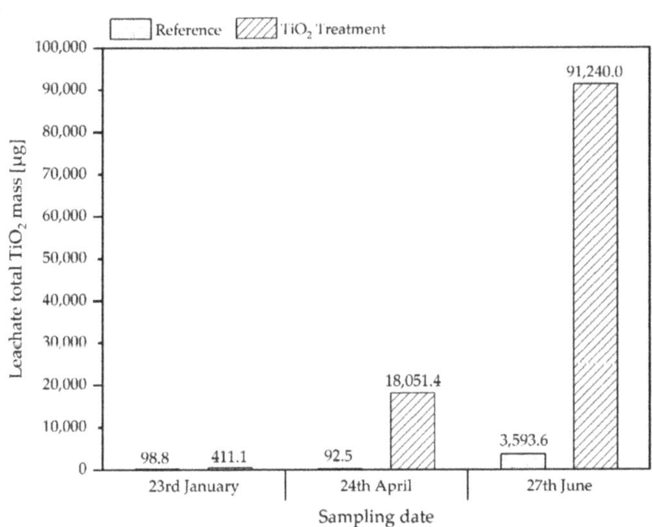

Figure 3. Total TiO$_2$ content (derived from Ti-concentrations) determined in leachates from pilot scale simulation reactors collected after watering of the soil at end of the respective preceding winter phase.

3.2.2. SpICP-MS/MS Analyses of Leachates

The significantly higher release of titanium in the leachates from the treatment reactor is also visible in the nanoparticle specific analysis of the leachates by spICP-MS, as shown in Figure 4. The higher TiO$_2$ particle discharge in leachates after the first winter period (23 January) is 3.7 times higher in the treated reactor and rises to 640 times after the second simulated winter phase (17 April) and drops to 8.2 times after the third winter phase (27 June) (Figure 4). Again, the differences in discharge are much higher than could be expected from the total Ti-content determined for the soil/ash mixtures applied to the two reactors.

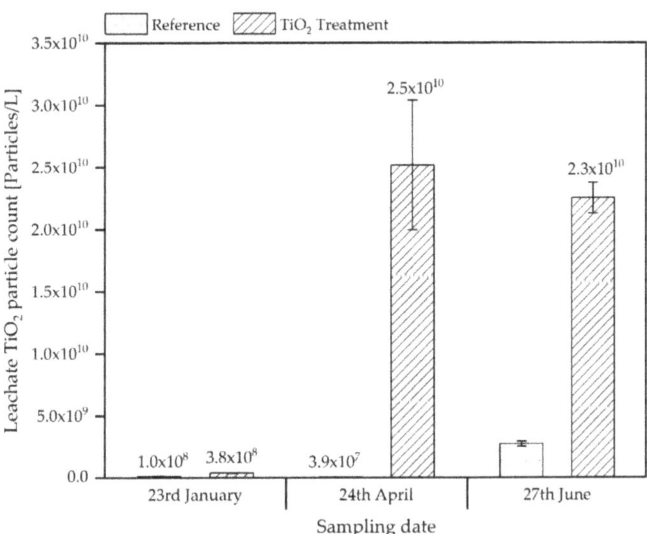

Figure 4. TiO$_2$ particle concentrations (Particles/L) determined in leachates from pilot scale simulation reactors collected after watering of the soil at end of the respective preceding winter phase.

3.3. Column Elution Experiments

The percolate samples obtained from standard column eluate experiments were analysed analogously to the leachates from the pilot scale experiments for their total TiO_2 content (ICP-OES analysis after HF digestion) and particle size distribution (spICP-MS).

Figure 5 shows the TiO_2 content in the percolates of soil/ash-material taken at end of the three simulated annual seasonal cycles from the treatment and reference reactor to investigate the behaviour during soil-related use. In the percolates, an increased TiO_2 load was detected for the treated SSA/soil mixture compared to the reference mixture. In contrast to the analyses of the reactor leachates, the ratio between TiO_2 released from $nTiO_2$-treated material and the reference material was found to be more or less stable during all three sampling times. Already, percolates from the first sampling (17 January) showed increased TiO_2 contents in the $nTiO_2$ treated SSA/soil mixture. This was most likely due to the very high water:solid ratios used in the column elution compared to the pilot scale experiment. However, the maximum TiO_2 concentrations determined are in the same range as the maximum concentration found in the pilot scale experiment.

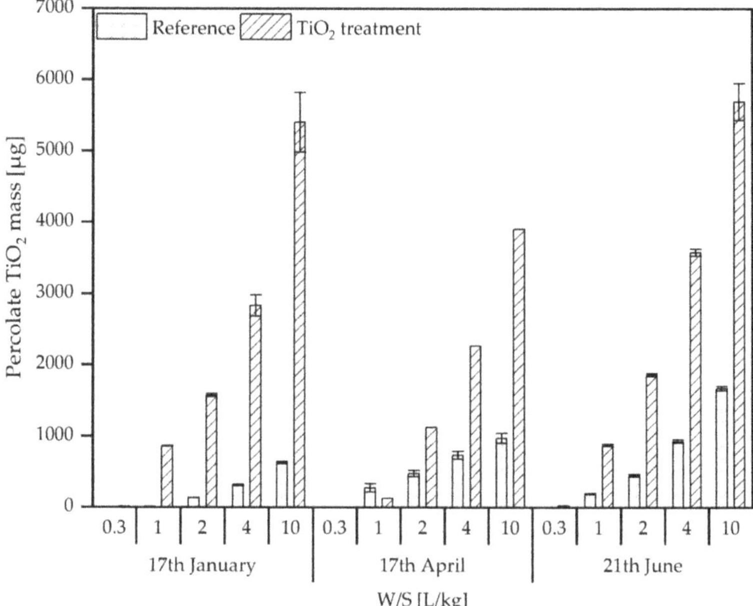

Figure 5. Total TiO_2 content (µg) in column percolates from reference and $nTiO_2$ treated SSA/soil mixtures out of pilot scale simulation reactors collected at three different sampling times.

The spICP-MS/MS analyses of the column percolates (reference and $nTiO_2$ treatment) show a higher $nTiO_2$ particle discharge from the $nTiO_2$ treated SSA/soil mixture (Figure 6). The particle number in the percolates of the reference and TiO_2 treated SSA/soil mixtures increases in both cases with the water/solid ratio (W/S). In the TiO_2-treated percolates, the particle concentration at a W/S of 10 L/kg at the first sampling (17 January) extend concentration in the reference percolates by a factor 10. During the following samplings (17 April and 21 June), particle counts in the $nTiO_2$ percolates extend the reference percolates by a factor of 4. In general, the $nTiO_2$ analysis confirms the findings of the total TiO_2 analysis, indicating that most of the TiO_2 determined was due to leached $nTiO_2$.

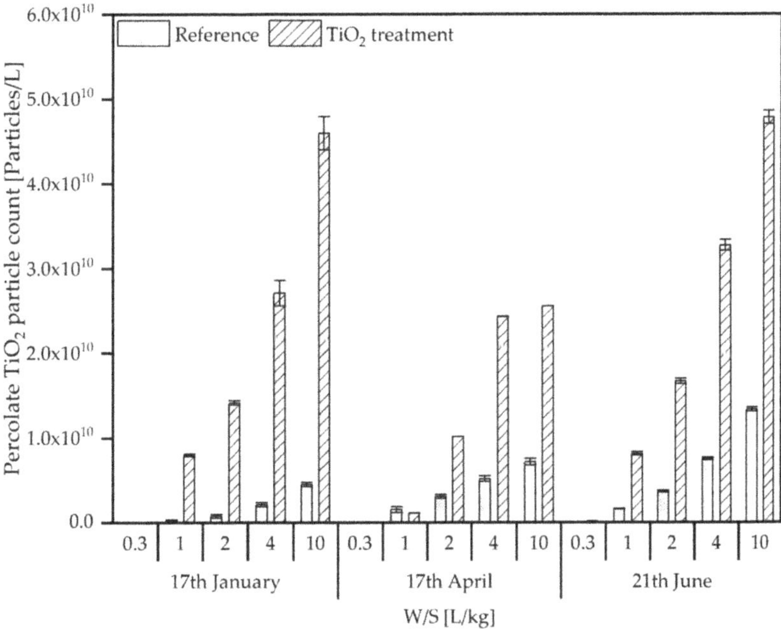

Figure 6. $nTiO_2$ particle concentrations (Particles/L) determined in column percolates from reference and $nTiO_2$ treated SSA/soil mixtures out of pilot scale simulation reactors collected at three different sampling times.

4. Discussion

In this study, $nTiO_2$ was used as an exemplary ENM for the preparation of ENM-containing SSA produced for the soil simulation experiments. In the simulation of agricultural use of this $nTiO_2$ amended SSA, a clearly increased release of $nTiO_2$ was found in the leachates of the pilot scale simulation reactors after a certain time compared to a reference ash from non-amended sewage sludge. The amount of release was significantly higher than could be expected due to the differences in the total TiO_2 contents in the soil-ash mixture relative to the reference ash. A cause for the comparatively high releases could not be determined within the project. After three freezing and thawing cycles, the amount of TiO_2 released from the reference was increased significantly, indicating mobilization by changes of the soil structure due to frost wedging. This is also observed in increasing amounts of TiO_2 in the column experiments throughout the three samplings. A difference between the leaching from the simulation reactors and standard column elution is the significantly higher ratio between TiO_2 released from reference and treated SSA in the reactors compared to the ratios observed for the column experiments, which were nearer to the ratio of total Ti contents of both materials (Figure 2) at the end of the study. For instance, while, for the reactors, the release of TiO_2, expressed in terms of total concentrations, from treated samples compared to the reference, was elevated by a factor of approx. 25, the analogous ratio for the standard column experiments was found to be approx. 3 (Figures 3 and 5). The differences are not that pronounced if the particle counts from spICP-MS analyses are assessed with factors of approx. 10 (leachates of treated-against-reference reactors) and 4 (column percolates of treated-against-reference soil/ash-mixtures). Thus, the particle counts agree with the total titanium on a relative scale for the column experiments, whereas, there seems to be an underestimation for the reactor leachates. This might be explained by losses of, e.g., bigger particles, which would significantly contribute to the total titanium mass in the sample, due to sedimentation prior to measurement or the underestimation of the number of bigger particles due to the applied analysis time of 60 s. In contrast, the total Ti determination after digestion captures the complete amount of Ti in the sample. However, in principle, the spICP-MS analyses give similar results to the determination of the total concentrations and can possibly be a fast alternative analytical method that, additionally, provides also information on particle sizes.

Looking further and in more detail to the obtained data, some differences can be noticed not only between the elution behavior of the test materials but also between the elution procedures itself. The elution procedure in the column and the simulation experiment were very different. In the column elution, within 7 days, a water to solid ratio of 10 L per kg soil material was applied. In the simulation, the maximum water to solid ratio used was 0.5 L per kg of soil material over a time of 250 days. This results in very different flow velocities in both approaches; it is not surprising that these huge differences in the elution conditions give different results in eluate analysis.

This significant influence of different elution conditions is strikingly visible if the released TiO_2 amounts are related to the total amount of soil/ash mixture in the respective experiments. For instance, for the last sampling date, after three seasonal cycles, around 130 µg/kg (released amount (Figure 3) referred to 700 kg of soil/ash (dry weight) material per reactor) was released from the reactor, while, for the column elution experiments, 5172 µg/kg (TiO_2 content in percolate referred to 1.1 kg soil/ash per column) was released. Thus, the column leaching experiments are overestimating the leaching for this kind of particle in comparison to the more realistic simulation experiments.

The higher relative release of TiO_2 from reference samples in the column experiments indicates that the natural "background" TiO_2 might be attached differently to the soil matrix, as it is more easily mobilized under the experimental conditions compared to the simulation experiments.

Therefore, the main finding of the study is that the standard column experiment according to DIN 19528 might be suitable to show increased mobility and emissions of $nTiO_2$ during soil-related recycling. This finding is confirmed by simulation experiments under more realistic conditions. It proves the general suitability of the standardized column elution method for predicting an increased release of $nTiO_2$ from a soil-ash mixture.

However, due to the limitations of a pilot scale experiment, which does not allow to test a reasonable number of different soils that would be needed for a more general statement, results cannot be generalized at this point. However, they indicate that increased leaching of ENM measured in a standard soil column approach relative to a reference material might serve as reasonable prediction of increased leaching of ENM under realistic environmental conditions. As a next step, a larger data set should be determined on both SSAs using a set of different soils for standard column elution to obtain the influence of different soil matrices.

5. Conclusions

The results of this study demonstrate that a standard soil column experiment might be suitable to reliably indicate an increased mobility of $nTiO_2$ from SSA mixed with agricultural soil relative to a respective reference material containing only "background" TiO_2. However, due to the limited data set of so far just one soil general conclusions cannot be made at this point. Still, the result is important for future exposure assessment, as it might offer an experimental laboratory tool to determine the risk of leaching from a recycling material.

The use of an accepted standard (DIN 19528) for the determination of the mobility offers the chance to get a new standard for testing ENM in soil material just by extension of the scope of an existing standard. In order to establish effective thresholds and to make them available for regulatory practice, respective standards for laboratory use are required. However, further research is necessary to verify current results and to base findings on a broader database before standardization processes might start. Laboratory elution experiments could give a useful contribution to the classification and safer use of incineration ashes, which contain not only potential pollutants but also many components that are too valuable for landfill disposal. Even if no direct transfer from the laboratory elution tests on the simulation experiments can be derived, the general concept is confirmed.

Author Contributions: Conceptualization, B.M., D.H. and K.H.-R.; methodology, B.M., D.H., K.H.-R.; validation, B.M., N.S. and D.H.; formal analysis, N.S. and B.M.; investigation, N.S. and K.W; resources, K.W., J.O.; data curation, B.M. and N.S.; writing—original draft preparation, B.M. and N.S.; writing—review and editing, B.M., D.H. and K.W.; visualization, N.S.; supervision, D.H. and B.M.; project administration, D.H. and B.M.; funding acquisition, D.H., J.O., K.H.-R. and B.M. All authors have read and agreed to the published version of the manuscript.

Funding: This research was funded by the German Federal Environment Agency (UBA) and was part of the project "Investigations on the possible release of nanoparticles in the deposition and soil-related application of mineral wastes" (FKZ 3712 33 327).

Data Availability Statement: Restrictions apply to the availability of these data. Data was obtained from the project funded by the German Environment Agency (FKZ 3712 33 327) and are available from the corresponding authors with the permission of the German Environment Agency.

Acknowledgments: The authors acknowledge the plant operator at ZVK Steinhäule for his support during the experiments.

Conflicts of Interest: The authors declare no conflict of interest.

References

1. Kaegi, R.; Sinnet, B.; Zuleeg, S.; Hagendorfer, H.; Mueller, E.; Vonbank, R.; Boller, M.; Burkhardt, M. Release of silver nanoparticles from outdoor facades. *Environ. Pollut.* **2010**, *158*, 2900–2905. [CrossRef] [PubMed]
2. Gondikas, A.P.; von der Kammer, F.; Reed, R.B.; Wagner, S.; Ranville, J.F.; Hofmann, T. Release of TiO$_2$ nanoparticles from sunscreens into surface waters: A one-year survey at the old Danube recreational Lake. *Environ. Sci. Technol.* **2014**, *48*, 5415–5422. [CrossRef] [PubMed]
3. Keller, A.A.; McFerran, S.; Lazareva, A.; Suh, S. Global life cycle releases of engineered nanomaterials. *J. Nanopart. Res.* **2013**, *15*, 1692–1709. [CrossRef]
4. Westerhoff, P.; Song, G.; Hristovski, K.; Kiser, M.A. Occurrence and removal of titanium at full scale wastewater treatment plants: Implications for TiO$_2$ nanomaterials. *J. Environ. Monit.* **2011**, *13*, 1195–1203. [CrossRef] [PubMed]
5. Kaegi, R.; Voegelin, A.; Sinnet, B.; Zuleeg, S.; Hagendorfer, H.; Burkhardt, M.; Siegrist, H. Behavior of metallic silver nanoparticles in a pilot wastewater treatment plant. *Environ. Sci. Technol.* **2011**, *45*, 3902–3908. [CrossRef] [PubMed]
6. Wiechmann, B.; Dienemann, C.; Kabbe, C.; Brandt, S.; Vogel, I.; Roskosch, A. Sewage Sludge Treatment in Germany; Umweltbundesamt, Dessau-Roßlau. 2015. Available online: https://www.umweltbundesamt.de/sites/default/files/medien/378/publikationen/sewage_sludge_management_in_germany.pdf (accessed on 10 December 2021).
7. DüMV, Verordnung über das Inverkehrbringen von Düngemitteln, Bodenhilfsstoffen, Kultursubstraten und Pflanzenhilfsmitteln (Düngemittelverordnung–DüMV) (German Fertilizer Ordinance); Bundesgesetzblatt (Federal Law Gazette) Jahrgang 2012 Teil I Nr. 58. 2012. Available online: https://www.gesetze-im-internet.de/d_mv_2012/ (accessed on 10 December 2021).
8. Wielinski, J.; Gogos, A.; Voegelin, A.; Müller, C.R.; Morgenroth, E.; Kaegi, R. Release of gold (Au), silver (Ag) and cerium dioxide (CeO$_2$) nanoparticles from sewage sludge incineration ash. *Environ. Sci. Nano* **2021**, *8*, 3220–3232. [CrossRef] [PubMed]
9. BBodSchV, Bundes-Bodenschutz- und Altlastenverordnung vom 12. Juli 1999 (BGBl. I S. 1554), die Zuletzt durch Artikel 126 der Verordnung vom 19. Juni 2020 (BGBl. I S. 1328) Geändert Worden ist (German Soil Protection Ordinance); Bundesgesetzblatt (Federal Law Gazette) Jahrgang 1999 Teil I, Nr. 36. 1999. Available online: https://www.gesetze-im-internet.de/bbodschv/ (accessed on 10 December 2021).
10. Piccinno, F.; Gottschalk, F.; Seeger, S.; Nowack, B. Industrial production quantities and uses of ten engineered nanomaterials in Europe and the world. *J. Nanopart. Res.* **2012**, *14*, 1109. [CrossRef]
11. Foss Hansen, S.; Heggelund, L.R.; Revilla Besora, P.; Mackevica, A.; Boldrin, A.; Baun, A. Nanoproducts–what is actually available to European consumers? *Environ. Sci. Nano* **2016**, *3*, 169–180. [CrossRef]
12. Landi, S.; Carneiro, J.; Soares, O.S.G.P.; Pereira, M.F.R.; Gomes, A.C.; Ribeiro, A.; Fonseca, A.M.; Parpot, P.; Neves, I.C. Photocatalytic performance of N-doped TiO$_2$nano-SiO$_2$-HY nanocomposites immobilized over cotton fabrics. *J. Mater. Res. Technol.* **2019**, *8*, 1933–1943. [CrossRef]
13. Noireaux, J.; López-Sanz, S.; Vidmar, J.; Correia, M.; Devoille, L.; Fisicaro, P.; Loeschner, K. Titanium dioxide nanoparticles in food: Comparison of detection by triple-quadrupole and high-resolution ICP-MS in single-particle mode. *J. Nanopart. Res.* **2021**, *23*, 4. [CrossRef]
14. BioAbfV, Bioabfallverordnung in der Fassung der Bekanntmachung vom 4. April 2013 (BGBl. I S. 658), die Zuletzt durch Artikel 3 Absatz 2 der Verordnung vom 27. September 2017 (BGBl. I S. 3465) Geändert Worden ist. Bundesgesetzblatt (Federal Law Gazette) Jahrgang 1998 Teil I, S. 2955. 1998. Available online: https://www.gesetze-im-internet.de/bioabfv/BJNR295500998.html (accessed on 10 December 2021).

15. Praetorius, A.; Tufenkji, N.; Goss, K.-U.; Scheringer, M.; von der Kammer, F.; Elimelech, M. The road to nowhere: Equilibrium partition coefficients for nanoparticles. *Environ. Sci. Nano* **2014**, *1*, 317–323. [CrossRef]
16. Börner, R.; Meiller, M.; Oischinger, J.; Daschner, R. *Untersuchung Möglicher Umweltauswirkungen bei der Entsorgung Nanomaterialhaltiger Abfälle in Abfallbehandlungsanlagen*; Umweltbundesamt: Dessau-Roßlau, Germany, 2016.
17. Oischinger, J.; Meiller, M.; Daschner, R.; Hornung, A.; Warnecke, R. Fate of nano titanium dioxide during combustion of engineered nanomaterial-containing waste in a municipal solid waste incineration plant. *Waste Manag. Res.* **2019**, *37*, 1033–1042. [CrossRef] [PubMed]
18. Schäfer, E. Beispiel einer zukunftsorientierten kommunalen Abwasserreinigung. In *Verwertung von Klärschlamm*; Holm, O.E., Thomé-Kozmiensky, E., Quicker, P., Kopp-Assenmacher, S., Eds.; Thomé-Kozmiensky Verlag GmbH: Neuruppin, Germany, 2018; pp. 121–130.
19. Ländergemscinaft Abfall, LAGA PN98 Richtlinie für das Vorgehen bei Physikalischen, Chemischen und Biologischen Untersuchungen im Zusammenhang mit der Verwertung/Beseitigung von Abfällen. Grundregeln für die Entnahme von Proben aus Festen und Stichfesten Abfällen Sowie Abgelagerten Materialien; Ländergemeinschaft Abfall. 2001. Available online: https://www.laga-online.de/documents/m32_laga_pn98_1503993280.pdf (accessed on 10 December 2021).
20. *Länderarbeitsgemeinschaft Abfall, LAGA-Mitteilung 19, Merkblatt für die Entsorgung von Abfällen aus Verbrennungsanlagen für Siedlungsabfälle*; ESV: Berlin, Germany, 1994.
21. Chen, G.; Liu, X.; Su, C. Distinct effects of humic acid on transport and retention of TiO_2 rutile nanoparticles in saturated sand columns. *Environ. Sci. Technol.* **2012**, *46*, 7142–7150. [CrossRef] [PubMed]
22. Wang, Y.; Gao, B.; Morales, V.L.; Tian, Y.; Wu, L.; Gao, J.; Bai, W.Y.L. Transport of titanium dioxide nanoparticles in saturated porous media under various solution chemistry conditions. *J. Nanopart. Res.* **2012**, *14*, 1095–1104. [CrossRef]
23. *DIN 19528:2009-01*; Leaching of Solid Materials-Percolation Method for the Joint Examination of the Leaching Behaviour of Inorganic and Organic Substances. Beuth: Berlin, Germany, 2009.
24. Montano, M.D.; Olesik, J.W.; Barber, A.G.; Challis, K.; Ranville, J.F. Single Particle ICP-MS: Advances toward routine analysis of nanomaterials. *Anal. Bioanal. Chem.* **2016**, *408*, 5053–5074. [CrossRef] [PubMed]
25. Meermann, B.; Nischwitz, V. ICP-MS for the analysis at the nanoscale–A tutorial review. *J. Anal. At. Spectrom.* **2018**, *33*, 1432–1468. [CrossRef]
26. Oischinger, J.; Meiller, M.; Daschner, R.; Hennecke, D.; Hund-Rinke, K.; Meisterjahn, B.; Schröder, N. *Untersuchungen zur Möglichen Freisetzung von Nanopartikeln bei der Ablagerung und Bodenbezogenen Anwendung von Mineralischen Abfällen-Abschlussbericht*; Umweltbundesamt: Dessau-Roßlau, Germany, 2020.

Article

Mass Transfer Principles in Column Percolation Tests: Initial Conditions and Tailing in Heterogeneous Materials

Binlong Liu, Michael Finkel and Peter Grathwohl *

Center for Applied Geoscience, University of Tübingen, Schnarrenbergstraße 94-96, 72076 Tübingen, Germany; binlong.liu@uni-tuebingen.de (B.L.); michael.finkel@uni-tuebingen.de (M.F.)
* Correspondence: grathwohl@uni-tuebingen.de

Abstract: Initial conditions (pre-equilibrium or after the first flooding of the column), mass transfer mechanisms and sample composition (heterogeneity) have a strong impact on leaching of less and strongly sorbing compounds in column percolation tests. Mechanistic models as used in this study provide the necessary insight to understand the complexity of column leaching tests especially when heterogeneous samples are concerned. By means of numerical experiments, we illustrate the initial concentration distribution inside the column after the first flooding and how this impacts leaching concentrations. Steep concentration gradients close to the outlet of the column have to be expected for small distribution coefficients ($K_d < 1$ L kg^{-1}) and longitudinal dispersion leads to smaller initial concentrations than expected under equilibrium conditions. In order to elucidate the impact of different mass transfer mechanisms, film diffusion across an external aqueous boundary layer (first order kinetics, FD) and intraparticle pore diffusion (IPD) are considered. The results show that IPD results in slow desorption kinetics due to retarded transport within the tortuous intragranular pores. Non-linear sorption has not much of an effect if compared to K_d values calculated for the appropriate concentration range (e.g., the initial equilibrium concentration). Sample heterogeneity in terms of grain size and different fractions of sorptive particles in the sample have a strong impact on leaching curves. A small fraction (<1%) of strongly sorbing particles (high K_d) carrying the contaminant may lead to very slow desorption rates (because of less surface area)—especially if mass release is limited by IPD—and thus non-equilibrium. In contrast, mixtures of less sorbing fine material ("labile" contamination with low K_d), with a small fraction of coarse particles carrying the contaminant leads to leaching close to or at equilibrium showing a step-wise concentration decline in the column effluent.

Keywords: leaching test; equilibrium condition; non-equilibrium condition; modelling; sorption kinetics; non-linear sorption; heterogeneity

Citation: Liu, B.; Finkel, M.; Grathwohl, P. Mass Transfer Principles in Column Percolation Tests: Initial Conditions and Tailing in Heterogeneous Materials. *Materials* **2021**, *14*, 4708. https://doi.org/10.3390/ma14164708

Academic Editor: Saeed Chehreh Chelgani

Received: 18 July 2021
Accepted: 16 August 2021
Published: 20 August 2021

Publisher's Note: MDPI stays neutral with regard to jurisdictional claims in published maps and institutional affiliations.

1. Introduction

Leaching tests are widely used for the determination of contaminant release rates from soils [1–4], recycling materials [5–11], construction products [12–16], radioactive and other waste materials [17,18]. Compared to traditional batch shaking tests, column tests are preferred for assessing the risk of release of potential pollutants into groundwater or surface waters because they are closer to natural conditions [19,20]. Different mechanisms controlling desorption kinetics may result in complex leaching behaviors. While initially the observed column effluent concentration often reflects equilibrium conditions between the solid phase (incl. intraparticle pores) and the mobile aqueous phase [21,22], the concentrations decrease and often an extended tailing is observed due to slow desorption processes such as intraparticle diffusion [23–25].

Although most leaching test procedures aim at equilibrium conditions in the column before the leaching starts, the true concentration distribution before the start of the percolation depends not only on the test procedure (contact time, pre-equilibration time, flow velocity during first flooding) but also on the properties of both the solid material and the pollutant of interest [26]. Equilibration and long-term leaching are further complicated if the test sample consists of a heterogeneous mixture of different material types and grain sizes [26–28], which is the common case if waste materials such as demolition waste or soils are tested.

Finkel and Grathwohl (2017) evaluated the role of initial conditions for column leaching tests with intraparticle pore diffusion models by comparing the hypothetical scenarios, a perfectly equilibrated column vs. a column that wasn't equilibrated at all [26]. They could show that in many practical cases, peak and cumulative leachate concentrations are rather independent of the initial conditions. However, if release kinetics are slow due to large grain size or small intragranular porosity, the sensitivity to initial conditions is relevant, in particular for initial peak and early cumulative leachate concentrations.

The shortcomings of all previous studies [26,29,30], is that only uniform initial concentrations in the columns were considered in the leaching models. However, due to the specific conditions during the flooding of the column filled with initially dry material, the true initial conditions at the start of the leaching test may considerably deviate from this ideal, i.e., uniform distribution.

Against this background, the objectives of this study are (i) to illustrate the possibly non-uniform initial conditions that may be achieved after the first flooding of the column, (ii) to show the impact of these initial conditions on the temporal development of the effluent concentrations, and (iii) to investigate how heterogeneous mixtures of particles having different properties affect both the initial conditions in the column and the leaching of the solutes. To achieve that, we used numerical solutions for flow and transport in a column coupled to two kinetic models: (i) solute diffusion through an aqueous boundary layer and (ii) intraparticle pore diffusion. The implementation of the numerical models is described in detail in the appendices.

2. Theory and Background

2.1. Local Equilibrium: The Advection—Dispersion Equation

In order to facilitate the understanding of mass transfer-limited cases of contaminant release in a column, we briefly introduce the equilibrium case for which the advection-dispersion model is commonly used:

$$\frac{\partial}{\partial t}(n\,C_w + \rho_b\,C_s) + \frac{\partial}{\partial x}\left(n\,v\,C_w - n\,D_L\frac{\partial C_w}{\partial x}\right) = 0 \tag{1}$$

where v [L T^{-1}], n [-] and D_L ($= \alpha v + D_p$) [L^2 T^{-1}] denote the seepage velocity of the water, the intergranular porosity and the longitudinal dispersion coefficient. α [L], D_p ($=nD_{aq}$) [L^2 T^{-1}] and D_{aq} [L^2 T^{-1}] denote the dispersivity, the pore diffusion coefficient and the aqueous diffusion coefficient of the solute. x [L] and t [T] are the length of the column and time. ρ_b ($= (1-n)\rho_s$) [M L^{-3}] is the dry bulk density of the packed bed in the column (porous media; ρ_s is the solids density). For local equilibrium conditions the concentration in the solid phase (C_s) is in equilibrium with the solute concentration in water (C_w) and the distribution coefficient K_d ($= C_s/C_w$) allowing for the calculation of the respective concentrations. Under these conditions, Equation (1) can be simplified as:

$$\frac{\partial C_w}{\partial t} = \frac{D_L}{R_d}\frac{\partial^2 C_w}{\partial x^2} - \frac{v}{R_d}\frac{\partial C_w}{\partial x} \tag{2}$$

where R_d [-] represents the retardation factor, defined as:

$$R_d = 1 + K_d\frac{\rho_b}{n} \tag{3}$$

Assuming equilibrium conditions and an initially uniform distribution of the solute in the column, leaching may be described by the analytical solution of Equation (2) [31]:

$$\frac{C_w}{C_{w,eq}} = 1 - 0.5\left[\text{erfc}\left(\frac{x - \frac{v}{R_d}t}{2\sqrt{\frac{D_L}{R_d}t}}\right) + \exp\left(\frac{v\,x}{D_L}\right)\text{erfc}\left(\frac{x + \frac{v}{R_d}t}{2\sqrt{\frac{D_L}{R_d}t}}\right)\right] \tag{4}$$

where erfc denotes the complementary error function. The aqueous concentration at equilibrium, $C_{w,eq}$, can be calculated from the initial solid concentration ($C_{s,ini}$) accounting for the mass balance when the contaminant mass in the water is equilibrated with the mass in the solid:

$$C_{w,eq} = \frac{C_{s,ini}}{K_d + \frac{n}{\rho_b}} \tag{5}$$

The ratio n/ρ_b [L^3 M^{-1}] equals the liquid to solid ratio within the column, which in most cases is much smaller than in a batch leaching test (e.g., 0.25 L kg^{-1} for a column test with a porosity of $n = 0.4$ and a solid density of $\rho_s = 2.65$ g cm^{-3}, compared to e.g., 10 L kg^{-1} in a batch test). Since leaching tests start for practical reasons with material packed more or less dry into the column, a uniform initial concentration is not necessarily achieved during the first flooding of the column. Initial conditions as assumed in Equation (4) (uniform concentration distribution), would only be achieved if the material is first mixed with water, equilibrated and then packed into the column, which is not practical. During the first flooding of the column, especially less sorbing solutes are displaced from the inlet and higher concentrations occur towards the outlet, as illustrated in Figure 1 (see also Appendix E). This may be accounted for by subtracting the distance of the solute displaced initially (x/R_d with $R_d > 1$) in Equation (4):

$$\frac{C_w}{C_{w,eq}} = 1 - 0.5\left[\text{erfc}\left(\frac{\left(x - \frac{x}{R_d}\right) - \frac{v}{R_d}t}{2\sqrt{\frac{D_L}{R_d}t}}\right)\right.$$
$$\left. + \exp\left(\frac{v\left(x - \frac{x}{R_d}\right)}{D_L}\right)\text{erfc}\left(\frac{\left(x - \frac{x}{R_d}\right) + \frac{v}{R_d}t}{2\sqrt{\frac{D_L}{R_d}t}}\right)\right] \tag{6}$$

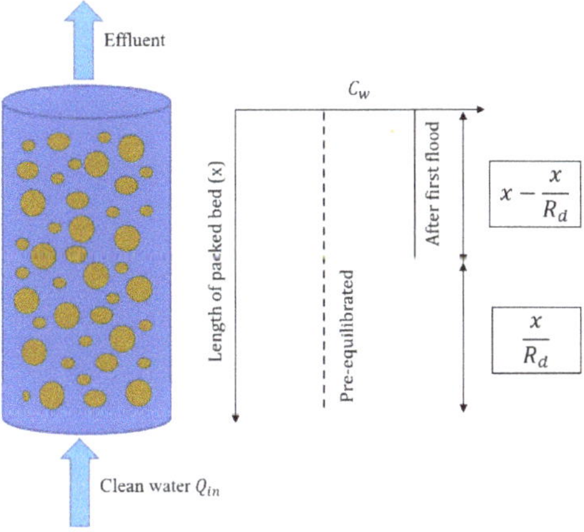

Figure 1. Initial concentration distribution in a column of length x for the "pre-equilibrated" case (dashed line) and after the first flooding of an initially dry column from the bottom (solid line); no dispersion, $R_d = 2$, after Grathwohl and Susset, 2009 [21].

In this case the initial solute concentration ($C_{w,eq}$) in the column effluent is in equilibrium with the initial concentration in the solids ($C_{w,eq} = C_{s,ini}/K_d$) and higher than calculated in Equation (5), especially if K_d values are low.

In order to explore the influence of mass transfer limitations on initial and long-term solute concentrations in column tests, two relevant mass transfer mechanisms are compared: (1) film diffusion where diffusion from the solid-water interface occurs through an aqueous boundary layer with a given thickness and (2) intraparticle pore diffusion where diffusion inside the porous particle limits mass transfer.

2.2. Desorption Kinetics Limited by Film Diffusion

The simplest model for the kinetic release of a solute from solids considers diffusion through an aqueous boundary layer surrounding spherical particles (Figure 2). Such film diffusion models are also widely used for the dissolution of minerals with high solubilities (e.g., salts). The solute release from the solid surface into the bulk water phase can be described by a linear driving force with constant mass transfer coefficient $k = D_{aq}/\delta_f$:

$$\frac{\partial C_w}{\partial t} = k\,A^o\left(C_w' - C_w\right) = \frac{D_{aq}}{\delta_f}\,\frac{m_d\,6}{\rho_s\,d\,V_w}\left(C_w' - C_w\right) \tag{7}$$

where δ_f [L], V_w [L^3], m_d [M] and d [L] denotes the thickness of the external film, the volume of water, the dry mass of the solids in the column and the particle diameter, respectively. A^o ($= 6\,m_d/(V_w\,\rho_s\,d)$) is the specific surface area of the particles per unit volume of water in the column [m^2 m^{-3} = m^{-1}] (the term $6/\rho_s\,d$) represents the specific surface area of spherical particles per dry mass, e.g., in m^2 g^{-1}). C_w' is the concentration at the solid-water interface where local equilibrium conditions apply ($C_w' = C_s/K_d$). The external film thickness (δ_f) can be estimated from empirical Sherwood numbers (Sh) and the particle diameter (d):

$$Sh = \frac{kd}{D_{aq}} = \frac{d}{\delta_f} \;\rightarrow\; \delta_f = \frac{d}{Sh} \tag{8}$$

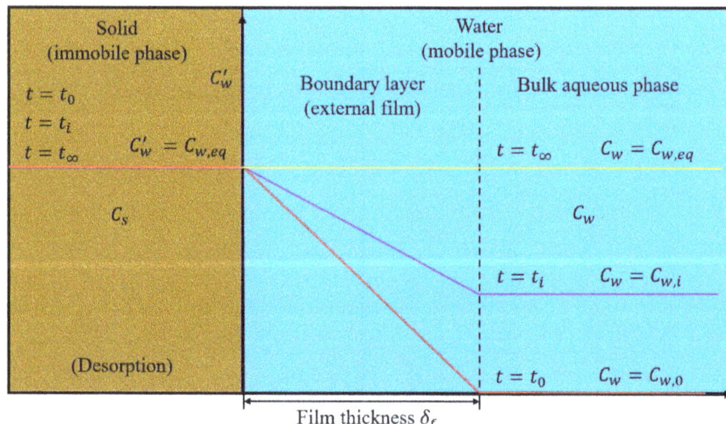

Figure 2. Scheme of mass transfer limited by film diffusion during the first flooding with fixed concentration at the interface (because the infiltrating water is always contacting fresh material as it advances).

For an overview on empirical relationships for the estimation of Sherwood numbers see Appendix A. The mass balance in such two-phase systems expressed by their respective rates is:

$$V_w\frac{\partial C_w}{\partial t} = -m_d\frac{\partial C_s}{\partial t} \tag{9}$$

Thus, the solute mass gained (or lost) by the water phase equals the solute mass lost (or gained) from the solid phase. If V_w and m_d in a packed bed (porous media) are replaced by n and ρ_b, the density of the solids (ρ_s) in Equation (7) drops out. Film diffusion coupled to the one-dimensional advection-dispersion equation (Equation (1)) yields:

$$\frac{\partial C_w}{\partial t} = D_L \frac{\partial^2 C_w}{\partial x^2} - v \frac{\partial C_w}{\partial x} + \frac{D_{aq}}{\delta_f} \frac{6\,(1-n)}{n\,d} \left(\frac{C_s}{K_d} - C_w \right) \tag{10}$$

Using the finite volume method, the column is spatially discretized by a number of cells (see Figure A1) and the governing equation (Equation (10)) is solved iteratively by employing the Newton–Raphson scheme. Details of the numerical solution of the film diffusion model are presented in Appendix B.

2.3. Desorption Limited by Intraparticle Pore Diffusion

If the release of compounds from the solid phase is governed by intra-granular diffusion, e.g., within a porous grain (Figure 3), then mass transfer is described by Fick's second law in radial coordinates:

$$\frac{\partial}{\partial t} \left(\varepsilon\, C_{w,intra} + \rho_p C_s \right) = D_e \left[\frac{\partial^2 C_{w,intra}}{\partial r^2} + \frac{2}{r} \frac{\partial C_{w,intra}}{\partial r} \right] \tag{11}$$

with the boundary conditions

$$C_{w,intra}(r = R, t) = C_w \tag{12}$$

$$\frac{\partial C_{w,intra}}{\partial r}(r = 0, t) = 0 \tag{13}$$

r [L] is the radial coordinate in the sphere and D_e [L^2 T^{-1}] the effective diffusion coefficient. $C_{w,intra}$ [M L^{-3}] is the concentration of solute in the intra-granular pore water. ε [-] denotes the intraparticle porosity. R [L] and ρ_p [M L^{-3}] $(= \rho_s(1 - \varepsilon))$ denote the radius and bulk density of the particle (sphere).

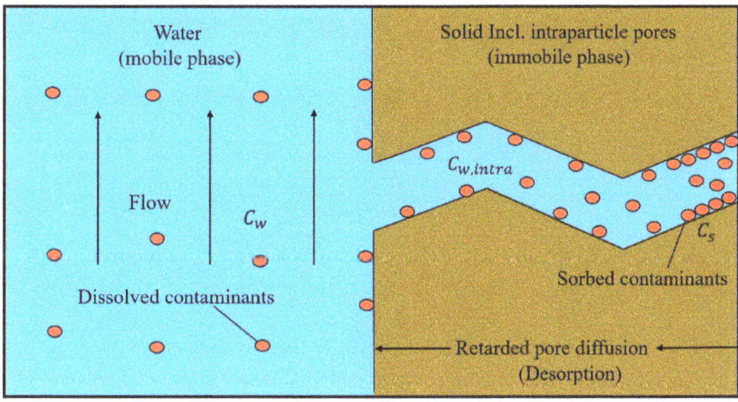

Figure 3. Scheme of mass transfer limited by intraparticle pore diffusion.

For linear sorption with concentration independent distribution coefficients (i.e., $C_s = K_d\, C_{w,intra}$) Equation (11) becomes:

$$\frac{\partial C_{w,intra}}{\partial t} = D_a \left[\frac{\partial^2 C_{w,intra}}{\partial r^2} + \frac{2}{r} \frac{\partial C_{w,intra}}{\partial r} \right] \tag{14}$$

where D_a [L^2 T^{-1}] is the apparent diffusion coefficient, defined as:

$$D_a = \frac{D_e}{\varepsilon + K_d \rho_p} = \frac{D_{aq}\varepsilon}{\tau(\varepsilon + K_d \rho_p)} \approx \frac{D_{aq}\varepsilon^2}{\varepsilon + K_d(1-\varepsilon)\rho_s} \tag{15}$$

Empirical studies showed that D_e increases approximately with the square of the porosity [32]. This corresponds to a tortuosity τ [-] of the intra-granular pores—if expressed as a function of intra-granular porosity—of $\tau \approx 1/\varepsilon$.

Considering intraparticle diffusion, the advection-dispersion model (Equation (1)) can be rewritten as:

$$\frac{\partial}{\partial t}\left(nC_w + (1-n)\left(\varepsilon C_{w,intra} + \rho_p C_s\right)\right) + \frac{\partial}{\partial x}\left(nvC_w - nD_L\frac{\partial C_w}{\partial x}\right) = 0 \tag{16}$$

The corresponding equilibrium concentration ($C_{w,eq}$) in water after first flooding can be rewritten as $C_{w,eq} = C_{s,ini}/(\varepsilon/\rho_p + K_d)$, which is slightly lower than for non-porous solids ($C_{w,eq} = C_{s,ini}/K_d$) because the intragranular pore space is assumed to be initially free of water. The deviation becomes insignificant with the increase of K_d ($\varepsilon/\rho_p + K_d \approx K_d$ when $K_d \gg \varepsilon/\rho_p$).

Coupling the intraparticle pore diffusion model (Equation (11)) to the one-dimensional advection-dispersion equation (Equation (16)) allows for the expression of the change of the solute concentration in the bulk water:

$$\frac{\partial C_w}{\partial t} = D_L\frac{\partial^2 C_w}{\partial x^2} - v\frac{\partial C_w}{\partial x} - \frac{(1-n)}{n}D_e\left[\frac{\partial^2 C_{w,intra}}{\partial r^2} + \frac{2}{r}\frac{\partial C_{w,intra}}{\partial r}\right] \tag{17}$$

The intraparticle pore diffusion model (Equations (11)–(13)) was implemented numerically using a finite volume method where the spherical particle was discretized by a number of spherical shells of equal volume (based on the method of Jäger and Liedl [28]). The column was spatially discretized by a number of cells (see Figure A1) and all the governing equations (Equations (11)–(13) and (17)) were solved iteratively applying the Newton–Raphson scheme. Compared to the 1D film diffusion case, the intraparticle pore diffusion case is more complicated and becomes a 2D problem. Details of the numerical solution of the intraparticle pore diffusion model are given in Appendix C.

2.4. Set-Up of "Numerical" Column Tests

The boundary conditions of the numerical experiments are based on the set-up of column tests in daily practice in Germany [21,33,34]. Table 1 lists the relevant material properties and the parameter ranges applied. The saturation time for the first pore volume of the column and the contact time (after the first flooding period) were set to 5 h. Initially, experiments with uniform materials were simulated where the intraparticle porosity, distribution coefficient, aqueous diffusion coefficient and tortuosity were set the same for each grain size fraction. The Sherwood number in packed beds was estimated based on the empirical formula proposed by Liu et al. (2014) (Equation (A3)) [35]. In order to illustrate the influence of longitudinal dispersion on the initial concentration distribution in the column after the first flooding, we used fine particles ($d_{p,\,fine} = 63$ μm) where kinetics are very fast, and the local equilibrium assumption is valid. The numerical model did not consider dispersion beyond the outlet of the column. Non-linear sorption was simulated using the Freundlich model ($C_s = K_{fr}C_w^{\frac{1}{n}}$ where K_{fr} and $1/n$ denote the Freundlich coefficient and Freundlich exponent, respectively).

Many factors may contribute to sample heterogeneity, such as grain size distribution and particle properties (sorption, porosities, etc.). To highlight the impact of particle size and properties we focused on two grain size classes and different fractions of sorptive/reactive particles in the sample. Distribution coefficients were varied in a range of 0.1–100 L kg^{-1}. Lower K_d values (<0.1 L kg^{-1}) were not considered here (this would have resulted in very high initial aqueous concentrations). If the K_d values become large ($K_d > 100$ L kg^{-1}), then the differences between the pre-equilibrated case and the "first flooding" scenario vanish and effluent concentrations are constant over long time periods. The K_d range chosen covers many frequent environmental contaminants, such as per- and polyfluoroalkyl substances (PFAS), chlorinated solvents, polycyclic aromatic hydrocarbons and some heavy metals.

Table 1. Summary of parameter values and ranges used to set up the numerical experiments.

Property	Symbol (Unit)	Reference and [Alternative Values]
Net column length	X_{col} (cm)	30
Inner column diameter	D_c (cm)	5.46
Total volume of column	V_{tot} (L)	0.70
Dry solid density	ρ_s (kg L^{-1})	2.60
Inter-granular porosity	n (-)	0.45
Intraparticle porosity	ε (-)	0.05
Solid mass in column	m_d (kg)	1
Liquid to solid ratio in column	LS_{col} (L kg^{-1})	0.31
Initial concentration in solid phase	$C_{s,ini}$ (µg kg^{-1})	1000
Contact time	t_c (h)	5
Dispersivity	α (m)	[0, 0.03]
Water flow velocity	v (m s^{-1})	1.67×10^{-5}
Aqueous diffusion coefficient	D_{aq} (m^2 s^{-1})	1×10^{-9}
Particle diameters	d (µm)	[63, 2000]
Distribution coefficients	K_d (L kg^{-1})	[0.1, 1, 10, 100]
Freundlich coefficients	K_{fr} (µg kg^{-1}:(µg L^{-1})$^{1/n}$)	[1.58, 7.94, 39.81]
Freundlich exponent	$1/n$	0.7
Sherwood number	$Sh = 2 + 0.1 Pe^{1/2}$ (-)	[2.1, 2.6]

3. Results and Discussion

3.1. Impact of Initial Conditions on Leaching

In order to investigate the impact of initial conditions on leaching behavior, we compared two scenarios: (1) a column filled with pre-equilibrated material where the initial concentration distribution in the column was uniform ($C_{w,eq} = C_{s,ini} / \left(K_d + \frac{n}{\rho_b} \right)$) and (2) columns with non-uniform concentration distributions after first flooding where concentrations increased towards the outlet (to a maximum of $C_{w,eq} = C_{s,ini} / K_d$) while at the inlet the solute was already depleted. To illustrate this, we used the film diffusion model with fine grain sizes, and thus, fast kinetics (local equilibrium conditions). Figure 4 shows the initial aqueous concentration distribution in the up-flow column test after the first flooding of the column compared to the pre-equilibrium case. The results show that the differences in the initial concentration profiles became smaller with increasing sorption. At high K_d values, the deviation of the initial concentration profiles only occurred at the inlet of the column. At low K_d values, very high concentrations are expected at the column outlet; in extreme cases this may lead to a density increase in the leachate and—especially if flow is stopped—to density driven flow within the column. This would cause dilution and lower leachate concentrations when the flow is restarted as was potentially observed by Naka et al. [36]. Density driven mixing may be caused by soluble materials, e.g., sulphate or other salts and not necessarily the target compounds. This phenomenon is quite similar to the case where the dispersion is taken into account (see bottom curves of Figure 4 and also Figures S1–S8 in SM), which leads to more "mixing" in the column and thus lower initial concentrations at the outlet, especially for low K_d values.

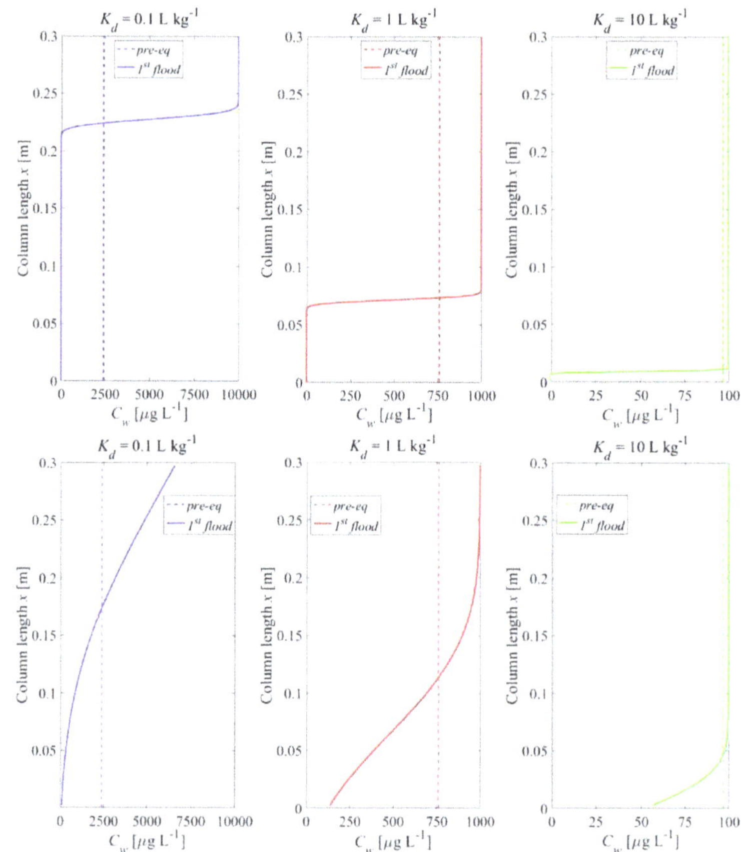

Figure 4. Initial aqueous concentration distributions in the column after the first flooding (solid lines) if mass transfer is controlled by film diffusion for three different distribution coefficients, K_d, compared to the pre-equilibrated case (dashed lines). **Top panel**: without dispersion; **bottom panel**: with dispersion; $n = 0.45$, $v = 1.67 \times 10^{-5}$ m s^{-1}, $\alpha/x = 0$ or 0.1, $C_{s,ini} = 1000$ µg kg^{-1}, $t_c = 5$ h, $d_{p,fine} = 63$ µm.

Figure 5 shows how the initial conditions (pre-equilibrated column and column after first flooding) influence the leaching curves. Compared to the pre-equilibrated case, a higher equilibrium concentration appeared at the outlet of the column after the first flooding period and more contaminant mass was released from the column at early times, followed by a rapidly decreasing concentration (see Figure 5, 2nd row). The deviations vanished with increasing K_d and became almost insignificant for $K_d \geq 10$ L kg^{-1}. Dispersion also reduces differences between the pre-equilibrated and the first flooding case. At high K_d values, the maximum concentrations were still achieved but the tailings became smoother. With the decrease of the K_d value, the concentration gradients at the inlet became steeper and the "back" dispersion fluxes towards the outlet increased as well. In extreme cases, the peak concentration at the column outlet was smaller than the maximum concentration expected (e.g., $K_d = 0.1$ L kg^{-1}). The effect of initial conditions on normalized concentrations looks like a phase shift (see Figure 5, 1st row). This would lead to an underestimation of K_d values derived from the pre-equilibrium analytical solution (Equation (4)) if the conditions in the column after the first flooding are not appropriately considered. The lower the K_d, the earlier the cumulative leachate concentration reaches its maximum value ($m_{cum,\,max} = 1000$ µg kg^{-1}). Dispersion shifts this point to later times (see Figure 5, 3rd row).

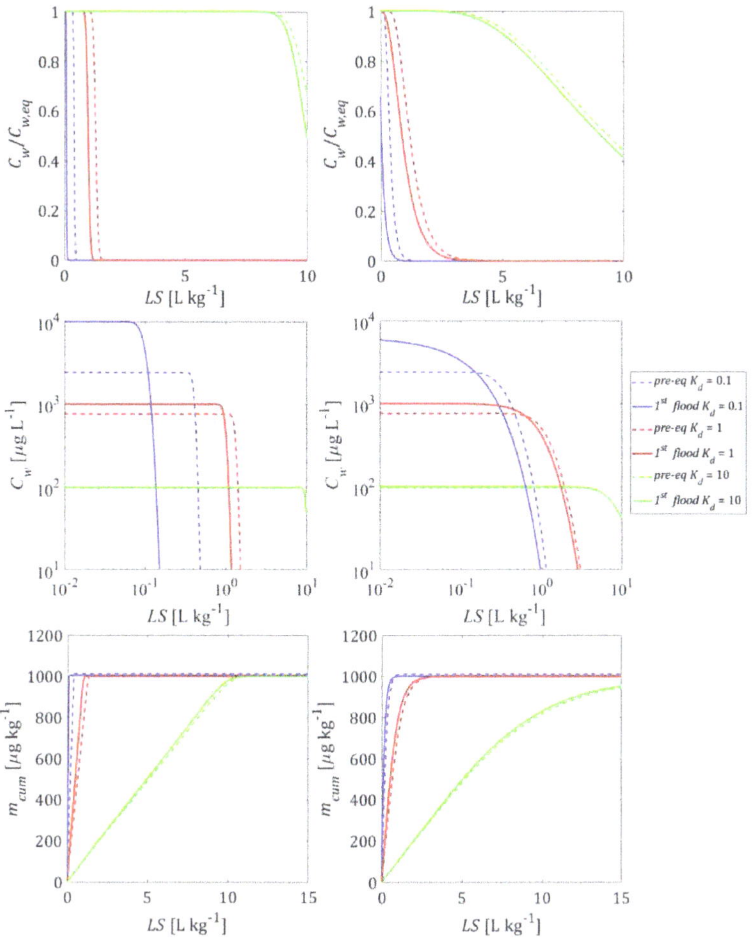

Figure 5. Normalized and absolute concentration ($C_w / C_{w,eq}$, C_w) as well as cumulative concentration (m_{cum}) in the column effluent vs. time (expressed as liquid to solid ratio: *LS*) for the initial conditions (depicted in Figure 4) established after the first flooding of the column (solid lines) compared to the pre-equilibrated case (dashed lines). **Left column**: without dispersion; **right column**: with dispersion.

3.2. Initial Conditions and Leaching with Mass Transfer Limitations

Mass-transfer limitations may change the picture considerably, with respect to initial conditions and the development of leachate concentrations over time. Figure 6 shows the influence of film diffusion (FD) compared to intraparticle pore diffusion (IPD) limited desorption on the initial concentration distribution in the column after the first flooding period. For large K_d values, equilibration is achieved after shorter distances in the column because of the retardation of the clean water front. The deviations between FD and IPD are due to different mass transfer zone lengths, $X_{s,63.2\%}$ (see Appendix D for a discussion of the concept and calculation of this length for FD and IPD). For FD, the mass transfer zone length is independent of the K_d value and proportional to the particle size (Equation (A28)). In contrast, for IPD the length of the mass transfer zone increases with particle size to the power of 3/2 ($d^{3/2}$) and decreases with $\sqrt{K_d}$ (Equation (A32)) (e.g., $X_{s,63.2\%} = 10$ cm, 3.5 cm, and 1.1 cm for K_d values of 0.1 L kg^{-1}, 1 L kg^{-1} and 10 L kg^{-1} (see Figure A2)). The length of the mass transfer zone for IPD is much longer than for FD, but differences decrease with increasing K_d values. For K_d values of 1 L kg^{-1} and 10 L kg^{-1}, the mass transfer zone lengths for IPD are much shorter than the column length (X_{col} = 30 cm), which indicates that the equilibrium concentration is achieved at the outlet of the column after the first flooding. For small K_d values (e.g., $K_d = 0.1$ L kg^{-1}), the equilibrium concentrations are not achieved at the outlet if dispersion is considered (see Figure 6, lower panel) although the mass transfer zone length ($X_{s,63.2\%} = 10$ cm) is still shorter than the column length. This is because the "clean" water front is close to the column outlet and dispersion "dilutes" the steep concentration gradients ("back dispersion"). The deviations between FD and fast kinetics almost vanish when dispersion is considered, indicating that with film diffusion, equilibrium is almost achieved. The development of the concentration distribution for IPD is also illustrated in animated graphs provided in the Supplementary Material (SM).

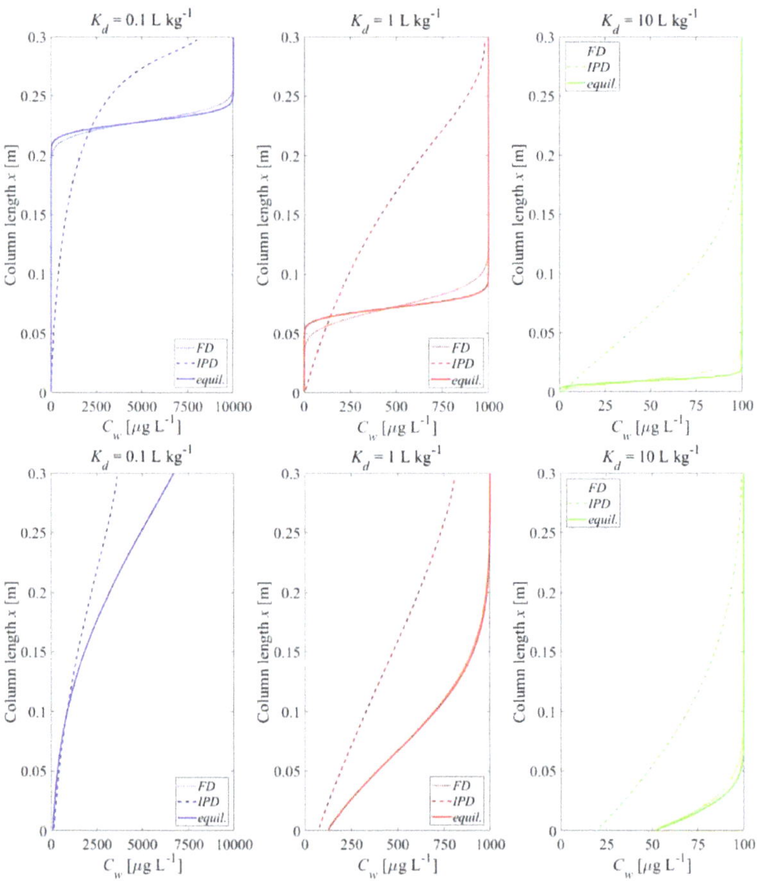

Figure 6. Initial aqueous concentration distributions in the column after the first flooding depending on the mass transfer limitation; dotted lines: film diffusion (FD), dashed lines: intraparticle diffusion (IPD); solid lines: fast kinetics (equilibrium, fine particles). **Top panel**: without dispersion; **bottom panel**: with dispersion; $n = 0.45$, $v = 1.67 \times 10^{-5}$ m s^{-1}, $\alpha/x = 0$ or 0.1, $C_{s,ini} = 1000$ µg kg^{-1}, $t_c = 5$ h, $D_{aq} = 1 \times 10^{-9}$ m^2 s^{-1}, $\varepsilon = 0.05$, $d_{p,coarse} = 2000$ µm, $d_{p,fine} = 63$ µm.

Figure 7 shows concentrations in aqueous leachates that correspond to the initial conditions shown in Figure 6. If leaching is limited by IPD, then the leaching process is slower and concentrations at early times are considerably lower than in the FD or the equilibrium model. This is due to the retarded diffusive transport within the tortuous intragranular pores and the correspondingly small effective diffusion coefficients (D_e). The contaminant release rate becomes lower and lower over time. Leachate concentrations decrease first with the square root of time (typical for transient diffusion) and then exponentially (see Figure 7 without dispersion, and Appendix D for details about the development of the internal mass transfer resistance over time). Note, the cumulative concentration curves confirm the mass conservation of the numerical solutions.

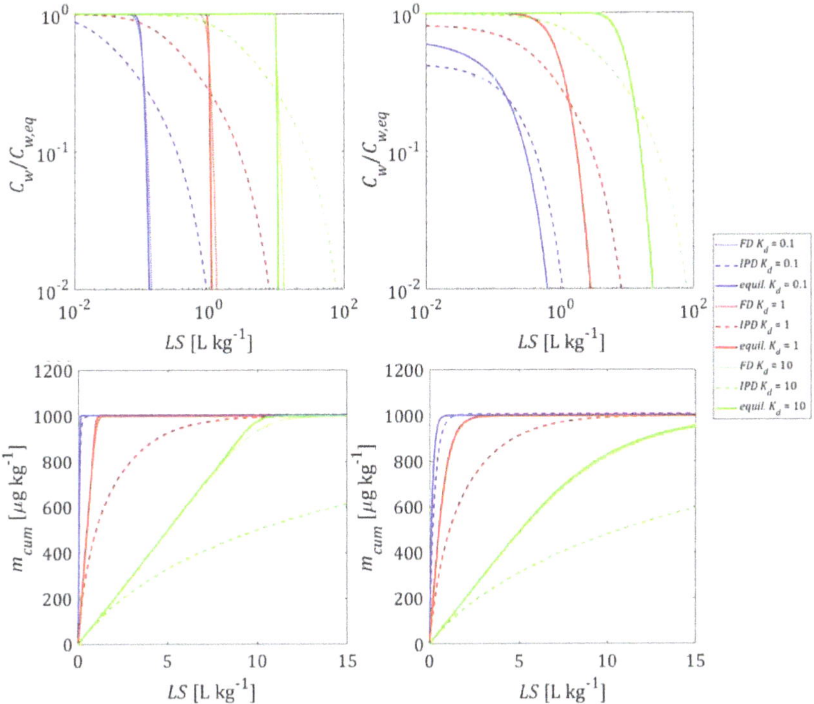

Figure 7. Normalized concentrations ($C_w/C_{w,eq}$) as well as cumulative concentrations (m_{cum}) in the column effluent vs. time (expressed as liquid to solid ratio: LS) for different mass-transfer processes, given the initial conditions depicted in Figure 6. **Left column**: without dispersion; **Right column**: with dispersion.

3.3. Nonlinear Sorption Isotherms

For many of the environmental contaminants and solid materials that are typically analyzed in column leaching tests, non-linear sorption isotherms describe the distribution of solutes between the aqueous and solid phases. The significance of this non-linearity for the development of the conditions in the column before the leaching starts has been analyzed exemplarily by defining Freundlich isotherms that result in the same "effective" K_d for the aqueous concentration at equilibrium: $K_{fr} = K_d/C_{w,eq}^{\frac{1}{n}-1}$.

Figure 8 shows the influence of nonlinear sorption on both the initial concentrations in the column and the leaching curves for the example of $K_d = 1$ L kg^{-1} when no dispersion is considered. The differences in the concentration distribution before percolation starts are moderate. Concentration profiles tend to be smoother with nonlinear sorption with a slightly lower maximum concentration at the column outlet for low to mid K_d values if dispersion is considered (see SM, Figure S1). Differences become more obvious in the tailing part of the leaching curves. Freundlich exponents smaller than 1 result in a longer tailing as is expected. The effect of nonlinear sorption looks similar to the dispersion effect, in both cases the leaching curves show more tailing (see SM, Figure S2). Nonlinearity of sorption is notably less significant than kinetic limitations in the mass transfer mechanisms.

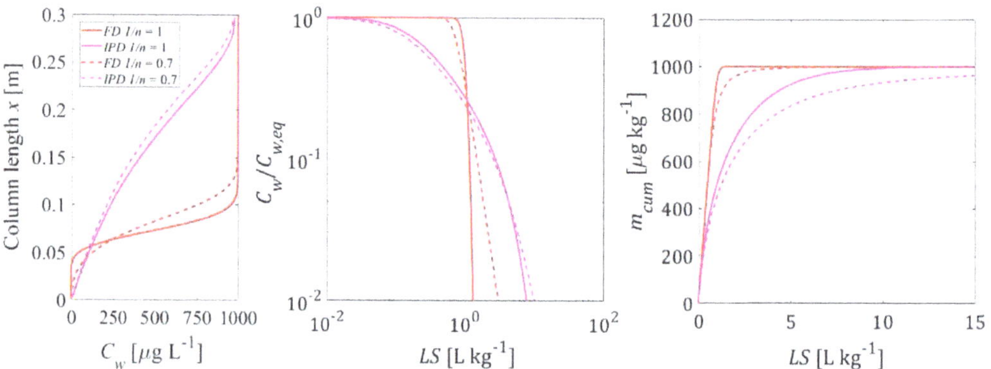

Figure 8. Influence of sorption non-linearity: initial aqueous concentration distribution in the column after the first flooding (**left graph**) and column effluent concentration (normalized: **mid graph**, cumulative: **right graph**) vs. time (expressed as liquid to solid ratio: *LS*); solid lines: linear sorption ($K_d = 1$ L kg^{-1}); dashed lines: non-linear sorption case (Freundlich coefficient $K_{fr} = 7.94$, exponent $1/n = 0.7$); $n = 0.45$, $v = 1.67 \times 10^{-5}$ m s^{-1}, $\alpha/x = 0$ or 0.1, $C_{s,ini} = 1000$ µg kg^{-1}, $t_c = 5$ h, $D_{aq} = 1 \times 10^{-9}$ m^2 s^{-1}, $\varepsilon = 0.05$, $d_{p,coarse} = 2000$ µm.

3.4. Impact of Heterogeneous Sample Composition

Real world materials that are typically investigated in leaching tests are not always homogeneous. Although the sample might be addressed as 'one material', its individual grains have different sizes and differ most likely also in other properties such as porosity, tortuosity, sorption capacity, etc., and may contain different amounts of the contaminants of interest. In order to illustrate the impact of material heterogeneity, we have carried out several numerical leaching experiments with hypothetical material compositions.

First, we consider three bi-modal material compositions. Each of these compositions consist of a fraction of contaminated particles (e.g., particulate organic carbon particles with high K_d) and another fraction of particles that neither contain contaminants nor possess any sorption capacity. If equilibrium conditions prevail during the first flooding and leaching, the heterogeneity of the sample does not matter, it is simply the average K_d ($K_{d,av}$) that rules. The situation changes if mass transfer between the solid and the aqueous phases is limited due to diffusion (FD or IPD). If only a small fraction of the particles in the sample carries the compounds of interest, the volume of particles released by the compound and thus the surface area available for mass transfer becomes much smaller. This may lead to pronounced non-equilibrium conditions after first flooding (see, e.g., Equations (A26) and (A30)) and during leaching. Figure 9 shows a comparison of the initial concentration profiles in the column after the first flooding, as well as the corresponding leaching curves that would develop for the three bi-modal material compositions (100/10/1% of the material is contaminated at a $K_d = 1/10/100$ L kg^{-1}, respectively). A small fraction of strong sorbents showed lower desorption rates compared to a large fraction of the weak sorbents. For this "exotic" case where only 1% of the particles carries all the contamination, initial nonequilibrium and long tailing was observed. This effect was very pronounced for intraparticle pore diffusion; the concentrations initially started on a plateau ("like equilibrium"), but then rapidly declined and showed a pronounced tailing and decrease with the square root of time (or *LS*). It may be noted, that longitudinal dispersion becomes less relevant if non-equilibrium conditions prevail at high K_d values (see Figures S3 and S4 in SM). If such pronounced initial nonequilibrium is observed, then extended periods of time would be needed to equilibrate the water in the column with the solids (e.g., a manifold of the contact time of 5 h).

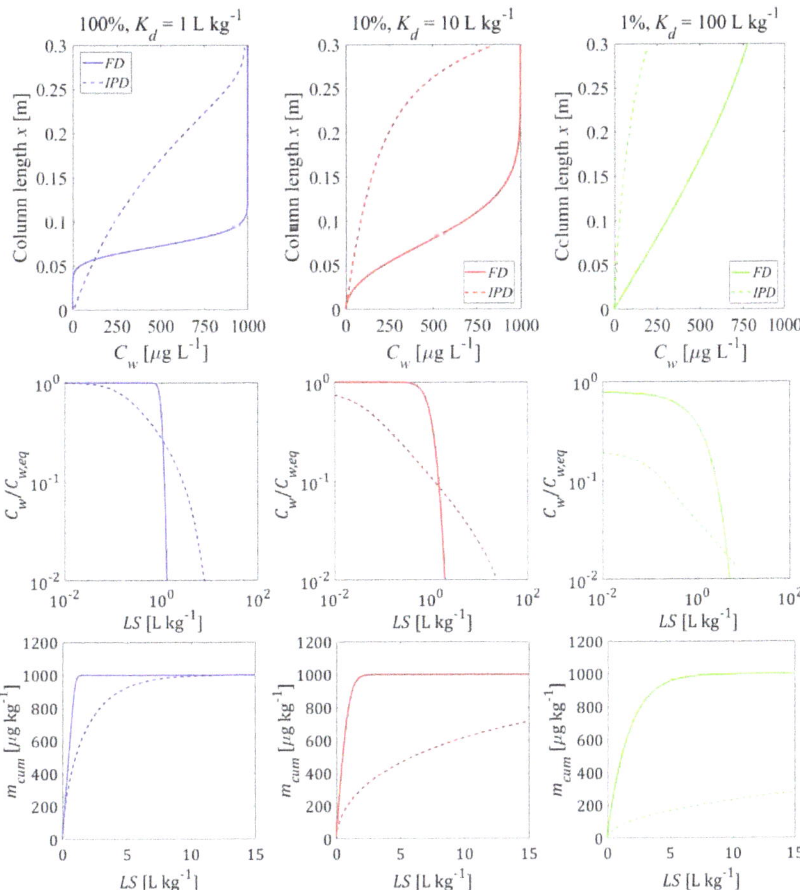

Figure 9. Behavior of bi-modal material compositions of sorbing and non-sorbing particles: initial concentration distribution in the column after the first flooding (**top panel**) and column effluent concentration (normalized: **mid panel**, cumulative: **bottom panel**) vs. time (expressed as liquid to solid ratio: LS). Left column: homogeneous case with average K_d ($= K_{d,av} = 1$ L kg^{-1}); mid column: only 10% of the particles carry the contaminant at $K_d = 10 \times K_{d,av}$; right column: only 1% of the particles carry the contaminant at $K_d = 100 \times K_{d,av}$; the average $K_{d,av}$ of the entire material is the same for all compositions; solid lines: film diffusion cases, dashed lines: intraparticle diffusion case; $n = 0.45$, $v = 1.67 \times 10^{-5}$ m s^{-1}, $\alpha = 0$ (no dispersion), $C_{s,ini} = 1000$ µg kg^{-1}, $t_c = 5$ h, $D_{aq} = 1 \times 10^{-9}$ m^2 s^{-1}, $\varepsilon = 0.05$, $d_{p,coarse} = 2000$ µm.

Samples consisting of mixtures of different particle sizes represent another typical and frequently occurring case of material heterogeneity. To illustrate the impact of such grain size heterogeneity, two bi-modal grain size distributions are considered here, introducing two grain sizes, coarse particles having a diameter of $d_{p,coarse} = 2000$ µm, and fine particles with $d_{p,fine} = 63$ µm. The 1st hypothetical grain size distribution consists of 10% fine particles and 90% coarse particles, the 2nd distribution of 90% fine particles and 10% coarse particles.

If mass transfer is limited by film diffusion, the establishment of the initial conditions as well as leaching (Figure 10) occurs under conditions close to equilibrium for both grain size distributions at all K_d values (0.1, 1, and 10 L kg^{-1}). While the shapes of all leaching curves are very similar, their locations are shifted in time according to the different K_d values by a factor of 10. If intraparticle pore diffusion is considered, tailing is observed if coarse particles predominate. This applies to both, the development of initial conditions in the column and leaching. If fine particles predominate, the leaching is close to equilibrium at early times; at later times, tailing is observed with the typical square root of time behavior. Considering the dispersion effect, non-equilibrium concentrations can be seen at the column effluent after first flooding especially at low K_d values ($K_d =0.1$ L kg^{-1}). Initial non-equilibrium conditions become more salient for intraparticle pore diffusion if coarse particles predominate (see Figures S5 and S6 in SM).

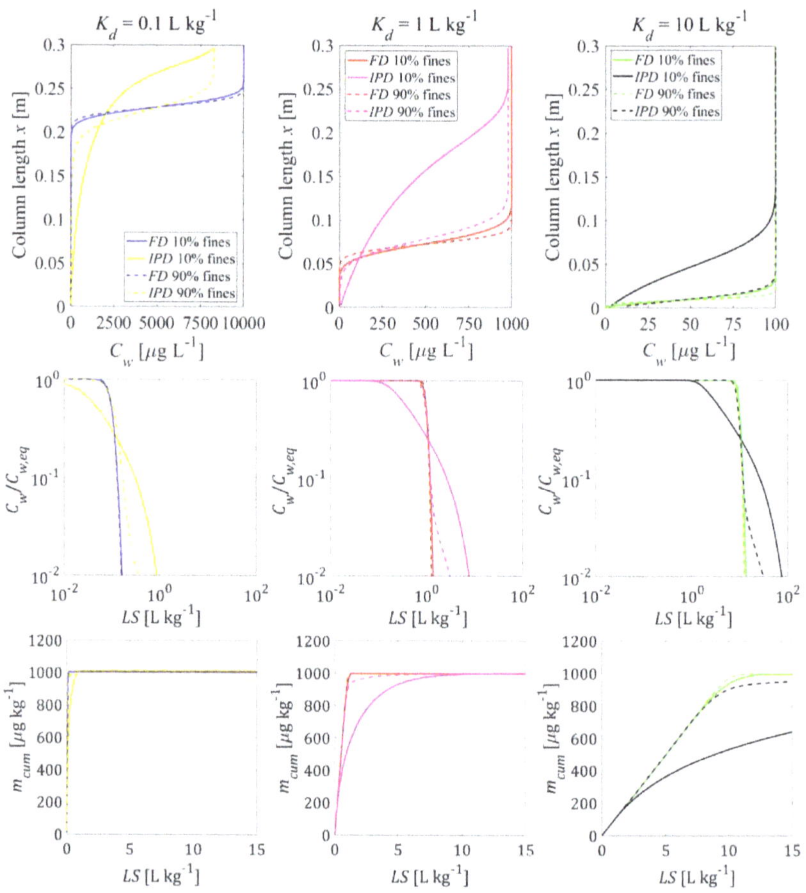

Figure 10. Behavior of the bi-modal material compositions of fine and coarse particles: initial concentration distribution in the column after the first flooding (**top panel**) and column effluent concentration (normalized: **mid panel**, cumulative: **bottom panel**) vs. time (expressed as liquid to solid ratio: *LS*); solid lines: fine particle mass fraction 10%; dashed lines: fine particle mass fraction 90%. ($n = 0.45$, $v = 1.67 \times 10^{-5}$ m s^{-1}, $\alpha = 0$ (no dispersion), $C_{s,ini} = 1000$ μg kg^{-1}, $t_c = 5$ h, $D_{aq} = 1 \times 10^{-9}$ m^2 s^{-1}, $\varepsilon = 0.05$, $d_{p,coarse} = 2000$ μm, $d_{p,fine} = 63$ μm).

Combining the heterogeneity of particle size (d) and sorption capacity (K_d), we consider three material compositions in the third case, which aims at showing circumstances where strong non-equilibrium conditions may be expected. For many materials this is probably not very realistic, but it may occur in material mixtures where a small particle fraction carries a "labile" contamination with a low K_d vs. just a few large particles carrying the contaminant with a large K_d. A hypothetical mixture containing 10% of fine particles with low sorption capacity ($K_d = 10$ L kg^{-1}) and 90% of coarse particles with high sorption capacity ($K_d = 100$ L kg^{-1}) is compared with two extreme cases where a hypothetical sample only contains pure fine particles with low sorption capacity and another hypothetical sample contains pure coarse particles with high sorption capacity. Figure 11 shows the initial concentration distribution for these three compositions after the first flooding period as well as the corresponding leaching curves. Sorption equilibrium is achieved rapidly if the sample consists of only fine particles with a small K_d or only coarse particles with a high K_d. Pure coarse material with a high K_d shows a low equilibrium concentration ($C_{w,eq} = C_{s,ini}/K_d = 1000$ µg kg^{-1}/100 L kg^{-1} = 10 µg L^{-1}) while pure fine material with a low K_d presents a much higher equilibrium concentration ($C_{w,eq} = C_{s,ini}/K_d = 1000$ µg kg^{-1}/10 L kg^{-1} = 100 µg L^{-1}) after a short flow distance. Interestingly, the mixed case where 10% of the column is fine material caused a high concentration which would be sorbed by the coarse materials leading to a slightly higher plateau concentration compared to pure coarse materials. The pollutants were redistributed between fine and coarse materials during the first flooding of the column. The concentration increase towards the outlet of the column in the mixed case is due to fast desorption from the fine material followed by slow sorption by the coarse material. The redistribution is almost complete at the inlet of the column because of the long residence time ($t_c = 5$ h). Since the fine particles make up only to 10% of the total mass, they are already depleted in contaminant concentrations inside the column and in equilibrium with the coarse particles (reflecting both extreme cases). The front of the high concentration caused by the fine particles is already close to the outlet, while the rest is in equilibrium with the 90% coarse particle fraction.

The leaching curve of the mixed case (red lines) reflects the properties of the two pure (homogeneous) cases with either fine or coarse particles. Ten percent of the fine particles with low sorption capacity led to a peak effluent concentration which was only slightly lower than the equilibrium concentration of the pure fine particles with low sorption capacity. However, because the fine particles made up only 10% of the total mass, this peak concentration leached out rapidly and the eluate concentrations followed the coarse particles with a high sorption capacity for long time periods (blue curves). Although this may appear to indicate non-equilibrium conditions (because of the rapid initial decline of the concentrations followed by a plateau or "tailing"), leaching from fine and coarse particles occurs at, or close to equilibrium. Compared to FD, the IPD in the mixed sample showed a slower concentration decline because of the desorption kinetics of the IPD of the coarse particles was slower than the case of FD and on the long-term control release kinetics. For the cumulative mass release the mixed case is close to the coarse material for both the FD and the IPD, whereas the fine-grained particles showed a much higher and faster release (see Figure 11: bottom panel).

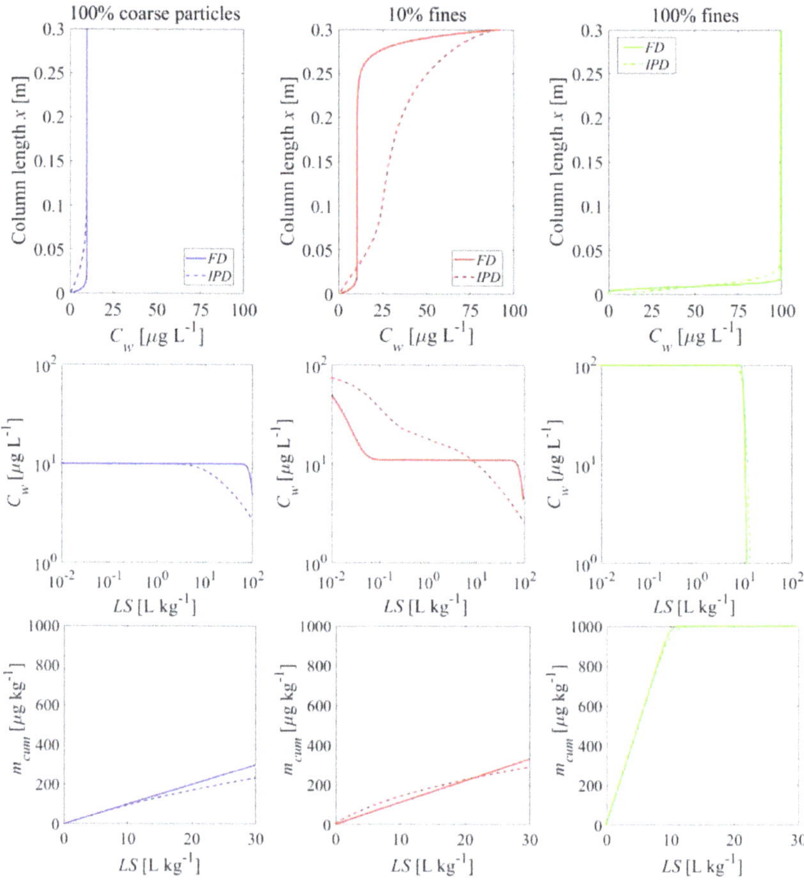

Figure 11. Behavior of bi-modal material compositions of fine particles with low sorption capacity (K_d =10 L kg^{-1}) and coarse particles with high sorption capacity ($K_d = 100$ L kg^{-1}): initial concentration distributions in the column after the first flooding (**top panel**) and the column effluent concentration (normalized: **mid panel**, cumulative: **bottom panel**) vs. time (expressed as liquid to solid ratio: *LS*). Left column: 100% coarse particles; mid column: mixed sample with 10% fine particles; right column: 100% fine particles; $n = 0.45$, $v = 1.67 \times 10^{-5}$ m s^{-1}, $\alpha = 0$ (no dispersion), $C_{s,ini} = 1000$ µg kg^{-1}, $t_c = 5$ h, $D_{aq} = 1 \times 10^{-9}$ m^2 s^{-1}, $\varepsilon = 0.05$, $d_{p,coarse} = 2000$ µm, $d_{p,fine} = 63$ µm.

4. Summary and Conclusions

We conducted numerical simulations to investigate the release characteristics of low to strongly sorbing compounds ($K_d = 0.1$–100 L kg^{-1}) in column leaching tests. Two different scenarios for the establishment of the initial conditions before the start of the leaching phase were considered: a fully pre-equilibrated column and a more realistic scenario where a column is flooded with water from the bottom. In order to highlight the effect of mass transfer limitations, two mechanisms are compared: film diffusion and intra-particle diffusion. Cases without and with dispersion illustrate how dispersive mixing may mask diffusion limited mass transfer. Furthermore, we looked into the impact of heterogeneous sample compositions in terms of reactive particle fractions and particle sizes. Since possible parameter combinations amount to almost infinite numbers, we have limited our analysis to just a few exemplary cases that illustrate the role of individual material properties. These few cases already show that virtually any leaching behavior can be produced with highly heterogeneous samples (depending on the mixing of different materials). The most important conclusions are:

Initial conditions have a significant impact on leaching at low K_d values ($K_d < 1$ L kg^{-1}). With increasing K_d, the differences between pre-equilibrium and non-equilibrium conditions gradually vanish for $K_d > 10$ L kg^{-1} (see Figure 5). Compounds with very low K_d ("salts") would reach extremely high concentrations ($K_d << 1$ L kg^{-1}) at the column outlet (see Figure 4) potentially leading to enhanced dispersion due to density fingering. The K_d values derived from retardation factors (R_d in Equation (4)) would be underestimated if the conditions in the column after the first flooding are not appropriately considered, due to a "phase shift" in normalized concentrations curves ($C_w/C_{w,eq}$ vs. LS).

Dispersion generally causes "smoothing" of concentrations gradients in the column and tends to "mask" film and intraparticle diffusion characteristics due to enhanced "mixing" of the solute within the column. It may lead to smaller initial concentrations at the column outlet after the first flooding period than expected for equilibrium; this is pronounced especially at low K_d values (see Figure 7 and Figure S2), which may be interpreted as non-equilibrium, but is just a consequence of dilution by dispersive mixing.

Intraparticle pore diffusion (IPD) generally shows slower desorption kinetics than **film diffusion** (FD) through an aqueous boundary layer. This is due to the much smaller effective diffusion coefficient in the intraparticle pores and the large diffusion distance that develops inside the particle over time, resulting in the typical square root of time decrease of concentrations (a slope of $1/2$ is observed in log-log plots of leaching curves, see Figures 7, 9 and 10). IPD is more sensitive to the variation of particle sizes than FD (see Figure 10). Mass transfer limitations in an aqueous boundary layer commonly exists for surface adsorbed compounds and easily soluble solids ("salts"). Elements such as heavy metals, which are slowly released from the solid phase, would require much lower solid state diffusion coefficients; if reaction fronts propagate into the particle releasing metals, intraparticle (solid) diffusion models apply again (shrinking core), which are very similar to the IPD approach used here.

Non-linear sorption has little influence on the leaching test results if the "right" effective K_d value is calculated for the proper concentration range (since for the nonlinear sorption the K_d depends on the concentration, large deviations may occur if just the K_{fr} is determined far away of the sample's concentration is used as "K_d"); nevertheless, as concentrations decrease nonlinear sorption causes more tailing (see Figure 8).

Heterogeneous samples with only a small fraction of strongly sorbing particles lead to much slower desorption rates (because of less surface area), especially if mass release is limited by intraparticle pore diffusion (see Figure 9). In extreme cases (just 1% of the material is contaminated at $K_d = 100 \times K_{d,av}$), leaching may start at a plateau (suggesting equilibrium), but far below equilibrium concentrations ($C_{w,peak} \ll C_{w,eq}$) and concentrations later decrease further; The K_d values derived from the initial aqueous concentration ($K_d = C_{s,ini}/C_{w,peak}$) would be overestimated while the K_d values calculated from retardation factors would be underestimated.

In contrast to that, already relatively small amounts of fine particles lead to initial equilibrium, but long-term tailing occurs and is dominated by the coarse particle fraction, especially for intraparticle pore diffusion. Since our FD simulations are close to equilibrium, results are not very affected by grain size heterogeneity (see Figure 10). Material mixtures of small amounts of fine particles (10%) with low sorption capacity ($K_d = 10$ L kg^{-1}) and large amounts of coarse particles with high sorption capacity ($K_d = 100$ L kg^{-1}), exhibit the respective characteristics of each of the individual components in different time periods (see Figure 11). Small amounts of fine particles with low sorption capacity dominate short term behavior of the mixtures and lead to a peak effluent concentration ($C_{w,peak}$) which approaches the equilibrium concentration expected for fine particles (see Figure 11). Since the mass fraction of fine particles is small (10%), the leachate concentrations drop rapidly and reach slightly higher equilibrium levels of 100% pure coarse particles due to the redistribution of pollutants between fine and coarse particles. Ten percent of fine particles with low sorption capacity causes a high equilibrium concentration which are sorbed by the coarse particles with high sorption capacity. K_d values derived from the initial aqueous concentration ($K_d = C_{s,ini}/C_{w,peak}$) would be underestimated, while K_d values derived from the following plateau concentration would be overestimated. Cumulative mass release, however, is often quite insensitive to mass transfer mechanisms (FD or IPD) especially for $LS < 5$ (see Figure 11).

Supplementary Materials: The following are available online at https://www.mdpi.com/article/10.3390/ma14164708/s1, Figure S1. Initial concentration profiles in the column after the first flooding (up-flow) without and with dispersion (top and bottom panel); solid lines: linear sorption; dashed lines: non-linear sorption cases (based on a Freundlich exponent $1/n = 0.7$), $n = 0.45$, $v = 1.67 \times 10^{-5}$ m s^{-1}, $\alpha/x = 0$ or 0.1, $C_{s,ini} = 1000$ µg kg^{-1}, $t_c = 5$ h, $D_{aq} = 1 \times 10^{-9}$ m^2 s^{-1}, $\varepsilon = 0.05$, $d_{p,coarse} = 2000$ µm, Figure S2. Normalized concentrations ($C_w/C_{w,eq}$) as well as cumulative concentrations (m_{cum}) in the column effluent vs. time (expressed as liquid to solid ratio: LS) for different initial conditions depicted in Figure S1; solid lines: linear sorption; dashed lines: nonlinear sorption. Left column: without dispersion; right column: with dispersion, Figure S3. Initial concentration distribution in the column after the first flooding (up-flow) for different bi-modal compositions of sorbing and non-sorbing particles; left column: homogeneous case with average K_d ($=K_{d,av} = 1$ L kg^{-1}); mid column: only 10% of the particles carry the contaminant at $K_d = 10 \times K_{d,av}$; right column: only 1% of the particles carry the contaminant at $K_d = 100 \times K_{d,av}$; the average $K_{d,av}$ of the entire material is the same for all compositions; solid lines: film diffusion case, dashed lines: intraparticle diffusion case. Top panel: without dispersion; bottom panel: with dispersion; $n = 0.45$, $v = 1.67 \times 10^{-5}$ m s^{-1}, $\alpha/x = 0$ or 0.1, $C_{s,ini} = 1000$ µg kg^{-1}, $t_c = 5$ h, $D_{aq} = 1 \times 10^{-9}$ m^2 s^{-1}, $\varepsilon = 0.05$, $d_{p,coarse} = 2000$ µm, Figure S4. Normalized concentrations ($C_w/C_{w,eq}$) as well as cumulative concentrations (m_{cum}) in the column effluent vs. time (expressed as liquid to solid ratio: LS) for different combinations of sorbing particles and distribution coefficients (initial conditions depicted in Figure S3); left: without dispersion; right: with dispersion; solid lines: film diffusion cases, dashed lines: intraparticle diffusion cases, Figure S5. Initial concentration distribution in the column after the first flooding (up-flow) for two different bi-modal grain size distributions of fine and coarse particles; solid lines: fine particle mass fraction 10%; dashed lines: fine particle mass fraction 90%. ($n = 0.45$, $v = 1.67 \times 10^{-5}$ m s^{-1}, $\alpha/x = 0$ or 0.1, $C_{s,ini} = 1000$ µg kg^{-1}, $t_c = 5$ h, $D_{aq} = 1 \times 10^{-9}$ m^2 s^{-1}, $\varepsilon = 0.05$, $d_{p,coarse} = 2000$ µm, $d_{p,fine} = 63$ µm); top panel: without dispersion; bottom panel: with dispersion, Figure S6. Influence of different grain size fractions and distribution coefficients on normalized concentrations ($C_w/C_{w,eq}$) as well as cumulative concentrations (m_{cum}) in the column effluent vs. time (expressed as liquid to solid ratio LS); left: without dispersion; right: with dispersion; solid lines: fine particle mass fraction 10%; dashed lines: fine particle mass fraction 90%; kinetic parameters are the same as Figure S5, Figure S7. Initial concentration distribution in the column after the first flooding (up-flow) for different bi-modal material compositions of fine particles with low sorption capacity ($K_d = 10$ L kg^{-1}) and coarse particles with high sorption capacity; left: 100% coarse particles ($K_d = 100$ L kg^{-1}); middle: mixed sample with 10% fine particles; right: 100% fine particles; solid lines: film diffusion (FD), dashed lines: intraparticle diffusion cases (IPD); $n = 0.45$, $v = 1.67 \times 10^{-5}$ m s^{-1}, $\alpha/x = 0$ or 0.1, $C_{s,ini} = 1000$ µg kg^{-1}, $t_c = 5$ h, $D_{aq} = 1 \times 10^{-9}$ m^2 s^{-1}, $\varepsilon = 0.05$, $d_{p,coarse} = 2000$ µm, $d_{p,fine} = 63$ µm; top panel: without dispersion; bottom panel: with dispersion, Figure S8. Leachate concentrations (C_w) as well as cumulative concentrations (m_{cum}) in the column effluent vs. time (expressed as liquid to solid ratio: LS) for different combinations of fine particles with low sorption capacity ($K_d = 10$ L kg^{-1}) and coarse particles with high sorption capacity ($K_d = 100$ L kg^{-1}); left: without dispersion; right: with dispersion; solid lines: film diffusion cases, dashed lines: intraparticle diffusion cases; kinetic parameters are the same as Figure S7.

Author Contributions: Conceptualization, B.L., M.F., P.G.; validation, B.L., M.F., P.G.; investigation, B.L., M.F., P.G.; writing—original draft preparation, B.L.; writing—review and editing, B.L., M.F., P.G.; visualization, B.L.; supervision, M.F., P.G. All authors have read and agreed to the published version of the manuscript.

Funding: This work was partly supported by the Collaborative Research Center 1253 CAMPOS, funded by the German Research Foundation (DFG, Grant Agreement SFB 1253/1 2017).

Institutional Review Board Statement: Not applicable.

Informed Consent Statement: Not applicable.

Data Availability Statement: The data presented in this study are available in this article.

Conflicts of Interest: The authors declare no conflict of interest.

Appendix A. Empirical Relationships for the Estimation of Sherwood Numbers

There are many studies available in the literature in which solid-liquid mass transfer in fluidized beds and flow through systems are investigated over a wide range of Reynolds numbers. Most of these correlations can be adequately described by the following equation:

$$Sh = A + BRe^{\theta}Sc^{\gamma} \tag{A1}$$

where Sh is the Sherwood number. A is a constant (theoretically $= 2$ for spherical particles in a stagnant infinite medium) and B is a constant to be determined by regression analysis of experimental data. Re and Sc denote the Reynolds and Schmidt number which are defined as:

$$Sc = \frac{\eta}{\rho_L D_{aq}} \qquad Re = \frac{\rho_L d\, v}{\eta} \tag{A2}$$

where η [M L^{-1} T^{-1}] denotes the dynamic viscosity of the fluid. ρ_L [M L^{-3}] is the density of the fluid and v [L T^{-1}] denotes the flow velocity.

The empirical exponents θ and γ in Equation (A1) may be determined experimentally or from theory. The Blasius (1908) solved the Navier-Stokes equation and continuity equation for laminar flow over a sharp leading edge and found that the ratio of fluid velocity boundary layer thickness to concentration boundary layer thickness is proportional to the Schmidt number with a power of $1/3$ ($= \gamma$ in Equation (A1)) which is widely used in literature [37–43]. Liu et al. (2014) showed a higher empirical exponent γ of $1/2$ based on penetration theory [35,44]. θ values depend on the experimental setup and are generally adapted from experimental data. Most of the empirical relationships show that θ values lie in the range of 0.5–0.75 [35,37,39–43].

Liu et al. (2014) proposed an empirical relationship for mass transfer in packed beds only based on the Peclet number ($Pe = Re \times Sc$) [35]:

$$Sh = 2 + 0.1Pe^{1/2} \tag{A3}$$

Equation (A3) is equivalent to Equation (A1) for $\theta = \gamma = 1/2$. This Sherwood number correlation was applied in the numerical models. Sh numbers obtained for the chosen column setup were close to 2 indicating slow mass transfer close to the theoretical limit (2).

Appendix B. Film Diffusion Coupled to Advective-Dispersive Transport

Equation (10) in the main text shows the governing equations of film diffusion coupled to advective-dispersive transport. These partial differential equations are solved numerically using the finite volume method (as illustrated in Figure A1a).

Discretizing the transport operator in space while keeping the time derivative yields the following system of ordinary differential equations:

$$\begin{aligned}
\frac{\partial C_{w,j}}{\partial t} &= D_L \frac{(C_{w,j-1} - 2C_{w,j} + C_{w,j+1})}{\Delta x^2} - v\frac{(C_{w,j} - C_{w,j-1})}{\Delta x} \\
&\quad + \frac{D_{aq}}{\delta_f} \frac{6(1-n)}{nd} \left(\frac{C_{s,j}}{K_d} - C_{w,j} \right) \\
\frac{\partial C_{s,j}}{\partial t} &= -\frac{D_{aq}}{\delta_f} \frac{6}{\rho_s\, d} \left(\frac{C_{s,j}}{K_d} - C_{w,j} \right)
\end{aligned} \tag{A4}$$

where $C_{w,j}$ [M L^{-3}], $C_{w,j-1}$ [M L^{-3}] and $C_{w,j+1}$ [M L^{-3}] denote the solute concentration in the water phase in volume j, $j-1$ and $j+1$, respectively. $C_{s,j}$ [M M^{-1}] denotes the solute concentration in the solid phase in volume j.

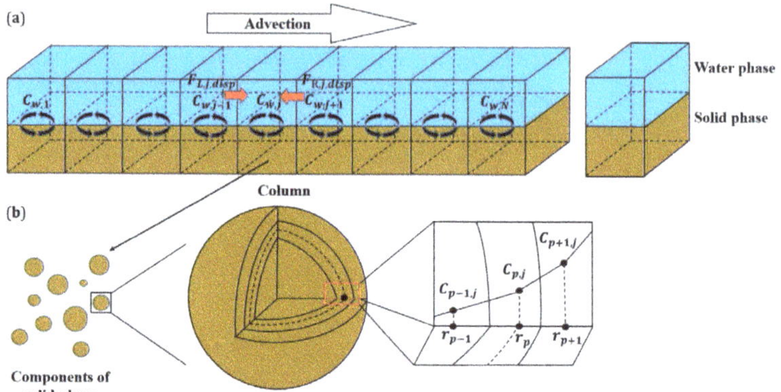

Figure A1. Discretization of the column into N parts (**a**). Representation of the solid phase as a composition of grains having different sizes and properties, each discretized by a number of L shells (**b**).

The approximation of the time derivative of Equation (A4) can be expressed as the concentration difference between the new and the previous time, divided by the time interval Δt. A time weighting factor φ was used to navigate between implicit and explicit time integration. For $\varphi = 0.5$, the Crank-Nicholson-scheme is realized, whereas for $\varphi = 0$ and $\varphi = 1$, the fully implicit and explicit scheme is used, respectively.

$$\frac{C_{w,j}^{k+1} - C_{w,j}^k}{\Delta t} = (1 - \varphi) \left(D_L \frac{\left(C_{w,j-1}^{k+1} - 2C_{w,j}^{k+1} + C_{w,j+1}^{k+1} \right)}{\Delta x^2} - v \frac{\left(C_{w,j}^{k+1} - C_{w,j-1}^{k+1} \right)}{\Delta x} + \frac{D_{aq}}{\delta_f} \frac{6(1-n)}{nd} \left(\frac{C_{s,j}^{k+1}}{K_d} - C_{w,j}^{k+1} \right) \right)$$

$$+ \varphi \left(D_L \frac{\left(C_{w,j-1}^k - 2C_{w,j}^k + C_{w,j+1}^k \right)}{\Delta x^2} - v \frac{\left(C_{w,j}^k - C_{w,j-1}^k \right)}{\Delta x} + \frac{D_{aq}}{\delta_f} \frac{6(1-n)}{nd} \left(\frac{C_{s,j}^k}{K_d} - C_{w,j}^k \right) \right)$$

$$\frac{C_{s,j}^{k+1} - C_{s,j}^k}{\Delta t} = -(1 - \varphi) \left(\frac{D_{aq}}{\delta_f} \frac{6}{\rho_s\, d} \left(\frac{C_{s,j}^{k+1}}{K_d} - C_{w,j}^{k+1} \right) \right)$$

$$- \varphi \left(\frac{D_{aq}}{\delta_f} \frac{6}{\rho_s\, d} \left(\frac{C_{s,j}^k}{K_d} - C_{w,j}^k \right) \right) \tag{A5}$$

where the indices k and $k + 1$ denote the corresponding concentration values at the previous time step and at the new time step.

In order to solve this system of equations, we may merge the two concentration vectors into a single one ($C = [C_w; C_s]$; with the semicolon being a line delimiter):

$$C = \begin{bmatrix} C_{w,1} \\ C_{s,1} \\ \vdots \\ C_{w,j} \\ C_{s,j} \\ \vdots \\ C_{w,N} \\ C_{s,N} \end{bmatrix}_{2N \times 1} = \begin{bmatrix} C_1 \\ C_2 \\ \vdots \\ C_{2j-1} \\ C_{2j} \\ \vdots \\ C_{2N-1} \\ C_{2N} \end{bmatrix}_{2N \times 1} \tag{A6}$$

with $j \epsilon [1, 2, \ldots, N]$.

A standard method of solving non-linear ordinary equations is the Newton–Raphson scheme [45]. It is based on linearizing the residual function $f(C^{k+1})$ at the current guess C^{k+1}_{guess} of C^{k+1}. The residual function $f(C^{k+1})$ is defined as:

$$
\begin{aligned}
f_{2j-1} = \frac{C^{k+1}_{w,j} - C^{k}_{w,j}}{\Delta t} \\
- (1 - \varphi)\left(D_L \frac{\left(C^{k+1}_{w,j-1} - 2C^{k+1}_{w,j} + C^{k+1}_{w,j+1}\right)}{\Delta x^2} - v\frac{\left(C^{k+1}_{w,j} - C^{k+1}_{w,j-1}\right)}{\Delta x} \right. \\
\left. + \frac{D_{aq}}{\delta_f}\frac{6(1-n)}{nd}\left(\frac{C^{k+1}_{s,j}}{K_d} - C^{k+1}_{w,j}\right)\right) \\
- \varphi\left(D_L \frac{\left(C^{k}_{w,j-1} - 2C^{k}_{w,j} + C^{k}_{w,j+1}\right)}{\Delta x^2} - v\frac{\left(C^{k}_{w,j} - C^{k}_{w,j-1}\right)}{\Delta x} \right. \\
\left. + \frac{D_{aq}}{\delta_f}\frac{6(1-n)}{nd}\left(\frac{C^{k}_{s,j}}{K_d} - C^{k}_{w,j}\right)\right)
\end{aligned}
\tag{A7}
$$

$$
\begin{aligned}
f_{2j} = \frac{C^{k+1}_{s,j} - C^{k}_{s,j}}{\Delta t} \quad + (1 - \varphi)\left(\frac{D_{aq}}{\delta_f}\frac{6}{\rho_s\, d}\left(\frac{C^{k+1}_{s,j}}{K_d} - C^{k+1}_{w,j}\right)\right) \\
+ \varphi\left(\frac{D_{aq}}{\delta_f}\frac{6}{\rho_s\, d}\left(\frac{C^{k}_{s,j}}{K_d} - C^{k}_{w,j}\right)\right)
\end{aligned}
$$

The residual function vector can be expressed as:

$$
f\left(C^{k+1}\right) = \begin{bmatrix} f_1 \\ f_2 \\ \vdots \\ f_{2j-1} \\ f_{2j} \\ \vdots \\ f_{2N-1} \\ f_{2N} \end{bmatrix}_{2N\times 1}
\tag{A8}
$$

The residual function vector becomes a zero vector if C^{k+1} is chosen right and a single step of the Newton–Raphson method can be denoted as:

$$
f\left(C^{k+1}\right) \approx f\left(C^{k+1}_{guess}\right) + \left(C^{k+1} - C^{k+1}_{guess}\right) \left.\frac{\partial f\left(C^{k+1}\right)}{\partial C^{k+1}}\right|_{C^{k+1} = C^{k+1}_{guess}}
$$

$$
C^{k+1} = C^{k+1}_{guess} - \frac{f\left(C^{k+1}_{guess}\right)}{J} = C^{k+1}_{guess} - \frac{f\left(C^{k+1}_{guess}\right)}{\left.\dfrac{\partial f\left(C^{k+1}\right)}{\partial C^{k+1}}\right|_{C^{k+1} = C^{k+1}_{guess}}}
\tag{A9}
$$

where J denotes the Jacobian matrix, which is the matrix of derivatives of all values of $f\left(C^{k+1}\right)$ with respective to all values of C^{k+1}. The residual $f\left(C^{k+1}\right)$ is reevaluated after updating C^{k+1}. If the resulting residual is not sufficiently close to zero, C^{k+1}_{guess} is set to the last solution of C^{k+1} and Equation (A9) is reapplied. In our case, the Jacobian matrix can be derived analytically:

$$
J = \begin{bmatrix}
\dfrac{\partial f_1}{\partial C_1} & \dfrac{\partial f_1}{\partial C_2} & \cdots & \dfrac{\partial f_1}{\partial C_{2N-1}} & \dfrac{\partial f_1}{\partial C_{2N}} \\[2ex]
\dfrac{\partial f_2}{\partial C_1} & \dfrac{\partial f_2}{\partial C_2} & \cdots & \dfrac{\partial f_2}{\partial C_{2N-1}} & \dfrac{\partial f_2}{\partial C_{2N}} \\[2ex]
\vdots & \vdots & \cdots & \vdots & \vdots \\[2ex]
\dfrac{\partial f_{2N-1}}{\partial C_1} & \dfrac{\partial f_{2N-1}}{\partial C_2} & \cdots & \dfrac{\partial f_{2N-1}}{\partial C_{2N-1}} & \dfrac{\partial f_{2N-1}}{\partial C_{2N}} \\[2ex]
\dfrac{\partial f_{2N}}{\partial C_1} & \dfrac{\partial f_{2N}}{\partial C_2} & \cdots & \dfrac{\partial f_{2N}}{\partial C_{2N-1}} & \dfrac{\partial f_{2N}}{\partial C_{2N}}
\end{bmatrix}_{2N\times 2N}
\tag{A10}
$$

In order to ensure the accuracy of the model, error control was employed at each time step and an error vector (ΔC^{k+1}) was used to monitor the difference between the old and new guess values, which is defined as:

$$\Delta C^{k+1} = \begin{bmatrix} \left| C^{k+1}_{1,\,guess,new} - C^{k+1}_{1,\,guess,old} \right| \\ \left| C^{k+1}_{2,\,guess,new} - C^{k+1}_{2,\,guess,old} \right| \\ \vdots \\ \left| C^{k+1}_{2j-1,\,guess,new} - C^{k+1}_{2j-1,\,guess,old} \right| \\ \left| C^{k+1}_{2j,\,guess,new} - C^{k+1}_{2j,\,guess,old} \right| \\ \vdots \\ \left| C^{k+1}_{2N-1,\,guess,new} - C^{k+1}_{2N-1,\,guess,old} \right| \\ \left| C^{k+1}_{2N,\,guess,new} - C^{k+1}_{2N,\,guess,old} \right| \end{bmatrix}_{2N \times 1} \tag{A11}$$

The iteration process stops if the maximum value of error vector ΔC^{k+1}_{max} is smaller than the tolerable error e (e.g., $e = 10^{-15}$).

Appendix C. Intraparticle Pore Diffusion Coupled to Advective-Dispersive Transport

Intraparticle pore diffusion is widely used to describe the sorptive uptake of pollutants in porous materials such as activated carbon, zeolites and many technical materials. Equations (11) and (17) describe intraparticle diffusion coupled to advective-dispersive transport. The intraparticle diffusion model approximates the solid grains as spherical particles. These spherical particles are discretized into a number of shells of equal volume. Mass transfer between solid and intra-granular water phases is assumed to be fast and local equilibrium is assumed. For sorption, the Freundlich isotherm model is employed for nonlinear and linear (exponent = 1) cases. Figure A1b shows the numerical grain model where the spherical grains are divided into L shells. For a specific shell p in volume j the corresponding difference-equations were used [46]:

$$\begin{aligned} \frac{1}{\Delta t} & \left(\varepsilon C^{k+1}_{p,j} + \rho_p K_{fr} \left(C^{k+1}_{p,j} \right)^{\frac{1}{n}} - \varepsilon C^k_{p,j} - \rho_p K_{fr} \left(C^k_{p,j} \right)^{\frac{1}{n}} \right) \\ &= \frac{D_e}{r_p^2} \frac{1-\varphi}{r_{p+0.5} - r_{p-0.5}} \left(r^2_{p+0.5} \frac{C^{k+1}_{p+1,j} - C^{k+1}_{p,j}}{r_{p+1} - r_p} - r^2_{p-0.5} \frac{C^{k+1}_{p,j} - C^{k+1}_{p-1,j}}{r_p - r_{p-1}} \right) \\ &+ \frac{D_e}{r_p^2} \frac{\varphi}{r_{p+0.5} - r_{p-0.5}} \left(r^2_{p+0.5} \frac{C^k_{p+1,j} - C^k_{p,j}}{r_{p+1} - r_p} - r^2_{p-0.5} \frac{C^k_{p,j} - C^k_{p-1,j}}{r_p - r_{p-1}} \right) \end{aligned} \tag{A12}$$

where the subscripts $p + 0.5$ and $p - 0.5$ represent the corresponding parameter value between shells (p and $p + 1$) and (p and $p - 1$), respectively. Subscript j denotes the corresponding parameter value in volume j. Subscripts k and $k + 1$ denote the "old" and "new" time levels.

Based on the boundary conditions (Equations (12) and (13), main text), the innermost shell and the outermost shell are treated specially. The solute concentration in the intra-granular water phase at the new time step ($C^{k+1}_{p,j}$) can be expressed as:

$$\begin{aligned} &\left[\varepsilon + \rho_p K_{fr} \left(C^{k+1}_{p,j} \right)^{\frac{1}{n}-1} + \frac{D_e \Delta t}{r_p^2} \frac{1-\varphi}{r_{p+0.5} - r_{p-0.5}} \left(\frac{r^2_{p+0.5}}{r_{p+1} - r_p} + \frac{r^2_{p-0.5}}{r_p - r_{p-1}} \right) \right] C^{k+1}_{p,j} \\ &= \left[\frac{D_e \Delta t}{r_p^2} \frac{1-\varphi}{r_{p+0.5} - r_{p-0.5}} \frac{r^2_{p+0.5}}{r_{p+1} - r_p} \right] C^{k+1}_{p+1,j} + \left[\frac{D_e \Delta t}{r_p^2} \frac{1-\varphi}{r_{p+0.5} - r_{p-0.5}} \frac{r^2_{p-0.5}}{r_p - r_{p-1}} \right] C^{k+1}_{p-1,j} \\ &+ \left[\frac{D_e \Delta t}{r_p^2} \frac{\varphi}{r_{p+0.5} - r_{p-0.5}} \frac{r^2_{p+0.5}}{r_{p+1} - r_p} \right] C^k_{p+1,j} + \left[\frac{D_e \Delta t}{r_p^2} \frac{\varphi}{r_{p+0.5} - r_{p-0.5}} \frac{r^2_{p-0.5}}{r_p - r_{p-1}} \right] C^k_{p-1,j} \\ &+ \left[\varepsilon + \rho_p K_{fr} \left(C^k_{p,j} \right)^{\frac{1}{n}-1} - \frac{D_e \Delta t}{r_p^2} \frac{\varphi}{r_{p+0.5} - r_{p-0.5}} \left(\frac{r^2_{p+0.5}}{r_{p+1} - r_p} + \frac{r^2_{p-0.5}}{r_p - r_{p-1}} \right) \right] C^k_{p,j} \end{aligned} \tag{A13}$$

for shell $p = 2$ to shell $p = L - 1$ and

$$\frac{1}{\Delta t}\left(\varepsilon C_{1,j}^{k+1} + \rho_p K_{fr}\left(C_{1,j}^{k+1}\right)^{\frac{1}{n}} - \varepsilon C_{1,j}^{k} - \rho_p K_{fr}\left(C_{1,j}^{k}\right)^{\frac{1}{n}}\right)$$
$$= \frac{D_e}{r_1^2}\frac{1-\varphi}{r_{1+0.5}-r_{1-0.5}}\left(r_{1+0.5}^2\frac{C_{2,j}^{k+1}-C_{1,j}^{k+1}}{r_2-r_1}\right) + \frac{D_e}{r_1^2}\frac{\varphi}{r_{1+0.5}-r_{1-0.5}}\left(r_{1+0.5}^2\frac{C_{2,j}^{k}-C_{1,j}^{k}}{r_2-r_1}\right)$$

After transformation :

$$\left[\varepsilon + \rho_p K_{fr}\left(C_{1,j}^{k+1}\right)^{\frac{1}{n}-1} + \frac{D_e\Delta t}{r_1^2}\frac{1-\varphi}{r_{1+0.5}-r_{1-0.5}}\frac{r_{1+0.5}^2}{r_2-r_1}\right]C_{1,j}^{k+1}$$
$$= \left[\frac{D_e\Delta t}{r_1^2}\frac{1-\varphi}{r_{1+0.5}-r_{1-0.5}}\frac{r_{1+0.5}^2}{r_2-r_1}\right]C_{2,j}^{k+1} + \left[\frac{D_e\Delta t}{r_1^2}\frac{\varphi}{r_{1+0.5}-r_{1-0.5}}\frac{r_{1+0.5}^2}{r_2-r_1}\right]C_{2,j}^{k}$$
$$+ \left[\varepsilon + \rho_p K_{fr}\left(C_{1,j}^{k}\right)^{\frac{1}{n}-1} - \frac{D_e\Delta t}{r_1^2}\frac{\varphi}{r_{1+0.5}-r_{1-0.5}}\frac{r_{1+0.5}^2}{r_2-r_1}\right]C_{1,j}^{k} \tag{A14}$$

for shell 1 (or the innermost shell, $p = 1$) and:

$$\frac{1}{\Delta t}\left(\varepsilon C_{L,j}^{k+1} + \rho_p K_{fr}\left(C_{L,j}^{k+1}\right)^{\frac{1}{n}} - \varepsilon C_{L,j}^{k} - \rho_p K_{fr}\left(C_{L,j}^{k}\right)^{\frac{1}{n}}\right)$$
$$= \frac{D_e}{r_L^2}\frac{1-\varphi}{R-r_{L-0.5}}\left(R^2\frac{C_{w,j}^{k+1}-C_{L,j}^{k+1}}{R-r_L} - r_{L-0.5}^2\frac{C_{L,j}^{k+1}-C_{L-1,j}^{k+1}}{r_L-r_{L-1}}\right)$$
$$+ \frac{D_e}{r_L^2}\frac{\varphi}{R-r_{L-0.5}}\left(R^2\frac{C_{w,j}^{k}-C_{L,j}^{k}}{R-r_L} - r_{L-0.5}^2\frac{C_{L,j}^{k}-C_{L-1,j}^{k}}{r_L-r_{L-1}}\right)$$

After transformation :

$$\left[\varepsilon + \rho_p K_{fr}\left(C_{L,j}^{k+1}\right)^{\frac{1}{n}-1} + \frac{D_e\Delta t}{r_L^2}\frac{1-\varphi}{R-r_{L-0.5}}\left(\frac{R^2}{R-r_L} + \frac{r_{L-0.5}^2}{r_L-r_{L-1}}\right)\right]C_{L,j}^{k+1}$$
$$= \left[\frac{D_e\Delta t}{r_L^2}\frac{1-\varphi}{R-r_{L-0.5}}\frac{R^2}{R-r_L}\right]C_{w,j}^{k+1} + \left[\frac{D_e\Delta t}{r_L^2}\frac{1-\varphi}{R-r_{L-0.5}}\frac{r_{L-0.5}^2}{r_L-r_{L-1}}\right]C_{L-1,j}^{k+1}$$
$$+ \left[\frac{D_e\Delta t}{r_L^2}\frac{\varphi}{R-r_{L-0.5}}\frac{R^2}{R-r_L}\right]C_{w,j}^{k} + \left[\frac{D_e\Delta t}{r_L^2}\frac{\varphi}{R-r_{L-0.5}}\frac{r_{L-0.5}^2}{r_L-r_{L-1}}\right]C_{L-1,j}^{k}$$
$$+ \left[\varepsilon + \rho_p K_{fr}\left(C_{L,j}^{k}\right)^{\frac{1}{n}-1} - \frac{D_e\Delta t}{r_L^2}\frac{\varphi}{R-r_{L-0.5}}\left(\frac{R^2}{R-r_L} + \frac{r_{L-0.5}^2}{r_L-r_{L-1}}\right)\right]C_{L,j}^{k} \tag{A15}$$

for shell L (or the outermost shell, $p = L$).

Based on the mass balance, solute mass change in the external water phase (M_w) equals the solute mass change in the spherical particles; for better understanding, the simple case of particles with uniform size is shown:

$$\frac{\partial M_w}{\partial t} = V_w\frac{\partial C_w}{\partial t} = 4\pi R^2 F N_p \tag{A16}$$

where F [M L^{-2} T^{-1}] denotes the solute flux density into the external water phase. R and N_p denote the radius and the total number of the spherical particles. The latter can be calculated by:

$$N_p = \frac{m_d}{\rho_p\left(\frac{4}{3}\pi R^3\right)} \tag{A17}$$

The solute flux density into the external water phase is given by:

$$F = D_e\frac{C_L - C_w}{R - r_L} \tag{A18}$$

Substituting F and N_p with Equations (A18) and (A17) in Equation (A16) and taking advection and dispersion into account, the solute concentration in the external water phase at the new time step $C_{w,j}^{k+1}$ can be expressed by:

$$\frac{C_{w,j}^{k+1}-C_{w,j}^{k}}{\Delta t} = (1-\varphi)\left(D_L\frac{\left(C_{w,j-1}^{k+1}-2C_{w,j}^{k+1}+C_{w,j+1}^{k+1}\right)}{\Delta x^2} - v\frac{\left(C_{w,j}^{k+1}-C_{w,j-1}^{k+1}\right)}{\Delta x}\right.$$
$$\left. +\frac{3D_e m_d}{\rho_p V_w R}\left(\frac{C_{L,j}^{k+1}-C_{w,j}^{k+1}}{R-r_L}\right)\right)$$
$$+\varphi\left(D_L\frac{\left(C_{w,j-1}^{k}-2C_{w,j}^{k}+C_{w,j+1}^{k}\right)}{\Delta x^2} - v\frac{\left(C_{w,j}^{k}-C_{w,j-1}^{k}\right)}{\Delta x}\right.$$
$$\left. +\frac{3D_e m_d}{\rho_p V_w R}\left(\frac{C_{L,j}^{k}-C_{w,j}^{k}}{R-r_L}\right)\right)$$

$$\text{(A19)}$$

After transformation :

$$\left[1+(1-\varphi)\ \left(\frac{3D_e m_d\Delta t}{\rho_p V_w R(R-r_L)} + \frac{2D_L\Delta t}{\Delta x^2} + \frac{v\Delta t}{\Delta x}\right)\right]C_{w,j}^{k+1}$$
$$= \left[(1-\varphi)\left(\frac{D_L\Delta t}{\Delta x^2} + \frac{v\Delta t}{\Delta x}\right)\right]C_{w,j-1}^{k+1} + \left[\frac{(1-\varphi)D_L\Delta t}{\Delta x^2}\right]C_{w,j+1}^{k+1}$$
$$+ \left[\frac{(1-\varphi)3D_e m_d\Delta t}{\rho_p V_w R(R-r_L)}\right]C_{L,j}^{k+1} + \left[\varphi\left(\frac{D_L\Delta t}{\Delta x^2} + \frac{v\Delta t}{\Delta x}\right)\right]C_{w,j-1}^{k}$$
$$+ \left[\frac{\varphi D_L\Delta t}{\Delta x^2}\right]C_{w,j+1}^{k} + \left[\frac{\varphi 3D_e m_d\Delta t}{\rho_p V_w R(R-r_L)}\right]C_{L,j}^{k}$$
$$+ \left[1 - \varphi\left(\frac{3D_e m_d\Delta t}{\rho_p V_w R(R-r_L)} + \frac{2D_L\Delta t}{\Delta x^2} + \frac{v\Delta t}{\Delta x}\right)\right]C_{w,j}^{k}$$

In order to solve this system of equations, we may merge the two concentration vectors to a single one ($C = [C_w; C_p]$; with the semicolon being a line delimiter):

$$C = \begin{bmatrix} C_{w,1} \\ C_{1,1} \\ \vdots \\ C_{p,1} \\ \vdots \\ C_{L,1} \\ \vdots \\ C_{w,j} \\ C_{1,j} \\ \vdots \\ C_{p,j} \\ \vdots \\ C_{L,j} \\ \vdots \\ C_{w,N} \\ C_{1,N} \\ \vdots \\ C_{p,N} \\ \vdots \\ C_{L,N} \end{bmatrix}_{(L+1)*N\times 1} = \begin{bmatrix} C_1 \\ C_2 \\ \vdots \\ C_{p+1} \\ \vdots \\ C_{L+1} \\ \vdots \\ C_{(L+1)*(j-1)+1} \\ C_{(L+1)*(j-1)+2} \\ \vdots \\ C_{(L+1)*(j-1)+p+1} \\ \vdots \\ C_{(L+1)*j} \\ \vdots \\ C_{(L+1)*(N-1)+1} \\ C_{(L+1)*(N-1)+2} \\ \vdots \\ C_{(L+1)*(N-1)+p+1} \\ \vdots \\ C_{(L+1)*N} \end{bmatrix}_{(L+1)*N\times 1} \qquad \text{(A20)}$$

with $p\epsilon[1, 2, \ldots, L]$ and $j\epsilon[1, 2, \ldots, N]$.

Using the Newton–Raphson scheme, the following residual function $f(C^{k+1})$ is linearized at the current guess C^{k+1}_{guess} of C^{k+1} [45]:

$$
\begin{aligned}
f_{(L+1)*(j-1)+1} =\ & \left[1+(1-\varphi)\left(\frac{3D_e m_d \Delta t}{\rho_p V_w R(R-r_L)}+\frac{2D_L \Delta t}{\Delta x^2}+\frac{v\Delta t}{\Delta x}\right)\right]C^{k+1}_{w,j}\\
&-\left[(1-\varphi)\left(\frac{D_L \Delta t}{\Delta x^2}+\frac{v\Delta t}{\Delta x}\right)\right]C^{k+1}_{w,j-1}+\left[\frac{(1-\varphi)D_L \Delta t}{\Delta x^2}\right]C^{k+1}_{w,j+1}\\
&-\left[\frac{(1-\varphi)3D_e m_d \Delta t}{\rho_p V_w R(R-r_L)}\right]C^{k+1}_{L,j}-\left[\varphi\left(\frac{D_L \Delta t}{\Delta x^2}+\frac{v\Delta t}{\Delta x}\right)\right]C^{k}_{w,j-1}\\
&-\left[\frac{\varphi D_L \Delta t}{\Delta x^2}\right]C^{k}_{w,j+1}-\left[\frac{\varphi 3D_e m_d \Delta t}{\rho_p V_w R(R-r_L)}\right]C^{k}_{L,j}\\
&-\left[1-\varphi\left(\frac{3D_e m_d \Delta t}{\rho_p V_w R(R-r_L)}+\frac{2D_L \Delta t}{\Delta x^2}+\frac{v\Delta t}{\Delta x}\right)\right]C^{k}_{w,j}\\[2ex]
f_{(L+1)*(j-1)+2} =\ & \left[\varepsilon+\rho_p K_{fr}\left(C^{k+1}_{1,j}\right)^{\frac{1}{n}-1}+\frac{D_e \Delta t}{r_1^2}\frac{1-\varphi}{r_{1+0.5}-r_{1-0.5}}\frac{r_{1+0.5}^2}{r_2-r_1}\right]C^{k+1}_{1,j}\\
&-\left[\frac{D_e \Delta t}{r_1^2}\frac{1-\varphi}{r_{1+0.5}-r_{1-0.5}}\frac{r_{1+0.5}^2}{r_2-r_1}\right]C^{k+1}_{2,j}-\left[\frac{D_e \Delta t}{r_1^2}\frac{\varphi}{r_{1+0.5}-r_{1-0.5}}\frac{r_{1+0.5}^2}{r_2-r_1}\right]C^{k}_{2,j}\\
&-\left[\varepsilon+\rho_p K_{fr}\left(C^{k}_{1,j}\right)^{\frac{1}{n}-1}-\frac{D_e \Delta t}{r_1^2}\frac{\varphi}{r_{1+0.5}-r_{1-0.5}}\frac{r_{1+0.5}^2}{r_2-r_1}\right]C^{k}_{1,j}\\[2ex]
f_{(L+1)*(j-1)+p+1} =\ & \left[\varepsilon+\rho_p K_{fr}\left(C^{k+1}_{p,j}\right)^{\frac{1}{n}-1}\right.\\
&\left.+\frac{D_e \Delta t}{r_p^2}\frac{1-\varphi}{r_{p+0.5}-r_{p-0.5}}\left(\frac{r_{p+0.5}^2}{r_{p+1}-r_p}+\frac{r_{p-0.5}^2}{r_p-r_{p-1}}\right)\right]C^{k+1}_{p,j}\\
&-\left[\frac{D_e \Delta t}{r_p^2}\frac{1-\varphi}{r_{p+0.5}-r_{p-0.5}}\frac{r_{p+0.5}^2}{r_{p+1}-r_p}\right]C^{k+1}_{p+1,j}-\left[\frac{D_e \Delta t}{r_p^2}\frac{1-\varphi}{r_{p+0.5}-r_{p-0.5}}\frac{r_{p-0.5}^2}{r_p-r_{p-1}}\right]C^{k+1}_{p-1,j}\\
&-\left[\frac{D_e \Delta t}{r_p^2}\frac{\varphi}{r_{p+0.5}-r_{p-0.5}}\frac{r_{p+0.5}^2}{r_{p+1}-r_p}\right]C^{k}_{p+1,j}-\left[\frac{D_e \Delta t}{r_p^2}\frac{\varphi}{r_{p+0.5}-r_{p-0.5}}\frac{r_{p-0.5}^2}{r_p-r_{p-1}}\right]C^{k}_{p-1,j}\\
&-\left[\varepsilon+\rho_p K_{fr}\left(C^{k}_{p,j}\right)^{\frac{1}{n}-1}-\frac{D_e \Delta t}{r_p^2}\frac{\varphi}{r_{p+0.5}-r_{p-0.5}}\left(\frac{r_{p+0.5}^2}{r_{p+1}-r_p}+\frac{r_{p-0.5}^2}{r_p-r_{p-1}}\right)\right]C^{k}_{p,j}\\[2ex]
f_{(L+1)*j} =\ & \left[\varepsilon+\rho_p K_{fr}\left(C^{k+1}_{L,j}\right)^{\frac{1}{n}-1}+\frac{D_e \Delta t}{r_L^2}\frac{1-\varphi}{R-r_{L-0.5}}\left(\frac{R^2}{R-r_L}+\frac{r_{L-0.5}^2}{r_L-r_{L-1}}\right)\right]C^{k+1}_{L,j}\\
&-\left[\frac{D_e \Delta t}{r_L^2}\frac{1-\varphi}{R-r_{L-0.5}}\frac{R^2}{R-r_L}\right]C^{k+1}_{w,j}-\left[\frac{D_e \Delta t}{r_L^2}\frac{1-\varphi}{R-r_{L-0.5}}\frac{r_{L-0.5}^2}{r_L-r_{L-1}}\right]C^{k+1}_{L-1,j}\\
&-\left[\frac{D_e \Delta t}{r_L^2}\frac{\varphi}{R-r_{L-0.5}}\frac{R^2}{R-r_L}\right]C^{k}_{w,j}-\left[\frac{D_e \Delta t}{r_L^2}\frac{\varphi}{R-r_{L-0.5}}\frac{r_{L-0.5}^2}{r_L-r_{L-1}}\right]C^{k}_{L-1,j}\\
&-\left[\varepsilon+\rho_p K_{fr}\left(C^{k}_{L,j}\right)^{\frac{1}{n}-1}-\frac{D_e \Delta t}{r_L^2}\frac{\varphi}{R-r_{L-0.5}}\left(\frac{R^2}{R-r_L}+\frac{r_{L-0.5}^2}{r_L-r_{L-1}}\right)\right]C^{k}_{L,j}
\end{aligned}
\tag{A21}
$$

The residual function vector can be organized as:

$$f\left(C^{k+1}\right) = \begin{bmatrix} f_1 \\ f_2 \\ \vdots \\ f_{p+1} \\ \vdots \\ f_{L+1} \\ \vdots \\ f_{(L+1)*(j-1)+1} \\ f_{(L+1)*(j-1)+2} \\ \vdots \\ f_{(L+1)*(j-1)+p+1} \\ \vdots \\ f_{(L+1)*j} \\ \vdots \\ f_{(L+1)*(N-1)+1} \\ f_{(L+1)*(N-1)+2} \\ \vdots \\ f_{(L+1)*(N-1)+p+1} \\ \vdots \\ f_{(L+1)*N} \end{bmatrix}_{(L+1)*N \times 1} \tag{A22}$$

Similar to the film diffusion case (see Appendix B), the C^{k+1} vector can be determined by Equation (A9) as well and the Jacobian matrix of intraparticle pore diffusion case can be expressed as:

$$J = \begin{bmatrix} \frac{\partial f_1}{\partial C_1} & \frac{\partial f_1}{\partial C_2} & \cdots & \frac{\partial f_1}{\partial C_{(L+1)*N-1}} & \frac{\partial f_1}{\partial C_{(L+1)*N}} \\ \frac{\partial f_2}{\partial C_1} & \frac{\partial f_2}{\partial C_2} & \cdots & \frac{\partial f_2}{\partial C_{(L+1)*N-1}} & \frac{\partial f_2}{\partial C_{(L+1)*N}} \\ \vdots & \vdots & \cdots & \vdots & \vdots \\ \frac{\partial f_{(L+1)*N-1}}{\partial C_1} & \frac{\partial f_{(L+1)*N-1}}{\partial C_2} & \cdots & \frac{\partial f_{(L+1)*N-1}}{\partial C_{(L+1)*N-1}} & \frac{\partial f_{(L+1)*N-1}}{\partial C_{(L+1)*N}} \\ \frac{\partial f_{(L+1)*N}}{\partial C_1} & \frac{\partial f_{(L+1)*N}}{\partial C_2} & \cdots & \frac{\partial f_{(L+1)*N}}{\partial C_{(L+1)*N-1}} & \frac{\partial f_{(L+1)*N}}{\partial C_{(L+1)*N}} \end{bmatrix}_{(L+1)*N \times (L+1)*N} \tag{A23}$$

In order to ensure the accuracy of the model, an error vector (ΔC^{k+1}) was used as described above for Equation (A11):

$$\Delta C^{k+1} = \begin{bmatrix} \left| C^{k+1}_{1,guess,new} - C^{k+1}_{1,guess,old} \right| \\ \left| C^{k+1}_{2,guess,new} - C^{k+1}_{2,guess,old} \right| \\ \vdots \\ \left| C^{k+1}_{p+1,guess,new} - C^{k+1}_{p+1,guess,old} \right| \\ \vdots \\ \left| C^{k+1}_{L+1,guess,new} - C^{k+1}_{L+1,guess,old} \right| \\ \vdots \\ \left| C^{k+1}_{(L+1)*(j-1)+1,guess,new} - C^{k+1}_{(L+1)*(j-1)+1,guess,old} \right| \\ \left| C^{k+1}_{(L+1)*(j-1)+2,guess,new} - C^{k+1}_{(L+1)*(j-1)+2,guess,old} \right| \\ \vdots \\ \left| C^{k+1}_{(L+1)*(j-1)+p+1,guess,new} - C^{k+1}_{(L+1)*(j-1)+p+1,guess,old} \right| \\ \vdots \\ \left| C^{k+1}_{(L+1)*j,guess,new} - C^{k+1}_{(L+1)*j,guess,old} \right| \\ \vdots \\ \left| C^{k+1}_{(L+1)*(N-1)+1,guess,new} - C^{k+1}_{(L+1)*(N-1)+1,guess,old} \right| \\ \left| C^{k+1}_{(L+1)*(N-1)+2,guess,new} - C^{k+1}_{(L+1)*(N-1)+2,guess,old} \right| \\ \vdots \\ \left| C^{k+1}_{(L+1)*(N-1)+p+1,guess,new} - C^{k+1}_{(L+1)*(N-1)+p+1,guess,old} \right| \\ \vdots \\ \left| C^{k+1}_{(L+1)*N,guess,new} - C^{k+1}_{(L+1)*N,guess,old} \right| \end{bmatrix}_{(L+1)*N \times 1} \tag{A24}$$

The iteration processes stop when the maximum value of the error vector ΔC^{k+1}_{max} is smaller than the tolerable error e (e.g., $e = 10^{-15}$).

Appendix D. Length of the Mass Transfer Zone (X_s) for the First Order Analytical Solution

Analytical solutions can be derived for the case of the first flooding of the column which are used here for verification of the numerical codes.

Appendix D.1. Analytical Solution Based on the Film Diffusion Model

During the first flooding of the column, the front water flow is always contacting fresh contaminant material. Therefore, the solute concentration at the particle-water boundary is constant and in equilibrium with the solids:

$$C_{w,eq} = \frac{C_{s,ini}}{K_d} \tag{A25}$$

Inserting Equation (A25) into Equation (7) gives:

$$\frac{\partial C_w}{\partial t} = k\frac{6(1-n)}{nd}\left(C_{w,eq} - C_w\right) \tag{A26}$$

which upon integration yields the following analytical solution for the initial condition $C_{w(t=0)} = 0$ (desorption):

$$\int_0^{C_w} \frac{\partial C_w}{(C_{w,eq} - C_w)} = \int_0^t k\frac{6(1-n)}{nd}\partial t$$

$$-\ln\left(C_{w,eq} - C_w\right) + \ln\left(C_{w,eq}\right) = -\ln\left(1 - \frac{C_w}{C_{w,eq}}\right) = k\frac{6(1-n)}{nd}t \qquad (A27)$$

$$\frac{C_w}{C_{w,eq}} = 1 - \exp\left(-k\frac{6(1-n)}{nd}t\right)$$

The contact time in Equation (A27) can be substituted with the ratio of the travel distance (x) and the flow velocity (v). The length of the mass transfer zone is defined by setting the argument of the exponential function to -1, referring to the location x where the solute concentration in the groundwater reaches 63.2% of the equilibrium concentration.

$$X_{s,63.2\%} = \frac{v\,n\,d}{6k(1-n)} \qquad (A28)$$

Equation (A28) shows that the length of mass transfer zone depends on the flow velocity, inter-granular porosity as well as particle size, but is independent of the distribution coefficient.

If the length of the mass transfer zone is shorter than the column length (X_{col}), a concentration higher than 63.2% of the equilibrium concentration will be observed in the column effluent until the mass transfer zone arrives at the column outlet. The time needed to reach 63.2% equilibrium concentration at the column outlet equals:

$$
\begin{aligned}
t_{63.2\%} &= \frac{X_s}{v} + \frac{(X_{col} - X_s)}{v}R_d \\
&= \frac{X_{col}}{v}\left(1 + K_d\frac{\rho_b}{n}\left(1 - \frac{X_s}{X_{col}}\right)\right)
\end{aligned}
\qquad (A29)
$$

Considering fast kinetics ($X_s \to 0$), $t_{63.2\%}(\approx X_{col}/(v/R_d))$ is mainly dominated by the retarded seepage velocity (v/R_d).

Appendix D.2. Analytical Solution Based on the Intraparticle Pore Diffusion Model

Expressing internal mass transfer resistance by means of intraparticle pore diffusion, with mass transfer coefficient $k = D_e/\delta_p$, where D_e is the effective intraparticle diffusion coefficient ($D_e = D_{aq}\varepsilon/\tau \approx D_{aq}\varepsilon^2$) and the mean square displacement δ_p ($\delta_p = \sqrt{\pi D_a\,t_c}$) representing the diffusion distance, which grows with the square root of contact time between particles and water (t_c) at early times, leads to:

$$\frac{\partial C_w}{\partial t} = k\,A^o\left(C_{w,eq} - C_w\right) = \frac{D_e}{\sqrt{\pi D_a\,t_c}}\frac{6(1-n)}{nd}\left(C_{w,eq} - C_w\right) \qquad (A30)$$

The contact time between water and dry particles can be estimated by the ratio of particle size and flow velocity ($t_c = d/v$).

For the initial condition $C_{w(t=0)} = 0$, integration of Equation (A30) yields the following analytical solution:

$$\int_0^{C_w} \frac{\partial C_w}{(C_{w,eq} - C_w)} = \int_0^t \frac{D_e}{\sqrt{\pi D_a\,t_c}}\frac{6(1-n)}{nd}\partial t$$

$$-\ln\left(C_{w,eq} - C_w\right) + \ln\left(C_{w,eq}\right) = -\ln\left(1 - \frac{C_w}{C_{w,eq}}\right) = \frac{D_e}{\sqrt{\pi D_a\frac{d}{v}}}\frac{6(1-n)}{nd}t \qquad (A31)$$

$$\frac{C_w}{C_{w,eq}} = 1 - \exp\left(-\frac{D_e}{\sqrt{\pi D_a\frac{d}{v}}}\frac{6(1-n)}{nd}t\right)$$

The length of the mass transfer zone of intraparticle pore diffusion can be calculated by:

$$X_{s,63.2\%} = \frac{v\,n\,d}{6\dfrac{D_e}{\sqrt{\pi Da\frac{d}{v}}}(1-n)}$$

$$= \sqrt{\frac{\pi d^3 v}{D_e\left(\varepsilon + K_d\rho_p\right)}\frac{n}{6(1-n)}} \tag{A32}$$

$X_{s,63.2\%}$ based on intraparticle pore diffusion increases with particle size to the power of 3/2 ($d^{3/2}$) and decreases with the square root of the distribution coefficient ($\sqrt{K_d}$). The time required to observe the corresponding concentration of 63.2% of the equilibrium concentration at the column outlet can be determined with Equation (A29).

Appendix D.3. Comparison of Analytical and Numerical Solution and Estimation of Mass Transfer Zone Length (X_s)

In Figure A2, analytical solutions for the increase of the concentration in the first water parcel over distance (which here represents time: $t = x/v$) during the first flooding of the column are shown for FD (Equation (A28)) and IPD (Equation (A31)). The numerical solutions are described in Appendices B and C.

A comparison reveals that the analytical solutions and the numerical solutions overlap almost perfectly for both FD and IPD (see Figure A2). This verifies the accuracy of the numerical model. The length of the mass transfer zone for FD is 0.35 cm and independent of K_d, and much shorter than for IPD with X_s = 10 cm, 3.5 cm and 1.1 cm for K_d values of 0.1 L kg^{-1}, 1 L kg^{-1} and 10 L kg^{-1}, respectively. The deviations between FD and IPD gradually vanish with increasing K_d values. If the initial concentration in the column leachate is close to equilibrium, it may be used for the determination of K_d ($K_d = C_{s,ini}/C_{w,peak}$); K_d is overestimated if the initial effluent concentration does not reach equilibrium ($C_{w,peak} < C_{w,eq}$). The length of the mass transfer zone (Equations (A28) and (A32)) may be used to assess equilibrium at the beginning of the column test.

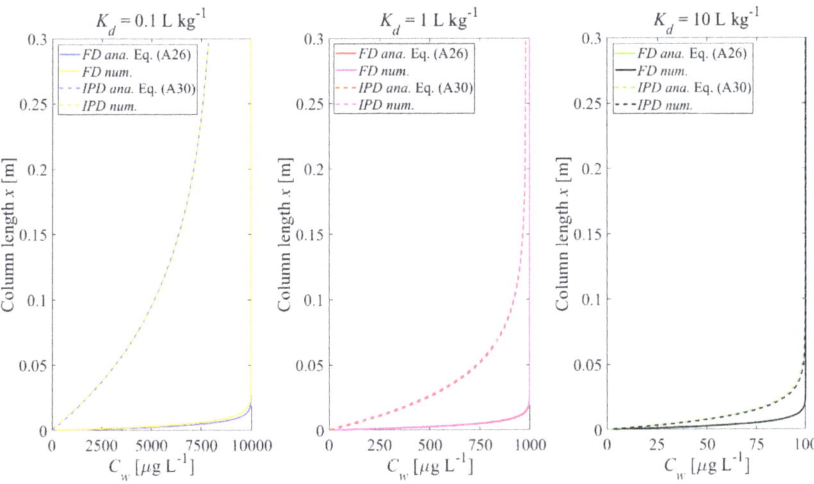

Figure A2. Concentration increase in a water parcel ($C_{w,peak}$) in the column during the first flooding (up-flow); solid lines: film diffusion; dashed lines: intraparticle diffusion; comparison between analytical (ana.) and numerical (num.) solutions. $n = 0.45$, $v = 1.67 \times 10^{-5}$ m s^{-1}, $\alpha /x = 0$ (no dispersion), $C_{s,ini} = 1000$ µg kg^{-1}, $t = 5$ h, $D_{aq} = 1 \times 10^{-9}$ m^2 s^{-1}, $\varepsilon = 0.05$, $d_{p,coarse} = 2000$ µm.

Appendix E. Comparison of Analytical and Numerical Solution (Code Verification)

In order to further confirm the accuracy of the numerical solution, the initial concentration distribution in the column after first flooding, as well as leaching curves are compared with the analytical solution (Equation (6)). The analytical solution is only valid for equilibrium sorption conditions and to compare it with the numerical solution, fine particles ($d_{p,\,fine}$ = 63 μm) are used to get close to equilibrium (to fast FD kinetics). Figures A3 and A4 show the good agreement of both solutions. The slight deviations between analytical and numerical solutions, especially at low K_d values, are due to kinetics in the numerical solution. Deviations gradually vanish with the increase of K_d.

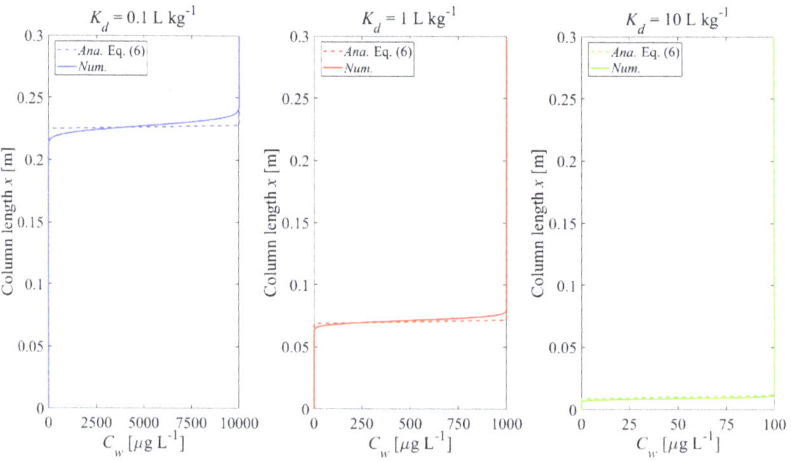

Figure A3. Concentration vs. distance in the up-flow column test after the first flooding of the column (initial condition). Comparison of the analytical solution (Equation (6), dashed lines) and numerical solution (solid lines); $n = 0.45$, $v = 1.67 \times 10^{-5}$ m s^{-1}, $\alpha = 0$ (no dispersion), $C_{s,ini} = 1000$ μ g kg^{-1}, $t_c = 5$ h, $d_{p,fine} = 63$ μ m.

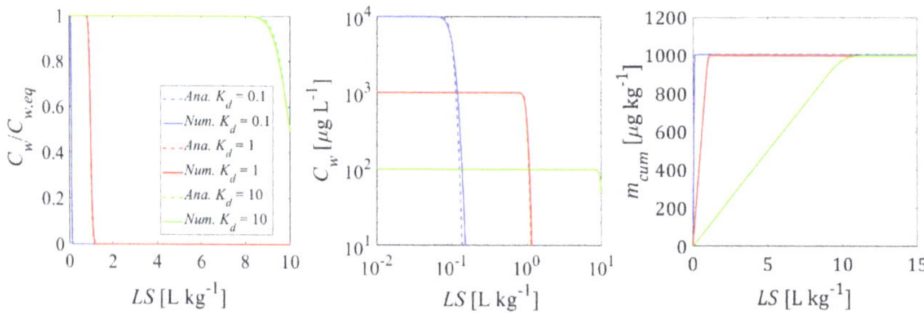

Figure A4. Normalized and absolute concentration ($C_w/C_{w,eq}, C_w$) as well as cumulative concentration (m_{cum}) in the column effluent vs. time (expressed as liquid to solid ratio: LS) for initial conditions shown in Figure A3; comparison of analytical solution (Equation (6), dashed lines) and numerical solution (solid lines).

References

1. Röhler, K.; Haluska, A.A.; Susset, B.; Liu, B.; Grathwohl, P. Long-term behavior of PFAS in contaminated agricultural soils in Germany. *J. Contam. Hydrol.* **2021**, *241*, 103812. [CrossRef] [PubMed]
2. Löv, Å.; Larsbo, M.; Sjöstedt, C.; Cornelis, G.; Gustafsson, J.P.; Kleja, D.B. Evaluating the ability of standardised leaching tests to predict metal(loid) leaching from intact soil columns using size-based elemental fractionation. *Chemosphere* **2019**, *222*, 453–460. [CrossRef] [PubMed]

3.   Inui, T.; Hori, M.; Takai, A.; Katsumi, T. Column percolation tests for evaluating the leaching behavior of marine sediment containing non-anthropogenic arsenic. In Proceedings of the 8th International Congress on Environmental Geotechnics (from 28th October to 1st November 2018 in Hangzhou, China); Zhan, L., Chen, Y., Bouazza, A., Eds.; Springer: Singapore, 2019; Volume 1. ICEG 2018, Environmental Science and Engineering. [CrossRef]

4.   Kalbe, U.; Bandow, N.; Bredow, A.; Mathies, H.; Piechotta, C. Column leaching tests on soils containing less investigated organic pollutants. *J. Geochem. Explor.* **2014**, *147*, 291–297. [CrossRef]

5.   Zhu, Y.; Hu, Y.; Guo, Q.; Zhao, L.; Li, L.; Wang, Y.; Hu, G.; Wibowo, H.; Francesco, D.M. The effect of wet treatment on the distribution and leaching of heavy metals and salts of bottom ash from municipal solid waste incineration. *Environ. Eng. Sci.* **2021**. ahead of print. [CrossRef]

6.   Kumar, A.; Samadder, S.R.; Kumar, V. Assessment of groundwater contamination risk due to fly ash leaching using column study. *Environ. Earth Sci.* **2019**, *78*, 18. [CrossRef]

7.   Chan, W.P.; Ren, F.; Dou, X.; Yin, K.; Chang, V.W.C. A large-scale field trial experiment to derive effective release of heavy metals from incineration bottom ashes during construction in land reclamation. *Sci. Total Environ.* **2018**, *637*, 182–190. [CrossRef] [PubMed]

8.   Di Gianfilippo, M.; Costa, G.; Verginelli, I.; Gavasci, R.; Lombardi, F. Analysis and interpretation of the leaching behaviour of waste thermal treatment bottom ash by batch and column tests. *Waste Manag.* **2016**, *56*, 216–228. [CrossRef]

9.   Tsiridis, V.; Petala, M.; Samaras, P.; Sakellaropoulos, G.P. Evaluation of interactions between soil and coal fly ash leachates using column percolation tests. *Waste Manag.* **2015**, *43*, 255–263. [CrossRef]

10.   Lange, C.N.; Flues, M.; Hiromoto, G.; Boscov, M.E.G.; Camargo, I.M.C. Long-term leaching of As, Cd, Mo, Pb, and Zn from coal fly ash in column test. *Environ. Monit. Assess.* **2019**, *191*, 602. [CrossRef]

11.   Liu, B.; Li, J.; Wang, Z.; Zeng, Y.; Ren, Q. Long-term leaching characterization and geochemical modeling of chromium released from AOD slag. *Environ. Sci. Pollut. Res.* **2020**, *27*, 921–929. [CrossRef]

12.   Schwab, O.; Bayer, P.; Juraske, R.; Verones, F.; Hellweg, S. Beyond the material grave: Life Cycle Impact Assessment of leaching from secondary materials in road and earth constructions. *Waste Manag.* **2014**, *34*, 1884–1896. [CrossRef]

13.   Diotti, A.; Galvin, A.P.; Piccinali, A.; Plizzari, G.; Sorlini, S. Chemical and leaching behavior of construction and demolition wastes and recycled aggregates. *Sustainability* **2020**, *12*, 10326. [CrossRef]

14.   Liu, H.; Zhang, J.; Li, B.; Zhou, N.; Xiao, X.; Li, M.; Zhu, C. Environmental behavior of construction and demolition waste as recycled aggregates for backfilling in mines: Leaching toxicity and surface subsidence studies. *J. Hazard. Mater.* **2020**, *389*, 121870. [CrossRef] [PubMed]

15.   Bandow, N.; Finkel, M.; Grathwohl, P.; Kalbe, U. Influence of flow rate and particle size on local equilibrium in column percolation tests using crushed masonry. *J. Mater. Cycles Waste Manag.* **2019**, *21*, 642–651. [CrossRef]

16.   Butera, S.; Hyks, J.; Christensen, T.H.; Astrup, T.F. Construction and demolition waste: Comparison of standard up-flow column and down-flow lysimeter leaching tests. *Waste Manag.* **2015**, *43*, 386–397. [CrossRef]

17.   Chen, Z.; Zhang, P.; Brown, K.; Branch, J.; van der Sloot, H.; Meeussen, J.; Delapp, R.; Um, W.; Kosson, D. Development of a geochemical speciation model for use in evaluating leaching from a cementitious radioactive waste form. *Environ. Sci. Technol.* **2021**, *55*, 8642–8653. [CrossRef]

18.   Molleda, A.; López, A.; Cuartas, M.; Lobo, A. Release of pollutants in MBT landfills: Laboratory versus field. *Chemosphere* **2020**, *249*, 126145. [CrossRef]

19.   Grathwohl, P.; van der Sloot, H. Groundwater risk assessment at contaminated sites (GRACOS): Test methods and modelling approaches. In *Groundwater Science and Policy*; Quevauviller, P., Ed.; RSC Publishing: Cambridge, UK, 2007.

20.   Susset, B.; Grathwohl, P. Leaching standards for mineral recycling materials—A harmonized regulatory concept for the upcoming German Recycling Decree. *Waste Manag.* **2011**, *31*, 201–214. [CrossRef] [PubMed]

21.   Grathwohl, P.; Susset, B. Comparison of percolation to batch and sequential leaching tests: Theory and data. *Waste Manag.* **2009**, *29*, 2681–2688. [CrossRef] [PubMed]

22.   Grathwohl, P. On equilibration of pore water in column leaching tests. *Waste Manag.* **2014**, *34*, 908–918. [CrossRef]

23.   Grathwohl, P.; Reinhard, M. Desorption of Trichloroethylene in Aquifer Material: Rate Limitation at the Grain Scale. *Environ. Sci. Technol.* **1993**, *27*, 2360–2366. [CrossRef]

24.   Rügner, H.; Kleineidam, S.; Grathwohl, P. Long term sorption kinetics of phenanthrene in aquifer materials. *Environ. Sci. Technol.* **1999**, *33*, 1645–1651. [CrossRef]

25.   Wang, G.; Kleineidam, S.; Grathwohl, P. Sorption/desorption reversibility of phenanthrene in soils and carbonaceous materials. *Environ. Sci. Technol.* **2007**, *41*, 1186–1193. [CrossRef]

26.   Finkel, M.; Grathwohl, P. Impact of pre-equilibration and diffusion limited release kinetics on effluent concentration in column leaching tests: Insights from numerical simulations. *Waste Manag.* **2017**, *63*, 58–73. [CrossRef] [PubMed]

27.   Kleineidam, S.; Rügner, H.; Grathwohl, P. Impact of grain scale heterogeneity on slow sorption kinetics. *Environ. Toxicol. Chem.* **1999**, *18*, 1673–1678. [CrossRef]

28.   Jäger, R.; Liedl, R. Prognose der Sorptionskinetik organischer Schadstoffe in heterogenem Aquifermaterial (Predicting sorption kinetics of organic contaminants in heterogeneous aquifer material). *Grundwasser* **2000**, *5*, 57–66. [CrossRef]

29. Xiao, Y.; Feng, Z.; Huang, X.; Huang, L.; Chen, Y.; Wang, L.; Long, Z. Recovery of rare earths from weathered crust elution-deposited rare earth ore without ammonia-nitrogen pollution: I. leaching with magnesium sulfate. *Hydrometallurgy* **2015**, *153*, 58–65. [CrossRef]

30. Yin, K.; Chan, W.P.; Dou, X.; Lisak, G.; Chang, V.W.C. Kinetics and modeling of trace metal leaching from bottom ashes dominated by diffusion or advection. *Sci. Total Environ.* **2020**, *719*, 137203. [CrossRef] [PubMed]

31. Ogata, A.; Banks, R.B. A solution of the differential equation of longitudinal dispersion in porous media. In *Professional Paper*; Geological Survey (U.S.): Washington, DC, USA, 1961; pp. A1–A7.

32. Boving, T.; Grathwohl, P. Matrix diffusion coefficients in sandstones and limestones: Relationship to permeability and porosity. *J. Contam. Hydrol.* **2001**, *53*, 85–100. [CrossRef]

33. Kalbe, U.; Berger, W.; Eckardt, J.; Simon, F.G. Evaluation of leaching and extraction procedures for soil and waste. *Waste Manag.* **2008**, *28*, 1027–1038. [CrossRef]

34. Kalbe, U.; Berger, W.; Simon, F.G.; Eckardt, J.; Christoph, G. Results of interlaboratory comparisons of column percolation tests. *J. Hazard. Mater.* **2007**, *148*, 714–720. [CrossRef]

35. Liu, Y.; Illangasekare, T.H.; Kitanidis, P.K. Long-term mass transfer and mixing-controlled reactions of a DNAPL plume from persistent residuals. *J. Contam. Hydrol.* **2014**, *157*, 11–24. [CrossRef]

36. Naka, A.; Yasutaka, T.; Sakanakura, H.; Kalbe, U.; Watanabe, Y.; Inoba, S.; Takeo, M.; Inui, T.; Katsumi, T.; Fujikawa, T.; et al. Column percolation test for contaminated soils- Key factors for standardization. *J. Hazard. Mater.* **2016**, *320*, 326–340. [CrossRef]

37. Arters, D.C.; Fan, L.S. Experimental methods and correlation of solid-liquid mass transfer in fluidized beds. *Chem. Eng. Sci.* **1990**, *45*, 965–975. [CrossRef]

38. Blasius, H. Grenzschichten in Flüssigkeiten mit kleiner Reibung (The boundary layers in fluids with little friction). *J. Appl. Math. Phys.* **1908**, *56*, 1–37.

39. Calderbank, P.H.; Moo-Young, M.B. The continuous phase heat and mass transfer properties of dispersions. *Chem. Eng. Sci.* **1995**, *50*, 3921–3934. [CrossRef]

40. Kirwan, J.; Armenante, P.M. Mass transfer to microparticle systems. *Chem. Eng. Sci.* **1989**, *44*, 2781–2796. [CrossRef]

41. Levins, D.M.; Glastonbury, J.R. Application of Kolmogorofff's theory to particle-liquid mass transfer in agitated vessels. *Chem. Eng. Sci.* **1972**, *27*, 537–543. [CrossRef]

42. Mao, H.; Chisti, Y.; Moo-Young, M. Multiphase hydrodynamics and solid-liquid mass transfer in an external-loop airlift reactor- a comparative study. *Chem. Eng. Commun.* **1992**, *113*, 1–13. [CrossRef]

43. Ohashi, H.; Sugawara, T.; Kikuchi, K.I. Mass transfer between particles and liquid in solid-liquid two-phase flow through vertical tubes. *J. Chem. Eng. Jpn.* **1981**, *14*, 489–491. [CrossRef]

44. Sherwood, T.K.; Pigford, R.L.; Wilke, R.L. *"Mass Transfer"*; McGraw-Hill: New York, NY, USA, 1975.

45. Cirpka, O.A. Solute and heat transport. In *Lecture Notes "Environmental Modeling 2"*; 2020; (unpublished, University of Tübingen).

46. Liedl, R.; Ptak, T. Modelling of diffusion-limited retardation of contaminants in hydraulically and lithologically nonuniform media. *J. Contam. Hydrol.* **2003**, *66*, 239–259. [CrossRef]

materials

MDPI

Article

Determining Adsorption Parameters of Potentially Contaminant-Releasing Materials Using Batch Tests with Differing Liquid-Solid Ratios

Hirofumi Sakanakura [1,*], Kenichi Ito [2], Jiajie Tang [3], Mikako Nakagawa [1] and Hiroyuki Ishimori [1]

[1] Material Cycles Division, National Institute for Environmental Studies, 16-2 Onogawa, Tsukuba City 305-8506, Japan; mikakonakagawa55@gmail.com (M.N.); ishimori.hiroyuki@nies.go.jp (H.I.)
[2] Center for International Relations, University of Miyazaki, 1-1, Gakuen Kibanadai-nishi, Miyazaki City 889-2192, Japan; itoken@cc.miyazaki-u.ac.jp
[3] Graduate School of Global Environmental Studies, Kyoto University, Yoshida-honmachi, Sakyo ku, Kyoto City 606-8501, Japan; tangjiajie0422@gmail.com
* Correspondence: sakanakura@nies.go.jp; Tel.: +81-29-850-2185; Fax.: +81-29-850-2091

Abstract: Adsorption parameters such as the distribution coefficient are required to predict the release behavior of contaminants using advection-dispersion models. However, for potentially contaminant-releasing materials (PCMs) such as dredged sludge and coal ash, these parameters cannot be obtained by conventional adsorption tests. This study developed a method to determine adsorption parameters for PCMs from a set of batch tests conducted in parallel as a function of the liquid-solid ratio (LS-parallel test). This LS-parallel test was performed on sandy soil derived from marine sediment using liquid-solid ratios from 1 to 300 L/kg. The water-contact time was also changed from 10 min to 28 d to elucidate the kinetics or equilibrium of contaminants released from the sample. Adsorption parameters were successfully obtained if the substance was under adsorption control. A column percolation test was performed to confirm the effectiveness of the obtained parameters. Good agreements were observed for SO_4^{2-} and B, but discrepancies remained for other substances such as F^- and As suggesting that improvements are necessary in both the LS-parallel test procedure and the advection-dispersion model.

Keywords: batch leaching test; liquid-solid ratio; column percolation test; advection-dispersion model; adsorption–desorption equilibrium

Citation: Sakanakura, H.; Ito, K.; Tang, J.; Nakagawa, M.; Ishimori, H. Determining Adsorption Parameters of Potentially Contaminant-Releasing Materials Using Batch Tests with Differing Liquid-Solid Ratios. *Materials* **2021**, *14*, 2534. https://doi.org/10.3390/ma14102534

Academic Editors: Franz-Georg Simon and Ute Kalbe

Received: 27 April 2021
Accepted: 10 May 2021
Published: 13 May 2021

Publisher's Note: MDPI stays neutral with regard to jurisdictional claims in published maps and institutional affiliations.

1. Introduction

Solid materials such as dredged sludge, construction and demolition waste, steel slag, coal ash, and municipal solid waste incineration ash are anticipated to be recycled in construction works. In particular, these materials could be used to make features such as roadbeds, embankments, and landfill. However, these materials are also known to contain trace contaminants that might be released into the environment during their application. Therefore, the environmental impact of such potentially contaminant-releasing materials (PCMs) should be evaluated. Leaching is one of the most environmental important aspects, as contaminants might be transferred to downstream environments by contacting water, resulting in soil and groundwater pollution. Leaching tests can be used to evaluate this aspect of PCMs; the simplest types of leaching test are called single batch tests, which have been standardized by many organizations and countries (e.g., [1,2]). However, the test conditions are quite different from the real environment in that, for example, a single batch test does not have a water flow. The column percolation test (e.g., [3]) is closer to reality because it features water flowing through a column filled with a PCM sample. However, this test is more complicated and time-consuming than single batch testing. Therefore, efforts have been made to relate the results between batch and column percolation tests [4–7]. For example, comparable values were obtained between the amount eluted under a liquid-solid ratio (LS_{batch}) of 10 L/kg in a batch test and the total amount released until a cumulative liquid-solid ratio (LS_{cum}) of 10 L/kg was reached in a column percolation test (e.g., Reference [7]). However, when considering the realistic nature of results, column tests are also limited because their flow path length (e.g., 30 ± 5 cm [3]) is usually shorter than actual cases—in the case of an embankment the flow path length might reach several meters. Furthermore, the water-contact times of both batch tests (less than 24 h for most standards) and column tests (around or less than a few months for most standards) are shorter than the actual periods of application, which can sometimes last for decades.

Numerical model calculations can complement such discrepancies of leaching tests, for which the role of the leaching test is to provide the model parameters [8]. To this end, Reference [9] reviewed numerical models of the release of contaminants from PCMs. The single-mode first-order decay model [10–12] is the simplest model in which the outlet eluate concentration decreases exponentially. Non-precipitating and non-adsorbing substances can be applied to this model, assuming that the total amount of a substance is dissolved in a complete mixing box simulating pore water, and that it is gradually diluted and flows out along with freshwater inflow. However, the assumption that the entire system remains limited to one complete mixing box is not realistic when the flow path length increases. Thus, Reference [13] developed the continuously stirred tank reactor cascade model, based on double porosity. Similarly, Reference [14] developed the dual-mode first-order decay model that also covers non-precipitating and non-adsorbing substances, and can depict various decrease curve shapes. However, this model cannot scale up and extrapolate with consistent logic. The coupled chemistry transport model [15,16] considers solubility equilibrium with redox reactions and mass transfer by advection and dispersion. Reproducibility in blind simulations will be high if the model captures every chemical reaction appropriately. However, the reaction formula and the substances involved must be identified precisely for each PCM and for each application. Furthermore, the solubility equilibrium only occurs in very limited circumstances; for many PCMs such as dredged sludge, construction demolition waste, steel slag, and coal ash, the eluate concentration is often much lower than the solubility. In such cases, these substances are regarded as having precipitation-free and adsorptive characteristics.

A further study has showed the validity of the advection-dispersion model for describing the release behavior of contaminants from PCMs [17]. This model has been widely applied to capture the underground spreading of pollution [18]. The model assumes adsorption equilibrium, and can calculate the real scales and times of PCMs' applications logically. Currently, it is necessary to perform a column percolation test to obtain the model parameters through the fitting of contaminant-release curves. If these parameters could be obtained from a simpler batch test method, then combined with model calculations, this could be quite practical for estimating the longer-term behavior of contaminants from PCMs in large scale scenarios. However, conventional adsorption tests (e.g., [19]) cannot provide adsorption parameters because the substance in question is also be eluted from the PCM itself.

In this study, a procedure was developed to obtain the adsorption parameters of PCMs through a set of batch tests with different liquid-solid ratios (LS-parallel test). This LS-parallel test was performed on a sandy soil derived from marine sediment, using not only different liquid-solid ratios, but also different water-contact times. This permitted discussion of the release mechanism regarding adsorption-control. Furthermore, a column percolation test was performed to ensure the validities of the parameters obtained in the LS-parallel test.

2. Theory

2.1. Advection-dispersion Model

In the advection-dispersion model, the transfer of a substance is represented by Equation (1) for a one-dimensional case:

$$\theta \frac{\partial C}{\partial t} = -\theta v \frac{\partial C}{\partial x} + \theta D \frac{\partial^2 C}{\partial x^2} - \rho_d \frac{\partial M_A}{\partial t} \tag{1}$$

where θ is the effective porosity (−), C is the concentration in the leachate (mg/L), t is the elapsed time (s), v is the actual flow velocity (m/s), x is a coordinate (m), D is the dispersion coefficient (m^2/s), ρ_d is the dry density (kg/L), and M_A is the amount of the substance in question that is adsorbed on solids (mg/kg). M_A is a function of C; its linear type (or Henry type) adsorption isotherm is shown in Equation (2):

$$M_A = K_d C \tag{2}$$

where K_d is the distribution coefficient (L/kg), which is peculiar to the contaminant in question [20]. Equation (3) is obtained from Equations (1) and (2):

$$\frac{\partial C}{\partial t} + \frac{v}{R} \frac{\partial C}{\partial x} = \frac{D}{R} \frac{\partial^2 C}{\partial x^2} \tag{3}$$

where R is the retardation factor (-), which is represented by Equation (4):

$$R = 1 + \frac{\rho_d}{\theta} K_d \tag{4}$$

The analytical solution of Equation (3) in the case where clean water moves through a column filled with a PCM is:

$$C(x,t) = C_0 \left[1 - \frac{1}{2} \left\{ erfc \left(\frac{Rx - vt}{2\sqrt{DRt}} \right) + \exp \left(\frac{vx}{D} \right) erfc \left(\frac{Rx + vt}{2\sqrt{DRt}} \right) \right\} \right] \tag{5}$$

Equation (5) shows that the concentration at a focused point decreases with time from the initial concentration, C_0. Some studies only describe the initial high concentration as "equilibrium" and describe the decrease curve as "non-equilibrium" (e.g., [17]), but this represents a misunderstanding. The monotonic decrease is always governed by the adsorption equilibrium represented by Equation (1) and including Equation (2), which is assuming an instant equilibrium. As a characteristic of the decrease curve, the larger the K_d, the slower the decrease. C_0 is calculated using Equation (6) [17]:

$$C_0 = \frac{M_T}{K_d + \theta / \rho_d} \tag{6}$$

where M_T is the amount of a given substance taking part in the adsorption equilibrium (mg/kg).

In a previous study [17], it was assumed that all substances were present at the solid surface initially, and that they did not increase with time (e.g., through new dissolution at the solid surface). Thus, they represented the mass parameter in $C_{s,ini}$ (mg/kg), i.e., the solid-phase initial concentration. However, the present study expresses the parameter M_T instead of $C_{s,ini}$ because M_T often increases with time, which will be described further in Section 4.1.

2.2. Adsorption Isotherms of PCMs

Column percolation tests are considered necessary to obtain model parameters such as M_T and K_d [17]. The present study, however, aims to develop a procedure that only requires batch tests, which are simpler. The theory is as follows. The amount of a contaminant present in the liquid phase per mass of the solid, M_L (mg/kg), can be expressed as follows:

$$M_L = \frac{V}{m} \times C = LS_{batch} \times C \tag{7}$$

where V is the liquid volume (L), m is the mass of solid (kg), and LS_{batch} is the ratio of liquid to solid in the batch condition (L/kg). Equation (7) shows that M_L has a linear relationship with C with the proportional constant LS_{batch}. The total amount of a substance per mass of solid taking part in adsorption equilibrium, M_T (mg/kg), can be expressed as the sum of the substances adsorbed on the solid surface and those present in the liquid phase:

$$M_T = M_A + M_L \tag{8}$$

Equation (9) can be obtained by substituting Equation (2) into Equation (8); when K_d and M_T are constant, M_L can be represented as a function of C.

$$M_L = -K_d C + M_T \tag{9}$$

In addition to a Henry type isotherm (Equation (2)), the relationship between M_A (mg/kg) and C (mg/L) can also be represented by Freundlich type and Langmuir type isotherms, as shown in Equations (10) and (11), respectively [21]:

$$M_A = K_F C^N \tag{10}$$

$$M_A = \frac{M_{sat} K_L C}{1 + K_L C} \tag{11}$$

where K_F, N, and K_L represent constants ($N < 1$), and M_{sat} represents the saturated adsorption amount (mg/kg). In all isotherms (Equations (2), (10), and (11)), the higher C (mg/L), the higher M_A (mg/kg). Equations (12) and (13) can be obtained by substituting Equations (10) and (11), respectively into Equation (8):

$$M_L = -K_F C^N + M_T \tag{12}$$

$$M_L = -\frac{M_{sat} K_L C}{1 + K_L C} + M_T \tag{13}$$

Figure 1 shows the relationship between M_L and C in (a) Henry type, (b) Freundlich type, and (c) (d) Langmuir type isotherms. In panels (a–d), the Y-intercept represents M_T. In panels (a–c), the X-intercept represents the liquid phase concentration, C, when the liquid volume, V, is zero (i.e., when the minimum amount of water is in contact with the dry material). In case of panel (d), there is no X-intercept as M_T is larger than M_{sat}, so C will rise until solubility is reached.

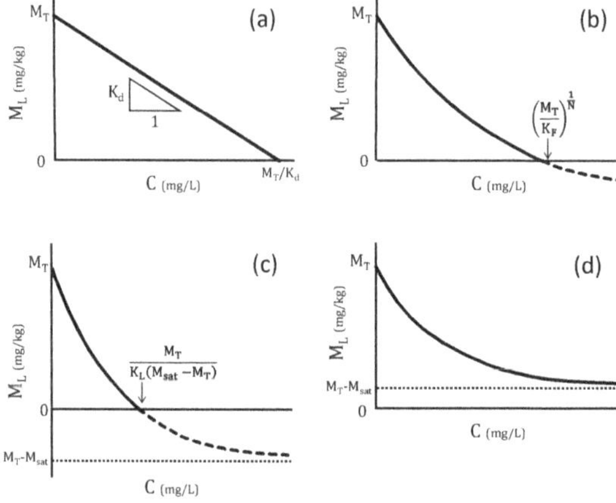

Figure 1. Schematic relationships between liquid phase concentration and eluted amount per mass in LS-parallel test. (**a**) Henry (linear) type, (**b**) Freundlich type, (**c**) Langmuir type of $M_T < M_{sat}$, and (**d**) Langmuir type of $M_T > M_{sat}$.

2.3. Application of LS-Parallel Test on PCMs

In a single batch test, one value for C is obtained, and M_L can be calculated from Equation (7). Therefore, by performing several batch tests in parallel on one PCM mother sample, while changing the liquid-solid ratio (LS-parallel test), various C and M_L values can be obtained. By plotting the relationship between C and M_L, as shown in Figure 1, and by approximation of any of Equations (9), (12), or (13), the adsorption parameters of the PCM can be obtained.

This procedure was developed under the assumption that the mass of substance taking part in the adsorption equilibrium, M_T, is constant. However, it should be noted that additional substances may be released from the solid phase over time and/or (co-)precipitated during a batch test. Another assumption is that the factors affecting adsorption equilibrium, such as pH, are the same among the batch tests conducted in parallel. The most important test condition would be duration, as these aspects might change dynamically with time. However, the adsorption isotherms (Equations (2), (10) and (11)) do not include time, indicating that the adsorption equilibrium is instantaneously achieved. Therefore, it is necessary to find the optimum LS-parallel test conditions, and especially the test time must be carefully examined.

3. Material and Methods

3.1. Material

A sandy soil derived from marine sediment in Japan, called a tsunami deposit, was used for analysis. This sample was obtained from a temporary stockpile from Miyagi prefecture after the 2011 East Japan Great Earthquake. As huge amounts (approximately 10 million tons) of tsunami deposits were generated after the disaster, it would be advantageous to recycle these tsunami deposits through civil engineering projects. However, the soil contains trace contaminants, and therefore evaluation of its environmental safety is necessary.

After sampling, the tsunami deposit was air-dried, sieved through a 2-mm sieve and stored at room temperature (25 °C) before the leaching tests. The particle density was 2.65 g/cm^3, and the water content was 0.6%. Table 1 shows the elemental composition as determined by aqua regia extraction and alkali melting, followed by inductively coupled plasma optical emission spectrometry (ICP-OES; Model 720, Agilent Technologies Inc., Santa Clara, CA, USA) and inductively coupled plasma mass spectrometry (ICP-MS; 7500CX, Agilent Technologies Inc., Santa Clara, CA, USA) except for Si, P, S, and Cl, which were measured by fluorescent X-ray spectroscopy (Primus II, Rigaku Corp., Matsubara-cho Akishima, Japan).

Table 1. Elemental composition of sandy soil.

	Content (% for Si-P, mg/kg for S-Sr)			Method		Content (mg/kg)			Method
Si	24.1			XRF	Zn	98.7	±	2.9	AD + AF
Al	7.20	+	0.26	AD + AF	Rb	66.0	±	1.6	AD + AF
Fe	3.72	±	0.09	AD + AF	Cr	52.9	±	8.6	AD + AF
K	3.19	±	0.18	AD + AF	Cu	25.6	±	0.5	AD + AF
Na	2.25	±	0.19	AD + AF	Ni	18.3	±	0.6	AD + AF
Ca	2.12	±	0.12	AD + AF	Pb	16.5	±	0.2	AD + AF
Mg	1.02	±	0.020	AD + AF	Se	14.8	±	0.5	AD + AF
Ti	0.331	±	0.023	AD + AF	As	12.7	±	1.3	AD + AF
P	0.120			XRF	Co	12.5	±	0.3	AD + AF
S	884			XRF	Cs	3.15	±	0.48	AD + AF
Mn	762	±	32	AD + AF	Mo	0.96	±	0.03	AD + AF
Cl	540			XRF	Sb	0.35	±	0.01	AD + AF
Ba	525	±	13	AD + AF	Cd	0.20	±	0.00	AD + AF
Sr	313	±	18	AD + AF					

AD + AF: acid digestion and alkali fusion.

3.2. LS-Parallel Test

The LS-parallel test comprises a series of batch test performed in parallel under different liquid-solid ratios. A LS-parallel test has been standardized in which the liquid-solid ratios are varied from 0.5 to 10 L/kg, the sample size is 2 mm or less, and the contact time is 48 h [22]. However, as shown in Table 2, here the test conditions were modified to accommodate wider ranges. The maximum liquid-solid ratio range was expanded to 300 L/kg, because the larger the liquid-solid ratio, the better the extrapolation prediction of M_T. The range of contact time was also expanded widely, from 10 min to 28 d, to allow for analysis of changes in the eluate conditions. The test times used here, such as 10 min and 28 days, are not suitable for future standardization, but it is meaningful to explore such extreme conditions to consider the release mechanisms that occur during a batch test system. The mass of sample, volume of solvent, and volume of vessel for each test are shown in Table 2. For each test condition, two or three subsamples were applied. In total, 70 single batch tests were executed. The vessels were made from high-density polyethylene. The solvent used was a 1 mmol/L calcium chloride solution, which was applied to reduce the generation of colloids. Mixing was performed with a tumbling shaker at approximately 5 rev/min. After each test, the solution was immediately filtered using a polytetrafluoroethylene (PTFE) membrane filter with a pore size of 0.45 μm. To check the effect of colloids on the 0.45 μm-filtrate, a part of the filtrate was further filtered with a 0.1 μm PTFE membrane filter for a contact time of only 6 h.

Table 2. Summary of LS-parallel test conditions.

	Liquid-Solid Ratio (L/kg)	Sample Amount (g)	Solution Volume (mL)	Vessel Volume (mL)	Replicates	Contact Time
LS 1	1	60	60	250	2	10 min, 6 h, 1 d, 7 d, 28 d
LS 3	3	20	60	250	2	10 min, 6 h, 1 d, 7 d, 28 d
LS 10	10	10	100	250	2	10 min, 6 h, 1 d, 7 d, 28 d
LS 30	30	5	150	250	2	10 min, 6 h, 1 d, 7 d, 28 d
LS 100	100	5	500	1000	3	10 min, 6 h, 1 d, 7 d, 28 d
LS 300	300	2.5	750	1000	3	10 min, 6 h, 1 d, 7 d, 28 d

3.3. Column Percolation Test

An acrylic column with an inner diameter of 5 cm was packed with 289 g of the soil sample, to obtain a thickness of 10 cm. This thickness deviates from the ISO 21268-3 up-flow percolation test, 30 ± 5 cm [3], due to the amount of the soil sample stored. As an eluent, a solution of 1 mmol/L calcium chloride was introduced from the bottom of the column using a peristaltic pump. A linear velocity of 15 ± 2 cm/d was applied, i.e., a flow rate of 288 ± 24 mL/d. Thus, an LS_{cum} of approximately 1.0 L/kg of eluent passed through the column per day. The flow was continued after the eluent level reached top of the column, which marks another deviation from ISO 21268-3; in the standard, after filling the column with eluent, the apparatus should be left for two days to achieve "equilibrium", and then the water flow should be restarted. In this study, non-stop flow was chosen to shorten the total test time; the LS-parallel test result justified that after two days the concentration was still changing, i.e., an equilibrium had not been reached. Under this non-equilibrium environment, the total time itself was important. Fourteen fractions of LS_{cum} of approximately 0.1, 0.3, 0.4, 0.6, 0.9, 1.2, 1.5, 1.8, 2.5, 3.1, 3.7, 5.2, 6.8, and 8.3 L/kg were taken to elucidate the release behavior of substances from the soil sample. The collected eluates were filtered using PTFE membrane filter with a pore size of 0.45 μm.

3.4. Measurement

The pH and electrical conductivity (EC) of the filtrate were measured immediately. Cl^-, F^-, PO_4^{3-}, and SO_4^{2-} were determined by ion chromatography (Dionex ICS 2100, Thermo Fisher Scientific Inc., Waltham, MA, USA). Al, As, B, Ba, Ca, Co, Cr, Cu, Fe, K, Mg, Mn, Mo, Na, Ni, Pb, Rb, Sb, Se, Si, Sr, Ti, and Zn were determined by ICP-OES (Model 720,

Agilent Technologies Inc.) or ICP-MS (7500CX, Agilent Technologies Inc.), depending on their concentration level.

4. Results and Discussion

4.1. Changes in pH with Time during the LS-Parallel Test

As pH significantly affects equilibria, such as the precipitation-dissolution and adsorption–desorption equilibria, it is desirable to restrict pH changes to a narrow range as possible among all eluates. This helps to obtain one parameter from a set of batch tests with differing conditions, such as liquid-solid ratio and contact time. Figure 2 shows the changes in pH with time during the LS-parallel test. Up until just after 10 min, pH remained in a narrow range (6.7–7.0). After this point, however, pH increased with time; the trends were clearer under lower LS conditions (hereinafter LS means LS_{batch} if there is no other notification). In LS 1, for example, pH reached 8.4 after 28 days. Larger LS conditions, such as 100 and 300 L/kg, showed relatively constant pH at approximately 7.0.

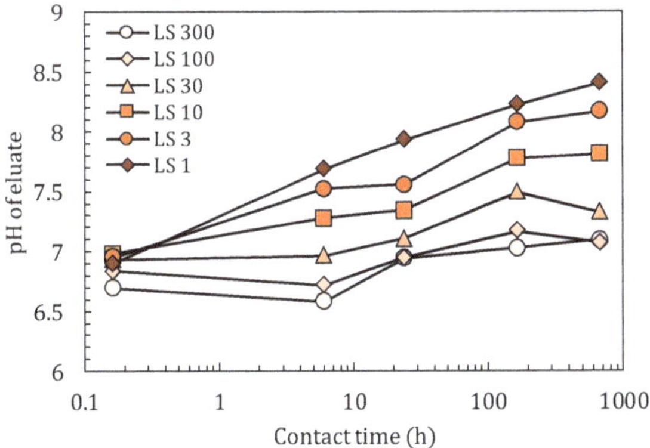

Figure 2. Change in pH in the LS-parallel test as a function of contact time.

It may be possible to control pH values by adding acid or alkali solutions. However, pH was not controlled in this study, because the concentrations of acid or alkali and the frequency of addition must be carefully adjusted to reduce fluctuations in pH. Furthermore, until the acid or alkali solution was sufficiently diluted in the mixture, high concentrations of the solution may have directly affected soil particles. Such results were experienced in preliminary batch tests. Therefore, in the results hereinafter it is necessary to consider that pH exhibited a range of approximately 1.8 during the batch test, which might have affected the equilibria of substances.

4.2. Approximation of Adsorption Parameters from LS-Parallel Test

Figure 3 shows the mutual relationships [C-t], [M_L-t], and [C-M_L], as obtained by LS-parallel tests on SO_4^{2-}, Na, B, Mg, F^-, and As. Regarding the relationship [C-t], the smaller the LS, the higher the eluate concentration, C. In Japan's soil environmental standard, B, F^-, and As are designated as regulated substances. They are judged by a batch test mixing with pure water with a liquid-solid ratio of 10 L/kg for 6 h. The reference values of B, F, and as are 1.0, 0.8, and 0.01 mg/L, respectively [23].

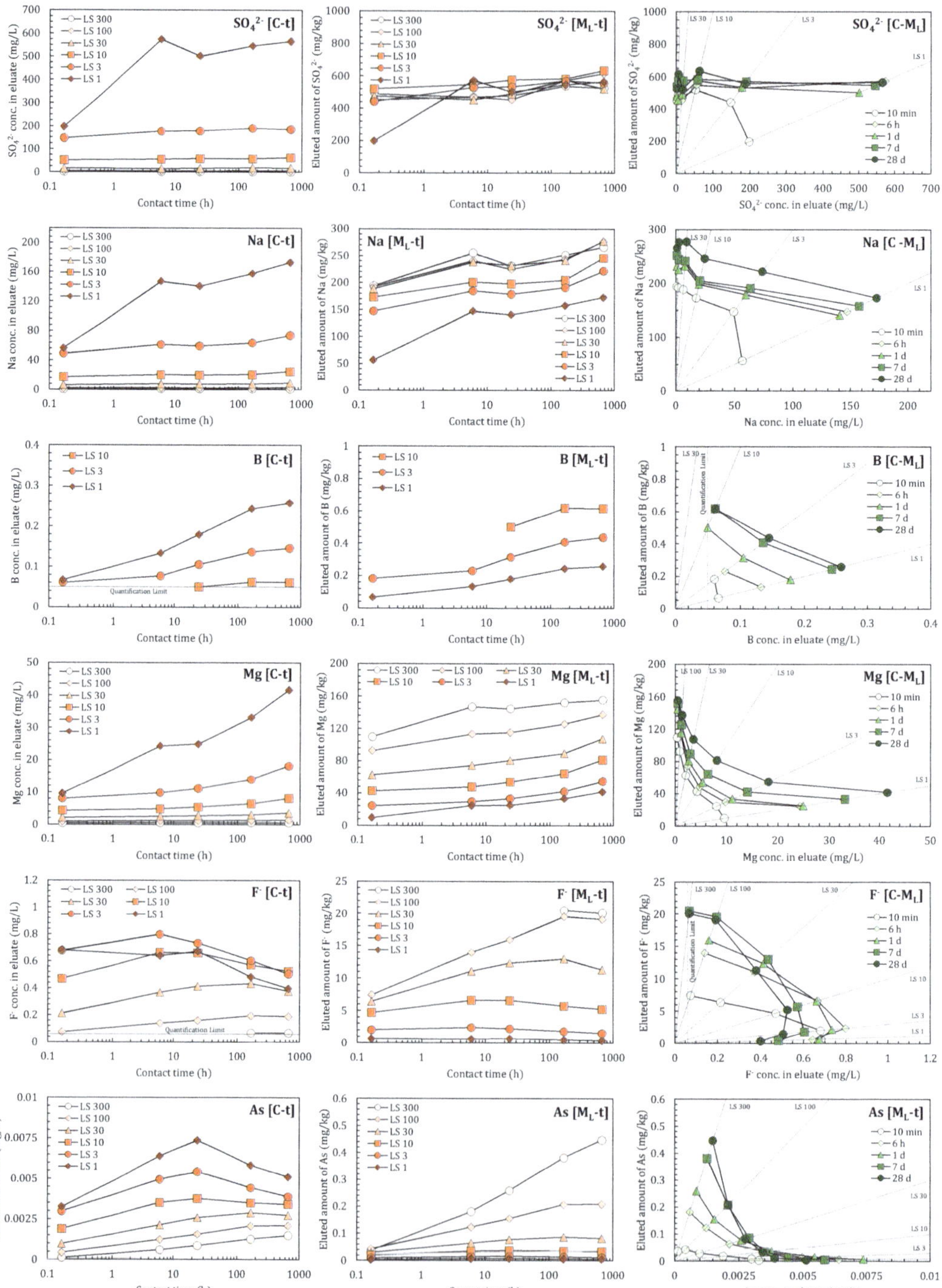

Figure 3. Changes in eluate concentration with time [*C-t*], changes in eluted amount with time [*M$_L$-t*], and relationship between eluate concentration and eluted amount [*C-M$_L$*]. In [*C-M$_L$*] panels, data of the same *LS$_{batch}$* condition are plotted on a straight-line passing through the origin. A lack of data means the eluate concentration was below the quantification limit.

The $[M_L\text{-}t]$ relationships were determined by converting C to M_L using Equation (9), as shown in the panels in the center column of Figure 3. Generally, larger LS values corresponded to higher M_L, with the exceptions of SO_4^{2-} and Na. The results coincided for SO_4^{2-} except for at 10 min in LS 1, whereas for Na they coincided above LS 30. The results obtained after 10 min in LS 1 appeared extremely low, presumably because even soluble substances could not be dissolved sufficiently at this time.

Using the results of C and M_L, the $[C\text{-}M_L]$ relationships were plotted as shown in the left-hand column of Figure 3. SO_4^{2-} showed almost horizontal, linear relationships. Referring to Figure 1, the slope representing K_d was almost zero, indicating that adsorption did not work for SO_4^{2-}. The parameters approximated from the $[C\text{-}M_L]$ relationships are summarized in Table 2. For SO_4^{2-}, K_d was almost zero and M_T remained almost unchanged until 28 days.

For Na, B, and Mg, $[C\text{-}t]$ and $[M_L\text{-}t]$ showed monotonous increases among all LS conditions (for B in LS 30, 100, and 300 all eluates were below the quantification limit). Such increases imply additional releases from the solid to the elution. The $[C\text{-}M_L]$ relationships of Na and B appeared to be linear in each LS data set, indicating that they exhibited Henry type adsorption isotherms. Mg exhibited a curved relationship that fitted well with a Langmuir type isotherm. All relationships gradually increased with time, indicating that the M_T of each substance increased gradually. The parameters obtained by fitting are summarized in Table 3. The observed gradual increases might represent intraparticle diffusion from inside the solid to the surface [24,25]. This could have proceeded with time due to differences in concentration inside and outside of the solid [26]. Besides, the adsorption isotherms shown in Equations (2), (10), and (11) feature no time-parameters, meaning that adsorption was considered to be in an instant equilibrium. Such differences in release mechanisms will be discussed in Section 4.4.

Table 3. Parameters obtained from LS-parallel tests.

	Duration Time in LS-Parallel Test	Referred LS Range (L/kg)	Referred Concentration Range (mg/L)	M_T (mg/kg)	Henry K_d (L/kg)	Langmuir M_{sat} (mg/kg)	Langmuir K_L (L/mg)	Coefficient of Determination R^2
SO_4^{2-}	6 h	1–10	50–600	530	0.06	-	-	0.604
	1 d	1–10	60–500	580	0.15	-	-	0.884
	7 d	1–300	2–500	570	0.03	-	-	0.153
	28 d	1–300	2–600	570	0.01	-	-	0.001
Na	10 min	3–300	0.6–50	190	0.96	-	-	0.988
	6 h	1–10	20–150	210	0.42	-	-	0.999
	1 d	1–10	20–140	210	0.47	-	-	1.000
	7 d	1–10	20–160	210	0.34	-	-	1.000
	28 d	1–10	25–170	260	0.49	-	-	1.000
B	6 h	1–3	0.08–0.1	0.36	1.7	-	-	-
	1 d	1–3	0.1–0.2	0.51	1.8	-	-	-
	7 d	1–10	0.06–0.2	0.72	2.0	-	-	0.971
	28 d	1–10	0.06–0.3	0.72	1.8	-	-	0.993
Mg	10 min	3–300	0.4–8.0	120	-	140	0.34	0.996
	6 h	1–100	1.1–25	170	-	160	0.65	0.994
	1 d	1–100	1.2–25	170	-	160	0.50	0.999
	7 d	1–100	1.3–33	170	-	150	0.37	0.995
	28 d	1–100	1.4–41	160	-	140	0.18	0.999
F^-	10 min	10–100	0.07–0.5	7.9	6.9	-	-	0.999
	6 h	30–100	0.1–0.4	16	14	-	-	-
	1 d	30–100	0.2–0.4	18	14	-	-	-
	7 d	30–300	0.07–0.4	23	21	-	-	0.947
As	10 min	1–300	0.0001–0.003	0.043	13	-	-	0.961
	6 h	1–300	0.0006–0.006	0.25	-	0.3	480	0.995
	1 d	30–300	0.0009–0.003	0.47	-	0.7	520	0.997
	7 d	30–300	0.001–0.003	1.3	-	1.6	980	0.932
	28 d	100–300	0.001–0.002	2.9	-	3.6	1500	-

In case of F^- and As, $[C\text{-}t]$ and $[M_L\text{-}t]$ initially increased, but after a certain period then began to decrease. Such behaviors were also observed for Co, Cu, Mo, Ni, Sb, and Se. Lower

LS conditions (higher *C* conditions) appeared to result in the observed decrease occurring earlier, which might have been due to precipitation and/or coprecipitation in batch test conditions [27]. Changes in the pH of the eluate (Figure 2) could also have affected the behaviors of pH-sensitive substances. SO_4^{2-}, Na, Mg, and B would not be so sensitive to precipitation, coprecipitation, or adsorption around the pH ranges observed in Figure 2. To confirm this assumption, however, a successful pH-adjusted batch test would be necessary. As a result, most parts of the [*C-M_L*] relationships of F^- and As appeared irregular due to concentration decreases. Therefore, such ranges should be removed when estimating adsorption parameters. In Table 3, the parameters of F^- and As were approximated using the data obtained before the observed decreases in concentration.

LS-parallel tests should be conducted under moderately short contact time (neither too short, such as 10 min, nor too long such as, >7 d) to obtain adsorption parameters. A contact time between 6 h and 1 d seemed to deliver the best results. Higher liquid-solid ratios could better maintain adsorption control and small changes in pH, although the concentration in the eluate could be lower than the quantification limit.

Although there are few studies on adsorption parameters for PCMs, the K_d of soluble salts such as Na, K, SO_4^{2-}, and Cl^- are estimated to be small, as in Table 3; for contaminated soils, the K_d of SO_4^{2-}, Cl^-, Cu, and ΣPAH are calculated to be 0.55, 0.50, 0.60, and 120, respectively, by fitting the advection-dispersion model of column test data [17]. Similarly, for APC residues, the K_d of Na, K, and Cl^- were calculated to be 1.3, 0.83–1.3, and 0.55 from column test data [7]. In addition, Reference [7] performed an LS-parallel test in the LS_{batch} range of 5–500 L/kg and calculated the K_d of Na, K, and Cl^- as 0.20–1.0, 0.29–1.3, and 0.38–6.0, respectively, by fitting the advection-dispersion model as a function of LS_{cum}. However, the handling of physical parameters such as dispersion length and porosity is unclear, so calculations using the procedure proposed in this study are expected.

4.3. Effect of Colloids Passing the Filter

Colloids could have passed through the filter during the liquid-solid separation step, and so may have affected the measured concentrations in the eluate [28–30]. To confirm the effect of colloids, eluates obtained using a 0.45 μm pore size filter (0.45 MF) were refiltered using a 0.1 μm pore size filter. This procedure was applied to eluates using only a 6 h-mixing time.

As shown in Figure 4, in *LS* 1 and 3, refiltration did not affect Fe concentration significantly, but under larger *LS* conditions concentrations clearly decreased in following refiltration. This suggests that in smaller *LS* conditions, colloids of 0.45 μm or smaller were removed at the first filtration by the cake that formed on 0.45 MF, but in larger *LS* conditions, such cake did not foam sufficiently to remove the colloids. Subsequent refiltration using 0.1 MF was able to remove colloids of 0.1–0.45 mm, resulting in significant reductions of their concentrations in the eluates [28]. Furthermore, in *LS* 300 it may have been difficult to remove colloids of 0.1 μm or smaller during refiltration, as the resulting concentration was higher and the observed variation was larger than those of *LS* 100. Ti and Al showed similar trends, and Pb and Zn also fluctuated significantly; these substances should therefore be excluded from the evaluation of adsorption parameters.

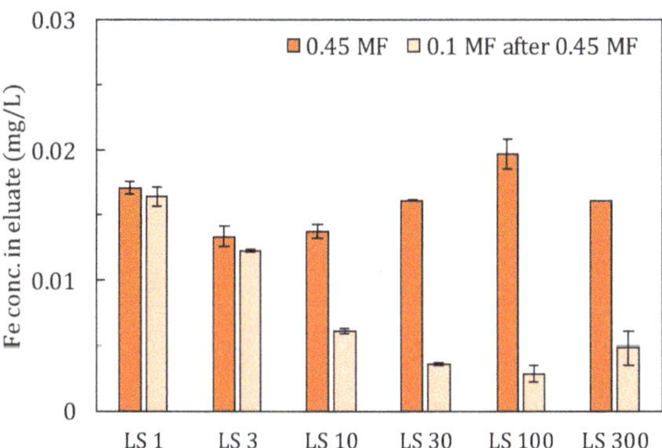

Figure 4. Concentration of Fe in filtrate using 0.45 μm membrane filter (MF) and in filtrate after re-filtration of 0.45 μm filtrate using the 0.1 μm membrane.

Care should be taken to properly remove colloids during the solid–liquid separation step, for example, by applying centrifugation with sufficient intensity and duration. Furthermore, considering the real environment, it would be necessary to develop a solid–liquid mixing method that minimizes the further generation of colloids through the abrasion of solids.

4.4. Reproductivity of Column Percolation Test by LS-Parallel Test

Figure 5 shows the changes in eluate concentration in the column percolation test, together with the results calculated using parameters from the LS-parallel test. In these calculations, the effective porosity, θ, and dispersion length, α, were estimated as 0.286 ($-$) and 0.0387 m, respectively, from the fitting of SO_4^{2-}. This was because K_d was almost zero and M_T did not increase over time during the LS-parallel test. In Figure 5, all parameters shown in Table 2 were used for calculations, because K_d and M_T certainly changed over time during not only the batch test, but also during column percolation tests (except for SO_4^{2-}). Therefore, the horizontal axis in Figure 5 shows the elapsed time. It should be noted that 1 day is almost equivalent to 1 LS_{cum}, as the sample volume of the column was 289 g and the water flow rate was 288 mL/d.

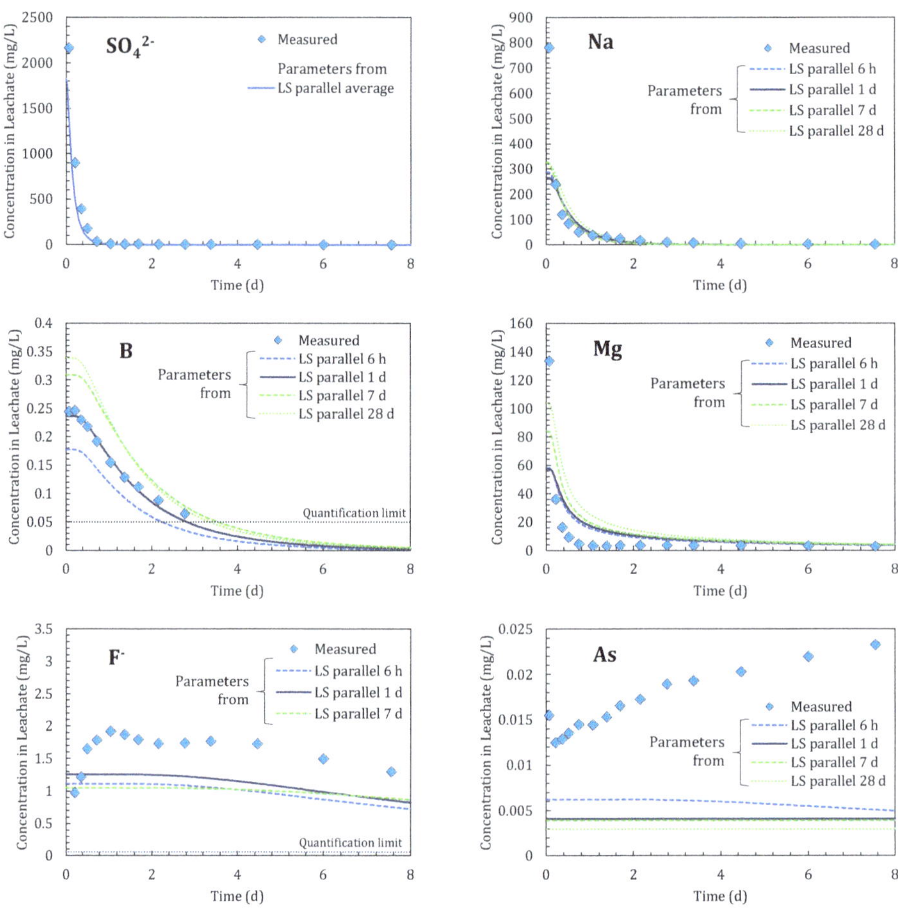

Figure 5. Column percolation test results and calculation results using parameters obtained in the LS-parallel tests.

The maximum concentration is one of the most important parameters in evaluating the environmental impact of a given substance [31]. In the advection-dispersion model, the eluate concentration decreased monotonically. Thus, the first eluate represented the maximum concentration, which can be calculated using Equation (6). In Figure 5, SO_4^{2-}, Na, B, and Mg showed their maximum concentrations in the first or the second eluate fractions. The calculated concentrations obtained by parameters from the LS-parallel test after one day were 67%, 38%, 96%, and 44% of measured SO_4^{2-}, Na, B, and Mg values in the column test, respectively; this timing would provide the best comparison to the measured results because the initial run of the eluate took 16.3 h. For SO_4^{2-}, Na, and B the decrease curves showed good agreement with the measured values. Ba, Co, Cr, Cu, K, Rb, Ni, Se, Si, Sr, and Zn also showed typical monotonic decreases. However, in the LS-parallel test, the concentrations of Co, Cr, Cu, Rb, Ni, Se, and Zn significantly fluctuated with time, probably due to the effects of colloids. In the column test, colloids originally contained in the soil would have been unlikely to spill out due to self-clogging. Furthermore, soil particles were not eroded during percolation. These results suggest that if the water-mixing procedure in the LS-parallel test were to be improved, the adsorption parameters of these substances could be obtained.

The eluate concentration of F^- increased from 0.97 to 1.9 mg/L, and then decreased gradually. Al, Mo, Sb, and Ti also showed peaks during the midpoint of their runs, and then decreased. The As concentration continued to rise until the end of the column test period. Fe and Mn also continued to rise from the first eluate to the end. However, the current advection-dispersion model could not simulate such partially or totally increasing trends because it only assumes an adsorption–desorption equilibrium. As seen in the LS-parallel tests conducted with different water-contact times, in the column test it could also be expected that dissolution and/or intraparticle diffusion was occurring. The advection-dispersion model should thus be developed to further consider these mechanisms.

4.5. Further Applications of LS-Parallel Test

The theory of obtaining the adsorption–desorption parameters of PCMs through the LS-parallel test is quite clear. Comparing the LS-parallel test and column test results confirmed that this theory is applicable to substances in which the adsorption–desorption equilibrium is dominant (SO_4^{2-}, Na, Mg, etc.). It is presumed that precipitation hardly occurred for these substances, and that the adsorption–desorption equilibrium was not significantly affected in the observed pH range.

The advantages of the proposed method are: (1) the parallel batch test is simpler and easier to conduct than the column percolation test, which means that the LS-parallel test could replace column percolation test to obtain the parameters; (2) as contaminants are released from the material itself, further addition of the contaminants is unnecessary, and the real chemical species from the PCM can be considered; and (3) the advection–related parameters are obtained at once for every substance released from the PCM, assuming the substance exhibits precipitation-free and adsorptive characters.

Regarding its further applications, the LS-parallel test can easily evaluate changes in the adsorption parameters of PCMs through specific conditions. For example, the mechanism could be investigated by LS-parallel test whether M_T decreases or K_d increases when a contaminant in a PCM is insolubilized with chemicals. The long-term stability of substances under weathering conditions could also be evaluated by analyzing a fewer amount of PCMs than would be required if using column percolation tests.

5. Conclusions

This study proposed a technique to determine the adsorption parameters using an LS-parallel test for materials that release contaminants. LS-parallel and column percolation tests were performed on a sandy soil derived from marine sediment.

In the LS-parallel test, adsorption parameters were successfully obtained if the substance was under adsorption control. Combining batch test conditions with the liquid-solid ratio and water contact time permits the investigation of leaching mechanisms from the inside of the solid phase, and of precipitation or coprecipitation reactions in the liquid phase after leaching. In conclusion, LS-parallel tests should be conducted under moderately short contact time; between 6 h and 1 d seemed to deliver the best results. Higher liquid-solid ratios could better maintain adsorption control and small changes in pH, although the concentration in the eluate could be lower than the quantification limit.

In the column percolation test, the behavior of SO_4^{2-} and B coincided well with the advection-dispersion model using adsorption parameters obtained from the LS-parallel test. However, for other substances, the initial concentration and release curves did not always fit, probably because these substances were continuously released from the soil, or because colloids affected differently between the LS-parallel tests and the column percolation test. These results suggest that improvements are necessary in the mixing method of the LS-parallel test procedure to suppress the release of colloids from the solids. Additionally, the advection-dispersion model should further be improved to express the release of substances from inside the solids.

As observed in the experiments, the adsorption parameters can change with time under different exposure conditions. Since the LS-parallel test is easy to apply, the combination of the LS-parallel test and the analysis method proposed in this study can be a very powerful tool to evaluate the changes in adsorption parameters and, moreover, the impact of PCMs on the environment.

Author Contributions: Conceptualization, H.S. and K.I.; methodology, H.S and H.I.; investigation, J.T. and M.N.; writing, H.S.; project administration, H.S.; funding acquisition, H.S. All authors have read and agreed to the published version of the manuscript.

Funding: This research was funded by JSPS KAKENHI Grant Numbers 19K04606.

Institutional Review Board Statement: Not applicable.

Informed Consent Statement: Not applicable.

Data Availability Statement: Data sharing is not applicable to this article.

Acknowledgments: In this section, you can acknowledge any support given, which is not covered by the author contribution or funding sections. This may include administrative and technical support, or donations in kind (e.g., materials used for experiments).

Conflicts of Interest: The authors declare no conflict of interest.

References

1. ISO 21268–1:2019(E). *Soil Quality—Leaching Procedures for Subsequent Chemical and Ecotoxicological Testing of Soil and Soil Materials—Part 1: Batch Test Using a Liquid to Solid Ratio of 2 L/kg Dry Matter*; International Standardization Organization: Geneva, Switzreland, 2019.
2. ISO 21268–2:2019(E). *Soil Quality—Leaching Procedures for Subsequent Chemical and Ecotoxicological Testing of Soil and Soil Materials—Part 2: Batch Test Using a Liquid to Solid Ratio of 10 L/kg Dry Matter*; International Standardization Organization: Geneva, Switzreland, 2019.
3. ISO 21268–3:2019(E). *Soil Quality—Leaching Procedures for Subsequent Chemical and Ecotoxicological Testing of Soil and Soil Materials—Part 3: Up-Flow Percolation Test*; International Standardization Organization: Geneva, Switzreland, 2019.
4. Kalbe, U.; Berger, W.; Eckardt, E.; Simon, F. Evaluation of leaching and extraction procedures for soil and waste. *Waste Manag.* **2008**, *28*, 1027–1038. [CrossRef]
5. Maszkowska, J.; Kołodziejska, M.; Białk-Bielińska, A.; Mrozik, W.; Kumirska, J.; Stepnowski, P.; Palavinskas, R.; Krüger, O.; Kalbe, U. Column and batch tests of sulfonamide leaching from different types of soil. *J. Hazard. Mater.* **2013**, *260*, 468–474. [CrossRef]
6. Meza, S.; Garrabants, A.; van der Sloot, H.; Kosson, D. Comparison of the release of constituents from granular materials under batch and column testing. *Waste Manag.* **2008**, *28*, 1853–1867. [CrossRef]
7. Quina, M.; Bordado, J.; Quinta-Ferreira, Q. Percolation and batch leaching tests to assess release of inorganic pollutants from municipal solid waste incinerator residues. *Waste Manag.* **2011**, *31*, 236–245. [CrossRef]
8. Schuwirth, N.; Hofmann, T. Comparability of and alternatives to leaching tests for the assessment of the emission of inorganic soil contamination. *J. Soils Sediments* **2006**, *6*, 102–112. [CrossRef]
9. Guyonnet, D.; Bodénan, F.; Brons-Laot, G.; Burnol, A.; Chateau, L.; Crest, M.; Méhu, J.; Moszkowicz, P.; Piantone, P. Multiple-scale dynamic leaching of municipal solid waste incineration ash. *Waste Manag.* **2008**, *28*, 1963–1976. [CrossRef]
10. Hjelmar, O.; van der Sloot, H.; Guyonnet, D.; Rietra, R.; Brun, A. Development of acceptance criteria for landfilling of waste: An approach based on impact modelling and scenario calculations. In *Sardinia 2001, Proceedings of the Eigth International Waste Management and Landfill Symposium, S. Margharita di Pula, Cagliari*; Christensen, T., Cossu, R., Stegmann, R., Eds.; CISA: Padova, Italy, 2001; Volume 3, pp. 711–721.
11. Lee, H.; Yu, G.; Choi, Y.; Jho, E.H.; Nam, K. Long-term leaching prediction of constituents in coal bottom ash used as a structural fill material. *J. Soils Sediments* **2017**, *17*, 2742–2751. [CrossRef]
12. Verginelli, I.; Baciocchi, R. Role of natural attenuation in modeling the leaching of contaminants in the risk analysis framework. *J. Environ. Manag.* **2013**, *114*, 395–403. [CrossRef] [PubMed]
13. Tiruta-Barna, L.; Barna, R.; Moszkowicz, P.; Hae-Ryong, B. Distributed mass transfer rate for modelling the leaching of porous granular materials containing soluble pollutants. *Chem. Eng. Sci.* **2000**, *55*, 1257–1267.
14. Kim, K.; Yang, W.; Nam, K.; Choe, J.K.; Cheong, J.; Choi, Y. Prediction of long-term heavy metal leaching from dredged marine sediment applied inland as a construction material. *Environ. Sci. Pollut. Res.* **2018**, *25*, 27352–27361. [CrossRef] [PubMed]
15. Parkhurst, D.L.; Appelo, C.A.J. User's guide to PHREEQC (version 2)—A computer program for speciation, batch-reaction, one-dimensional transport, and inverse geochemical calculation. *U.S. Geol. Surv. Water Resour. Investig.* **1999**, *312*. [CrossRef]
16. Astrup, T.; Rosenblad, C.; Trapp, S.; Christensen, T.H. Chromium release from waste incineration air-pollution-control residues. *Environ. Sci. Technol.* **2005**, *39*, 3321–3329. [CrossRef] [PubMed]
17. Grathowhl, P.; Susset, B. Comparison of percolation to batch and sequential leaching tests: Theory and data. *Waste Manag.* **2009**, *29*, 2681–2688. [CrossRef] [PubMed]
18. Ogata, A.; Banks, R.G. *A solution of the Differential Equation of Longitudinal Dispersion in Porous Media*; US Government Printing Office: Washington, DC, USA, 1961.
19. OECD Guideline for the Testing of Chemicals. Adsorption—Desorption Using a Batch Equilibrium Method. In *Adsorption—Desorption Using a Batch Equilibrium Method*; The Organisation for Economic Co-operation and Development: Paris, France, 2000.

20. Miller, C.; Weber, W. Sorption of Hydrophobic Organic Pollutants in Saturated Soil System. *J. Contam. Hydrol.* **1986**, *1*, 243–261. [CrossRef]
21. Fetter, C.; Boving, T.; Kreamer, D. *Contaminant Hydrology*, 2nd ed.; Prentice Hall: Hoboken, NJ, USA, 2017; ISBN 978-1478632795.
22. US EPA. *Method 1316, Liquid-Solid Partitioning as a Function of Liquid-T-Solid Ratio in Solid Materials Using a Parallel Batch Procedure, Revision 2017*; United States Environmental Protection Agency: Washington, DC, USA, 2017.
23. MOEJ. *Soil Environmental Standard*; Ministry of the Environment: Tokyo, Japan, 1991.
24. Grathwohl, P. On equilibration of pore water in column leaching tests. *Waste Manag.* **2014**, *34*, 908–918. [CrossRef] [PubMed]
25. Finkel, M.; Grathwohl, P. Impact of pre-equilibrium and diffusion limited release kinetics on effluent concentration in column leaching tests: Insights from numerical simulations. *Waste Manag.* **2017**, *63*, 58–73. [CrossRef]
26. Sakanakura, H. Diffusion test of 20 kinds of waste molten slags and competitive Materials. *J. Mater. Cycle Waste Manag.* **2005**, *7*, 71–77. [CrossRef]
27. Sakanakura, H.; Tanaka, N. Leaching mechanisms of waste molten slag in batch type experiments. *J. Jpn. Soc. Waste Manag. Experts* **1998**, *9*, 11–19. (In Japanese) [CrossRef]
28. Imoto, Y.; Yasutaka, T.; Someya, M.; Higashino, K. Influence of solid-liquid separation method parameters employed in soil leaching tests on apparent metal concentration. *Sci. Total Environ.* **2018**, *624*, 96–105. [CrossRef]
29. Someya, M.; Higashino, K.; Imoto, Y.; Sakanakura, H.; Yasutaka, T. Effects of membrane filter material and pore size on turbidity and hazardous element concentrations in soil batch leaching tests. *Chemosphere* **2021**, *265*, 128981. [CrossRef]
30. Yasutaka, T.; Imoto, Y.; Kurosawa, A.; Someya, M.; Higashino, K.; Kalbe, U.; Sakanakura, H. Effects of colloidal particles on the results and reproducibility of batch leaching tests for heavy metal-contaminated soil. *Soils Found.* **2017**, *57*, 861–871. [CrossRef]
31. Guyonnet, D. Discussion—Comparison of percolation test to batch and sequential leaching tests: Theory and data. *Waste Manag.* **2010**, *30*, 1746–1747. [CrossRef]

Article

Long-Term Leaching Behavior of Organic and Inorganic Pollutants after Wet Processing of Solid Waste Materials

Maria Prieto-Espinoza [1,2], **Bernd Susset** [1] **and Peter Grathwohl** [1,*]

[1] Center for Applied Geosciences, University of Tübingen, Schnarrenbergstr. 94-96, 72076 Tübingen, Germany; prietoespinoza@unistra.fr (M.P.-E.); bernd.susset@ifg.uni-tuebingen.de (B.S.)
[2] Institute Earth and Environment (ITES UMR 7063), University of Strasbourg, CNRS/EOST, 5 Rue Descartes, 67084 Strasbourg, France
* Correspondence: grathwohl@uni-tuebingen.de

Abstract: The recycling of mineral materials is a sustainable and economical approach for reducing solid waste and saving primary resources. However, their reuse may pose potential risks of groundwater contamination, which may result from the leaching of organic and inorganic substances into water that percolates the solid waste. In this study, column leaching tests were used to investigate the short- and long-term leaching behavior of "salts", "metals", and organic pollutants such as PAHs and herbicides from different grain size fractions of construction & demolition waste (CDW) and railway ballast (RB) after a novel treatment process. Specifically, silt, sand and gravel fractions obtained after a sequential crushing, sieving, and washing process ("wet-processing") of very heterogeneous input materials are compared with respect to residual contamination, potentially limiting their recycling. Concentrations in solid fractions and aqueous leachate were evaluated according to threshold values for groundwater protection to identify relevant substances and to classify materials obtained for recycling purposes according to limit values. For that, the upcoming German recycling degree was applied for the first time. Very good agreement was observed between short and extensive column tests, demonstrating that concentrations at L/S 2 ratios are suitable for quality control of recycling materials. Different solutes showed a characteristic leaching behavior such as the rapid decrease in "salts", e.g., SO_4^{2-} and Cl^-, from all solid fractions, and a slower decrease in metals and PAHs in the sand and silt fractions. Only the gravel fraction, however, showed concentrations of potential pollutants low enough for an unlimited re-use as recycling material in open technical applications. Sand fractions may only be re-used as recycling material in isolated or semi-isolated scenarios. Leaching from heterogeneous input materials proved harder to predict for all compounds. Overall, column leaching tests proved useful for (i) initial characterization of the mineral recycling materials, and (ii) continuous internal (factory control) and external quality control within the upcoming German recycling decree. Results from such studies may be used to optimize the treatment of mixed solid waste since they provide rapid insight in residual pollution of material fractions and their leaching behavior.

Keywords: mineral recycling material; leaching test; heterogeneity; compliance testing

Citation: Prieto-Espinoza, M.; Susset, B.; Grathwohl, P. Long-Term Leaching Behavior of Organic and Inorganic Pollutants after Wet Processing of Solid Waste Materials. *Materials* **2022**, *15*, 858. https://doi.org/10.3390/ma15030858

Academic Editor: Andrea Petrella

Received: 12 December 2021
Accepted: 19 January 2022
Published: 23 January 2022

1. Introduction

The largest solid waste stream in Germany with an annual volume of more than 275 million tones comprises 32% of construction and demolition waste (CDW), of which about 90% is reused [1]. Recycling mineral waste has a lot of advantages in terms of sustainability and economical aspects. To increase recycling potential, more and more companies start to treat excavated soil-stone mixtures, and demolition waste or railway ballast, combining crushing with dry and wet sieving and washing processes (wet processing). However, the reuse of mineral materials may pose potential risks of environmental pollution, resulting from leaching of organic and inorganic substances into percolating water and, ultimately, into groundwater [2–5].

In this context, the risk concerning potential contaminants in such materials must be addressed by their leaching potential into water rather than assessing total concentrations in the solid phase. Column leaching tests and batch shaking tests are frequently used to assess the transfer of contaminants into water [3,6,7]. Column leaching tests are preferred because they allow for assessing time-dependent behavior and simulating the flow of water through solid materials closer to natural conditions [8,9]. In Germany, the standard DIN 19528 (2009) is used for examining the leaching potential of inorganic and organic substances from solid materials [10]. Generally, "extensive column tests" are performed to characterize and evaluate the long-term leaching behavior of contaminants in which eluates are collected at different liquid-to-solid (L/S) ratios (e.g., 0.3, 1, 2, 4 and 10 L/kg). L/S ratios represent the time after which a certain volume of water has percolated through the solid material in the column (in L/kg dry matter). For compliance testing, "short column tests" may be employed that provide results of cumulative concentrations at one fixed L/S of 2 L/kg, which then are compared with threshold values set by special regulations [5,9,11]. Initially, equilibrium concentrations are often observed in column eluates [12]. The decrease in concentrations with increasing L/S ratios (or time) may be due to depletion of highly soluble substances, or a shift to non-equilibrium conditions because of mass transfer limitations (e.g., slow intraparticle diffusion) indicated by an extended tailing of the solute concentrations in the leachate [13]. The shift between equilibrium to non-equilibrium conditions may depend on initial conditions [12], flow velocities, grain-sizes, sorption capacity and contaminant release kinetics [8,9,12]. Typically, three basic leaching scenarios can be described for (i) fast leaching substances such as "salts" (e.g., sodium, potassium, chloride), where a rapid decline in concentrations in column effluents is observed (at L/S < 2 L/kg); (ii) intermediate compounds such as some metals, where mass release is governed by leaching parameters such as pH, redox conditions, ionic strength and DOC-complexations [14–17], and; (iii) for strongly sorbing compounds such as PAHs, where equilibrium concentrations prevail over extended periods of time [3].

Column leaching tests are thereby proposed as a common procedure for the evaluation of environmental qualities of solid waste recycling materials [5,9,18]. In Germany, the upcoming recycling directive [19] is based on improved methods for groundwater risk assessment to derive a new regulatory framework for the reuse of solid waste materials. For a given substance, the concentration level avoiding any significant alteration of the chemical status of groundwater is defined as the "insignificance threshold" concentration ("GFS", in German "Geringfügigkeitsschwelle") [20]. The GFS values are based on eco- and human toxicological tests and are not intended to set a quality goal for groundwater, but rather reflect a groundwater status unaffected by human activity [20].

Concentration limits at L/S 2 eluates are set depending on the type of mineral recycling material (e.g., CDW, RB, steel slag etc.), the type of technical application (open technical applications and isolated or semi-isolated technical applications protected from seepage water), the distance to the groundwater table, and the soil characteristics of the underground [19]. If the quality of the recycling material shows high variability, different "material classes" are defined by different sets of limit concentrations in eluates from the same mineral recycling material, so-called "material values" (e.g., material class RC–1, highest quality) [11]. The comparison of concentrations in L/S 2 eluates with GFS values and/or limit values of "material classes" will ultimately define permissible applications of the mineral recycling material. This concept is implemented in Germany within the upcoming recycling directive [19] with a quality assurance system and material-specific testing programs, where the quality of mineral recycling materials is assessed based on extensive column percolation tests [10]. Furthermore, short-term column percolation tests [10] are performed for internal and continuous external quality control [11].

In this study, short and extensive column tests were performed to examine the leaching of organic and inorganic substances from railway ballast (RB), and construction and demolition waste (CDW), which both underwent a sophisticated washing and grain size separation process, so-called wet processing. The input material ('In', highly heterogeneous

RB or CDW) as well as its separated grain-size fractions such as silt ('U', <0.063 mm), sand ('S', 0–2 mm) and gravel ('G', 2–8 mm) were examined. The aim of this study was (i) to characterize comprehensively the leaching behavior of organic and inorganic contaminants from input and recycled material fractions (i.e., RB and CDW); (ii) to assess the quality of the different grain size fractions with respect to threshold values for potential risks of groundwater contamination and material values; (iii) to compare results of the short and extensive tests, and; (iv) to examine the long-term leaching behavior of the investigated substances.

2. Materials and Methods

2.1. Materials

In total, 3 sets of construction and demolition waste (CDW1, CDW2 and CDW3) and 1 set of railway ballast (RB) were examined to reflect different sources of recycling materials and their variability. Samples were collected in December 2017 (RB and CDW1), March 2018 (CDW2) and May 2018 (CDW3). The original CDW material was a mixture of soil, demolition and construction waste. At the recycling plant, input materials were crushed and "cleaned" in a complex washing process. All solid waste materials were separated into different grain size fractions (Figure 1). Large fragments were sieved into different gravel-size fractions: 32–50 mm, 16–32 mm, 8–16 mm and 2–8 mm. Smaller particles in suspension underwent a centrifugation process, wherein the sand fraction (0–2 mm) was separated and further washed. Finally, the silt fraction (<0.063 mm) was separated from the suspension by centrifugation with the addition of flocculants and polymers. The input material as well as the silt, sand and gravel (2–8 mm) fractions were used for the column tests without further sieving or crushing (Figure 1). In addition, the aqueous solution (referred to as "washing water") used for separation and cleaning of the solid fractions at the recycling plant was analyzed.

Figure 1. Solid waste fractions of construction and demolition waste (CDW3) as received from the recycling plant: input material (In) as well as silt (U), sand (S) and gravel (G) fractions obtained after wet-treatment.

2.2. Experimental Setup

Prior to the column leaching tests, the gravimetric water content (w) of the solid fractions was determined by weighting and drying the wet material in an oven for 24 h at 105 °C. The dried material was further used to determine its volume and grain density using a gas pycnometer (micromeritrics/AccuPyc 1330). Quartz sand was used as a reference standard material (density: 2.65 g/cm^3). The final values were set by measuring 10 times the same material until reaching a standard deviation of less than 0.005 g/cm^3. In order to increase the permeability and to prevent mobilization of fine particles, the input material and silt fraction were mixed with clean quartz sand, as suggested elsewhere (e.g., [5]). From the input material, only particle fractions smaller than 32 mm were used (see Figure 1).

Column leaching tests were carried out according to the German standard DIN 19528 (2009) in a dark laboratory at a constant temperature of 20 °C [10]. The DIN 19528 has been validated for investigations on long-term leaching of salts and heavy metals from incineration bottom ash [9,21], comparisons with batch and lysimeters tests [3], and the effect of contact time in column percolation tests [9]. Furthermore, Lin et al. [5] recently proposed an optimization of the short column percolation tests (at L/S 2 eluates; DIN 19528,) by approving the use of sand admixtures in coarse grain fractions.

In total, 27 short column tests were performed for all fractions of the solid waste (i.e., RB, CDW1, CDW2 and CDW3), including 3 controls containing only a 3 cm layer of quartz sand. Short column eluates were collected until a L/S of 2 L/kg and analyzed for salts, metals, and organic substances such as BTEX, PCBs, herbicides and PAHs. In addition, 6 extensive column leaching tests were performed for sample CDW3 to examine the long-term leaching behavior of contaminants at L/S of 0.1, 0.3, 1, 2, 4 and 10 L/kg, including 1 control column. For the silt fraction, the earliest column eluate was collected at L/S 0.5 ratio. Glass columns with an inner diameter of 5 cm and a length of 30 cm were used for the sand and gravel fractions, whereas glass columns with an inner diameter of 7 cm were used for the input material and the silt fraction previously mixed with quartz sand. Before packing the samples into the columns, a 1 cm layer of quartz sand was placed at the bottom for better distribution of the water flow through the column inlet. A second quartz sand layer was placed at the top, at a filling height of about 28 cm, to prevent the release of fine particles. Additionally, glass wool was placed at the inlet and outlet openings. Teflon tubes were connected to the column inlets and the clean water reservoir consisting of a 50 L glass bottle containing Milli-Q water. The flow rate was set using a peristaltic pump (IPC 8, ISMATEC), and adjusted to allow a contact time of 5 h during the leaching tests. The initial flooding of the columns with clean water lasted approximately 2 h. Column eluates were collected in amber glass bottles at the corresponding L/S ratios and stored at a temperature of 20 °C until further analysis. Given that biodegradation and volatilization of organic compounds can occur, columns for PAHs were run in parallel—one for the analysis of ions and metals and the other for PAHs only. For PAH analysis, the collecting bottles previously contained 10 mL of cyclohexane (to avoid any biodegradation during sampling and storage), and an internal standard (10 μL, 5 perdeuterated PAHs according to DIN 38407–39 in toluene, each perdeuterated PAH 20 ng/μL).

2.3. Analytics

2.3.1. Turbidity, Electrical Conductivity, Ion Chromatography and DOC

All column eluates were analyzed for turbidity, electrical conductivity (EC, HACH LANGE), and pH (inoLab® pH 7110, WTW) within the first 2 h after collection. After filtration at 0.45 μm, major ions were analyzed by ion chromatography (DIONEX, DX-120). Total Organic Carbon (TOC) and Dissolved Organic Carbon (DOC) were measured via a TOC analyzer (Elementar, Vario TOC).

2.3.2. Metals, Phenols, EOX, PCBs, PHCs, Cyanide and Herbicides

Solid concentrations of heavy metals, petroleum hydrocarbons (PHCs, C10-C40), polychlorinated biphenyls (PCB), phenols, extractable organic halides (EOX) and cyanide were analyzed at the Gewerbliches Institut für Umweltanalytik GmbH (Industrial Institute for Environmental Analysis, Teningen, Germany). PHCs, PCBs, phenols, EOX and cyanide concentrations were measured by gas chromatography—tandem mass spectrometry (GC-MS/MS, Agilent).

Aqueous column eluates (aliquots of 20 mL) were filtered at 0.45 μm and acidified (HNO_3) prior to the analysis of heavy metals via inductively coupled plasma mass spectrometry (ICP-MS, Agilent). The herbicides atrazine, simazine, bromacil, desethylatrazine, hexazinone, dimefuron, diuron, flumioxazin, thiazafluron and ethidimuron were measured in 20 mL aliquots of column eluates by liquid chromatography—tandem mass spectrometry (LC-MS/MS, Agilent). All substances were measured according to protocols described in [20].

2.3.3. Polycyclic Aromatic Hydrocarbons (PAHs)

PAHs were measured in both solids and column eluates using GC-MS (Agilent/HP 5973). For solid concentrations, PAHs were extracted by Accelerated Solvent Extraction (ASE 300 DIONEX, Thermo Scientific), a technique that utilizes organic solvents at high temperature and pressures. Approximately 40 g of the solid samples were placed in the sample cell, along with 47 mm diameter filters on both ends of the extraction cell. Samples were extracted sequentially first with acetone and then with toluene (50 mL extracts) at a pressure of 100 bars and 100 °C [22]. Aqueous column eluates were extracted by liquid-liquid extraction. The bottles containing the column eluates along with 10 mL of cyclohexane (CH) and 10 µL of internal standard (10 µL, 5 perdeuterated PAHs according to DIN 38407–39, in toluene, each perdeuterated PAH 20 ng/µL) were horizontally shaken for 1 h (at 150 rpm), and subsequently filled with Milli-Q water until the solvent reached the bottleneck. The bottles were left overnight, and cyclohexane extracts were retrieved and treated with anhydrous sodium sulfate. All extracts were reduced to 200 µL by means of a nitrogen flow.

3. Results and Discussion

3.1. Pollutant Screening in Solid Fractions

Prior to the column tests, solids were analyzed for determination of initial concentrations and characterization of the materials according to precautionary values for soils [19]. Concentrations of PCBs were present in some solid fractions, but did not exceed precautionary values for soils (Appendix A, Table A1) [19]. Phenols, cyanide and EOX were not detected in the solid samples, except for the input material and the sand fraction of CDW2 with EOX concentrations of <0.5 µg/kg. PHCs (C10-C40) were present in concentrations below the limit value of material class BM-0* (<300 mg/kg; Table A1). For EOX and PHCs no precautionary values exist for soils [19]; therefore, limit values for material classes are used (Table A1). Metals exceeding precautionary values were detected in silt fractions of both RB and CDW materials, e.g., As (>10 mg/kg), Pb (>40 mg/kg), Cu (>20 mg/kg) and Zn (>60 mg/kg), while both, the silt and sand fractions exceeded the limit of solid concentrations for Cd (>0.4 mg/kg), Cr (>30 mg/kg) and Ni (>15 mg/kg) (Figure 2 and Table A1). Further metal solid concentrations are given in Table A1 for information.

Globally, the silt fraction was the most contaminated particularly with Cr and Cu in RB, and with the 16 PAHs in CDW samples (>3 mg/kg; Figure 2). The variability of solid concentrations in the different CDW samples is low for metals but high for PAHs. Moreover, RB shows different solid concentration patterns than CDW. The high variability of solid concentrations demonstrates that solid waste materials should preferably be examined as individual samples and according to grain size for intended use prior to recycling applications. While the washing process of the solid material into different grain size fractions should be considered as an important step for the separation of fractions suitable for recycling applications, the treatment is obviously not sufficient to clean up the materials to reach precautionary values for the sand and silt fractions. Only the gravel fraction reached concentrations below precautionary values (Figure 2), with the exception of PAHs in the gravel fraction of CDW3. The solid concentrations of PAHs in the sand and silt fraction of CDW 2 exceed the material value of RC-3 (20 mg/kg), and based on this, it cannot be reused and would have to be landfilled.

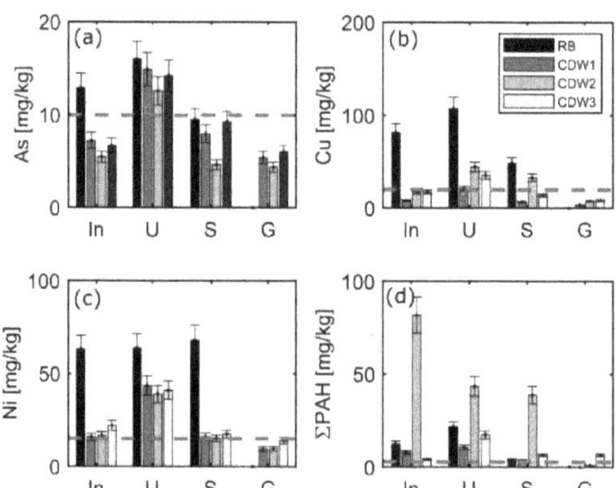

Figure 2. Concentrations of (**a**) As, (**b**) Cu, (**c**) Ni, and (**d**) the sum of the 16PAHs in solids: Input material (In), silt (U), sand (S) and gravel (G) of railway ballast (RB) and three sets of construction and demolition waste (CDW); input represents the material prior to separation into the different fractions (silt, sand and gravel), and the red dashed lines indicate precautionary values for sandy soils [19]. Error bars represent uncertainties in measurements.

3.2. Contaminant Concentrations in Eluates of Short Leaching Tests and Washing Water

Column eluates were examined at L/S 2 of the recycling materials RB, CDW1, CDW2 and CDW3 from 4 different solid fractions: input material (In), silt (U), sand (S) and gravel (G); column parameters are listed in Table 1. Further detailed concentrations are listed in the Supplementary Materials. The gravel fractions of both RB and CDW materials showed the lowest concentrations in the leachates, without exceeding GFS values, except for V (> 4 µg/L) and herbicides in RB (Appendix A, Table A2 and Supplementary Materials). Herbicides were only detected in RB (Table A2), where sand was the most contaminated fraction exceeding GFS values (>0.1 µg/L per herbicide). In general, leachates from CDW materials showed higher concentrations than those from RB, except for As and Mo (Figure 3). SO_4^{2-} showed the highest concentrations up to 270 mg/L, followed by Cl^- and NO_3^{2-}. Furthermore, the highest eluate concentrations at L/S 2 were observed in silt fractions of CDW, followed by the input material, and sand and gravel fractions (Figure 3). Overall, the gravel fractions of RB and CDW materials proved to be the least contaminated, and thus suitable for a free re-use as recycling material in open technical applications.

Concentrations of salts, metals and PAHs were highest in column eluates of CDW3, particularly in U and S fractions (Figure 3). Notably, the concentrations of some salts and metals increased in U and S fractions compared to In, indicating a possible redistribution and accumulation of contaminants in the fine-grained fractions during the washing process. While concentrations in the solids of silt fractions were 2–3 times higher than in the sand fraction (see Figure 2), leaching at L/S 2 resulted in much lower concentrations in U compared to S in most cases, suggesting that metals bind stronger to finer particles (Figure 3) [23]. From the PAHs, phenanthrene, fluoranthene and pyrene showed the highest concentrations in column eluates of the different solid fractions (Figure 3). The sum of the 16 PAHs exceeded the GFS values (>0.2 µg/L) in all column eluates except those of the gravel fractions (Figure 3). The input material (In) shows concentrations that fall in between those observed for U, S and G fractions (Figure 3). The results obtained for CDW reflects the impact of material heterogeneity on contaminant leaching [24], which is highly variable in all cases.

Table 1. Column parameters of short and extensive leaching tests of railway ballast (RB) and three sets of construction & demolition waste (CDW) materials. In: input material, U: silt fraction, S: sand fraction and G: gravel fraction.

Parameter	RB				CDW1				CDW2				CDW3				CDW3 Extensive Tests [a]		
	In [d]	U [e]	S	G	In	U	S	G	In	U	S	G	In	U	S	G	In	U	S
Water content [%] [b]	3.13	31.3	10.6	1.8	8.6	38.8	11.9	2.51	10.3	36.11	8.08	1.97	7.43	35.37	12.06	4.57	15	22	12.06
Dry sample [g]	1615	251	664	914	2417	182	676	882	1454	199	784	841	1326	193	873	840	927	172	839
Quartz sand [g]	1987	2225	-	-	2656	2208	-	-	2052	2192	-	-	2040	2208	-	-	1632	2507	-
Filling volume [cm^3] [b]	1193	1078	550	560	1116	1155	530	550	553	1116	550	530	550	1193	550	540	1155	1116	550
Porosity [%] [b]	51	48	56	41	30	40	52	41	31	38	46	41	34	40	41	42	34	41	44
Flow velocity [mL./min]	1.96	1.03	0.79	0.85	1.11	1.52	0.84	0.41	0.54	1.5	0.83	0.74	0.62	1.55	0.74	0.72	1.30	1.48	0.80
TOC [mg C/g] [c]	25.1	11.5	0.60	-	9.61	5.14	1.42	-	13.4	5.59	2.62	-	13.7	4.42	1.00	-	(see CDW3)		

[a] Extensive leaching tests of material CDW3 only. [b] For the input material (In) and silt fraction (U), values represent the mixture of sample with dry quartz sand. [c] Total organic carbon. TOC was not measured in the gravel fraction due to expected insignificant organic carbon content. [d] Column parameters of the input material (In) are related to the columns performed for ions and metals analyses. [e] Column parameters of the silt fraction (U) are related to the columns performed for ions and metals analyses.

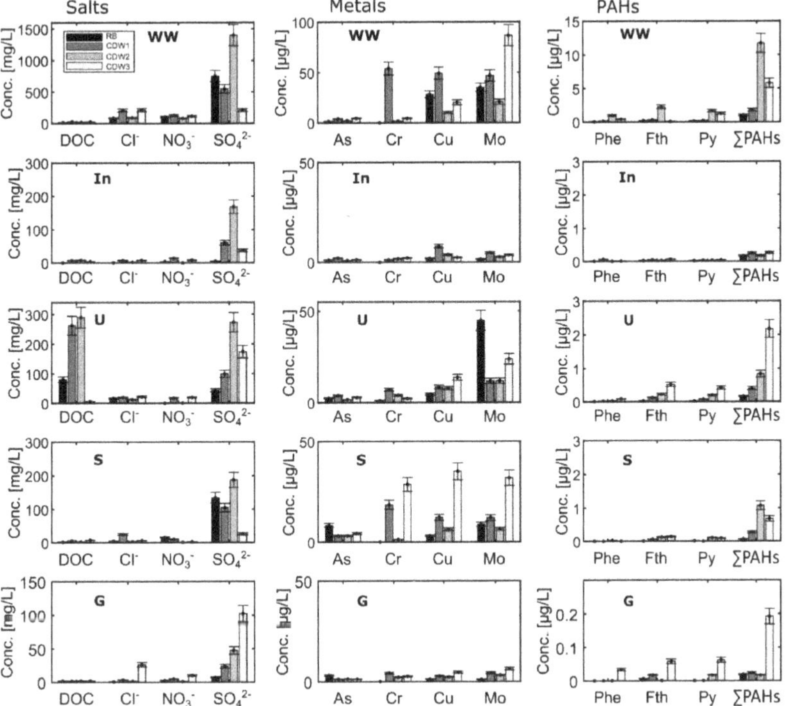

Figure 3. Concentrations of salts, metals and PAHs from short column tests at L/S 2 ratio of railway ballast (RB) and three sets of construction and demolition waste (CDW); WW represents concentrations in the washing water used for cleaning and separation of the input material (In) into the different size fractions silt (U), sand (S) and gravel (G). Error bars represent uncertainties in measurements.

In terms of usability of the solid waste material for different recycling purposes, the fine-solid fractions of RB and CDW are not suited for specific applications in technical constructions, which are sensitive with regard to groundwater protection (e.g., open applications with less than 1.5-m groundwater distance). These materials can be recycled only in isolated or semi-isolated applications with more than 1.5 m distance to the groundwater table and with suitable subsoil characteristics complying with the highest material classes (e.g., BM-F2 or BM-F3) [11,19]. As for the sand fraction of CDW, PAHs concentrations reached limit values for the best material class RC-1 of 4 µg/L. Overall, our results proved that the least contaminated fraction is the gravel (see Figures 2 and 3), which is suitable for free re-use in all open applications in all technical constructions. Sand and silt fractions can be re-used as recycling material only in isolated or semi-isolated technical applications. Concerning contamination with PAHs, limited applications are possible only if the solid concentration limits for PAHs are met additionally, which is not always the case (PAHs exceed limits for recycling even after wet processing, e.g., CDW-2).

Solute concentrations were also measured in the washing water (WW, Figure 3) used during the separation of the solid materials into different grain-size fractions (i.e., silt, sand and gravel) at the recycling plant. The washing water showed concentrations of metals such as As, Cr, Cu and Mo up to 4.5 µg/L, 54 µg/L, 50 µg/L and 86 µg/L, respectively. The most dominant anions were Cl^- and SO_4^{2-} with concentrations up to 212 mg/L and 550 mg/L, respectively. Aqueous concentrations of the sum of the 16 PAHs in WW reached up to 11.7 µg/L (Figure 3), particularly in CDW samples. Overall, the washing water (WW) showed concentrations exceeding the insignificance threshold values into groundwater (GFS values, [20]; see Table 2) and the limit values (methodological background values) for salts, and some of the metals and PAHs. These concentrations are in the range of material values of higher material classes as BM-F2 or RC-3 (Table 2) [19]. Therefore, the removal of contaminants during the washing process of solid waste material is essential to ensure adequate recycling fractions.

Table 2. "Insignificance threshold" concentrations (GFS) and material values of the examined organic and inorganic substances. GFS values are used to identify relevant substances in principle with regard to groundwater protection [20]. Material values are used for the classification of RB and CDW into material classes, which are linked with permissible applications in technical constructions, regulated in the upcoming German recycling degree [19].

Threshold and Material Values	F^- µg/L	Cl^- mg/L	SO_4^{2-} mg/L	As µg/L	Cr µg/L	Cu µg/L	Mo µg/L	Ni µg/L	Se µg/L	V µg/L	BaP µg/L	PAHs [f] µg/L	PCBs [g] µg/L	PHC [h] µg/L	Phenol µg/L
GFS values [a]	900	250	250	3.2	3.4	5.4	35	7	3	4	0.01	0.2	0.01	100	8
BM-F0* [b]	-	-	250	12	15	30	55	30	-	30	-	0.3	0.02	150	12
BM-F1 [c]	-	-	450	20	150	110	55	20	-	55	-	1.5	0.02	160	60
BM-F2 [d]	-	-	450	85	290	170	55	20	-	450	-	3.8	0.02	160	60
RC-1 [e]	-	-	600	-	150	110	-	-	-	120	-	4	-	-	-

[a] Insignificance threshold values for groundwater protection (GFS values) [20]. [b] Material values with regard to technical constructions of soil materials BM-F0* [19]. [c] Material values with regard to technical constructions of soil materials BM-F1 [19]. [d] Material values with regard to technical constructions of soil materials BM-F2 [19]. [e] Material values with regard to technical constructions of RC-1 defined as the highest quality construction and demolition waste [19]. [f] 15 PAHs, excluding naphthalene and methylnaphthalene. [g] Sum of PCBs (PCB-28, -52, -101, -138, -153, -180) and PCB-118. [h] Limit concentrations of petroleum hydrocarbons ranging from C10 to C40.

3.3. Comparison between Short and Extensive Column Tests: The Importance of Compliance Testing

Of the three sets of CDW (i.e., CDW1-CDW3), CDW3 material was selected to further examine the long-term leaching behavior of potential contaminants, as it proved to be the most contaminated solid material in L/S 2 eluates, particularly for the silt and sand fractions (Figure 3). The gravel fraction was not further examined as eluate concentrations in L/S 2 were lower than GFS values (see Figure 3 and Table 2). Figure 4 compares cumulative leaching in long-term to short-term tests at L/S 2 ratios. Figure 5 shows the grouping of salts, metals and PAHs in normalized leaching plots, and Figures 6–8 show the dynamics of the long-term leaching behavior in log-log plots.

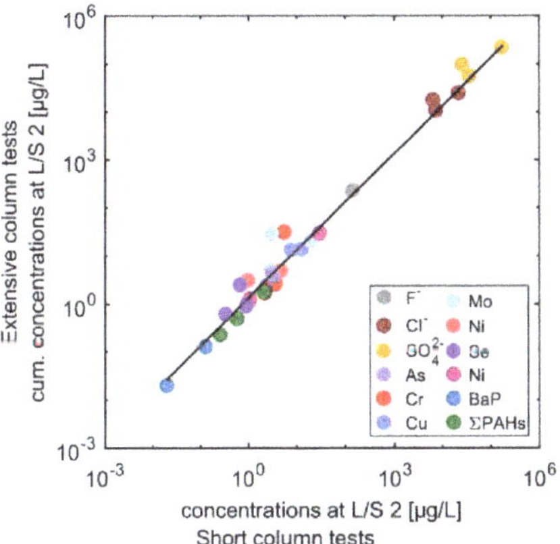

Figure 4. Comparison of short and extensive leaching test results for sample CDW3 (L/S 2); solid line represents the linear regression of the data (R^2 = 0.92, slope = 1.33).

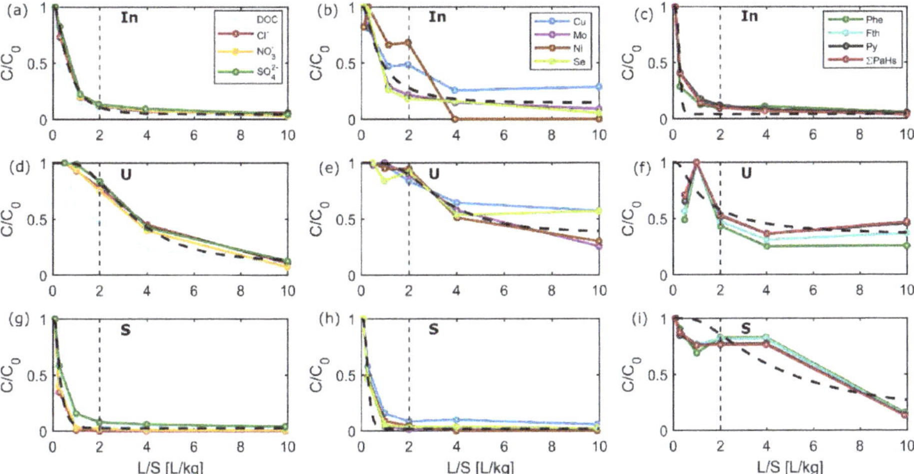

Figure 5. Normalized concentrations of selected groups of compounds and elements ("salts" left panel (**a,d,g**), "metals" middle panel (**b,e,h**), and PAHs right panel (**c,f,i**) vs. liquid-to-solid (L/S) ratio). Colored lines and symbols represent observations from extensive column test of different solid fractions of CDW 3: input material (In), silt (U), and sand (S); dashed lines represent the fitted results from the advection-dispersion model (with distribution coefficients K_d ranging from 0.28–3.64 L/kg and α/x ratios from 0.07–0.50 for salts and metals).

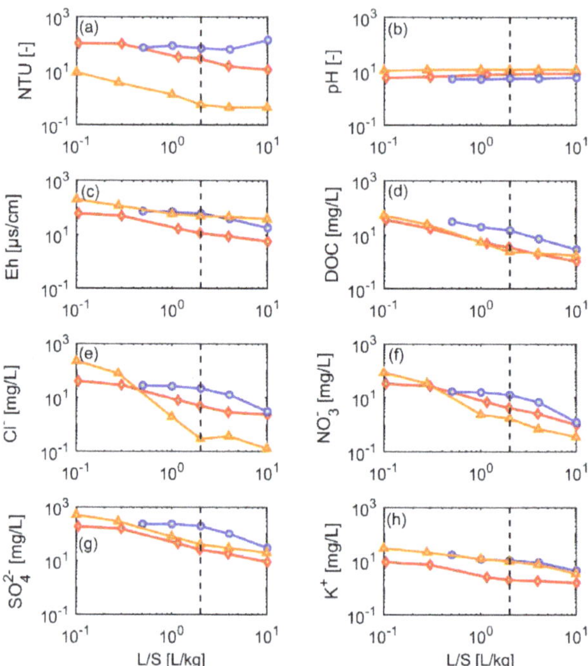

Figure 6. Leaching behavior of selected "salts" from different grain-size fractions of CDW material (CDW3) until a liquid-to-solid (L/S) ratio of 10 (dotted line L/S = 2); red diamonds: input material (mixture), blue circles: silt (<0.063 mm), orange triangles: sand (0–0.2 mm). (**a**) TSS, (**b**) pH, (**c**) Eh, (**d**) DOC, (**e**) Cl^-, (**f**) NO_3^-, (**g**) SO_4^{2-}, and (**h**) K^+.

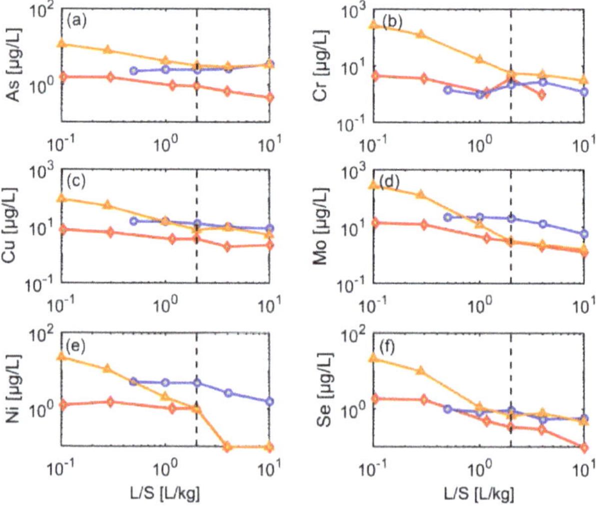

Figure 7. Leaching behavior of selected metals from different grain size fractions of CDW material (CDW3) until a liquid-to-solid (L/S) ratio of 10 (dotted line L/S = 2); red diamonds: input material (mixture), blue circles: silt (<0.063 mm), orange triangles: sand (0–0.2 mm). (**a**) As, (**b**) Cr, (**c**) Cu, (**d**) Mo, (**e**) Ni, and (**f**) Se.

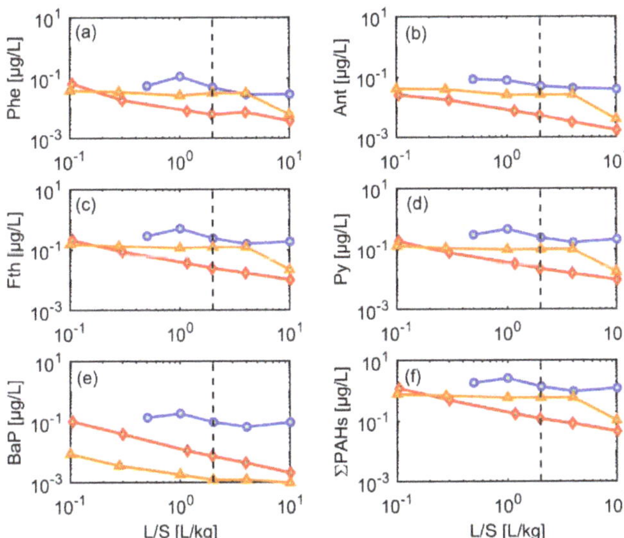

Figure 8. Leaching behavior of selected PAHs from different grain-size fractions of CDW material (CDW3) until a liquid-to-solid (L/S) ratio of 10 (dotted line L/S = 2); red diamonds: input material (mixture), blue circles: silt (<0.063 mm), orange triangles: sand (0–0.2 mm). (**a**) Phe, (**b**) Ant, (**c**) Fth, (**d**) Py, (**e**) BaP, and (**f**) the sum of 16PAHs.

Long-term column tests were sampled from LS 0.1 (0.5 for silt) to L/S 10 (extensive tests) and analyzed for "salts", metals and PAHs. Cumulative concentrations (C_{cum}) were calculated from the cumulative mass released up to L/S 2 divided by the total volume of water at L/S 2. As expected, very good agreement was observed between aqueous concentrations from the short and extensive column tests at L/S 2 ratios proving that one-step short column tests are sufficient for compliance testing, and thus reduce testing time (Figure 4). Short-term column percolation tests are thus suitable for continuous internal (facility control) and continuous external quality control. Some variability was observed in the sand fraction, particularly for metals, which may be due to more complex solubility behavior relative to pH and redox conditions [21,25].

3.4. Typical Release Patterns of Groups of Substances and Fitting of the Advection-Dispersion Transport Model

The different mass-release pattern of salts, metals and PAHs observed in eluates of the extensive column tests demonstrated that substances can be grouped into rapid ("salts"), intermediate ("metals") and slow leaching substances such as PAHs [3,4]. Figure 5 shows the similar leaching behavior of DOC, Cl^-, NO_3^- and SO_4^{2-} ("salts") based on normalized concentrations of the input material as well as silt and sand fractions. Metals such as Mo, Ni, Cu and Se may also be grouped, and showed a partially slower leaching than the "salts". In general, slower leaching was observed in the silt fraction, probably due to smaller grain size and higher sorption capacity. PAHs showed decreasing concentrations only in the input material, while partly stable concentrations were observed for the silt and sand fractions. These similar leaching patterns were observed for Phe, Fth, Py and most of the other 16 PAHs (Figure 5).

As suggested earlier [3], a simple parsimonious transport model may be used to describe leaching in column tests, and to obtain average K_d and longitudinal dispersivity (α) values by fitting to observed data (dashed line in Figure 5). A description of the model is provided in Appendix B. It should be noted that the fitting parameters K_d and dispersivity (here α/x - α as a function of the length of the pack column x) lump together all processes that are not accounted for in the analytical solution, such as non-equilibrium sorption/desorption, non-linear sorption and/or slow desorption, which all lead to extended tailing and thus increased "dispersivity" (e.g., $\alpha/x > 0.12$) [3]. While the approach worked reasonably well for "salts" and "metals", PAHs did not follow the model well, probably also due to artifacts in measurements (Figure 5).

The estimated (fitted) average K_d for salts and metals from eluates of the sand fraction were 0.28 to 0.32 L/kg, respectively, while eluates from the silt fraction resulted in the same K_d values of 3.6 L/kg. For the input material, K_d values of 0.7 and 1 L/kg were obtained for salts and metals, respectively. These results further support the high heterogeneity in the input material with a high fraction of silt leading to slower leaching (see also Figures 6 and 7). Dispersivities fitted as α/x were 0.31–0.50 in the input material and thus larger than in silt (0.14–0.15) and sand (0.07–0.16) fractions, possibly also due to the pronounced heterogeneity of the input material. Overall, the results of the fitting model indicate that in most cases leaching initially occurs at or reasonably close to equilibrium, as indicated by Grathwohl and Susset (2009), in particular for the homogeneous material fractions.

3.5. Dynamics of Contaminant Leaching

Figures 6–8 illustrate the long-term dynamics of contaminant leaching in log-log plots. With the exception of TSS and pH, the "salts" (SO_4^{2-}, NO^{3-}, NO^{2-}, Cl^- and Na^+) showed, initially, high concentrations (up to 400 mg/L, higher concentrations than in the short column tests but lower than in washing water), which decreased by 90% after L/S 2 in the sand fraction and input material (Figure 6). Silt showed the highest and most stable concentrations for PAHs and metals, which are attributed to high sorption capacity of the fine particles (Figure 7). In the sand fraction, a delayed decline in the metals was observed in some cases. The untreated input material showed mostly low and continuously decreasing concentrations, which likely results from the heterogeneity of the material largely composed by coarser fractions, which were least contaminated by PAHs and highly polluted fine materials (see Figures 2 and 3). This probably leads to a superposition of solute leaching from different material classes and the typical power-law behavior observed for the input material in Figures 6–8. Liu et al. (2021) [12] showed that heterogeneous mixtures of materials may result in very complex contaminant release characteristics in column leaching tests, especially if materials with different degrees of contamination are concerned. For example, a rapid initial decline in concentrations followed by concentration "tailing" maybe be explained by a heterogeneous material in which a small portion of less sorbing material (low K_d, high $C_{w,eq}$, low retardation) is mixed with a more strongly sorbing material (high K_d, low $C_{w,eq}$, high retardation) [12].

The leaching of metals typically varies from sample to sample and likely depends on several other parameters that change over time, such as pH, redox potential and ionic strength [3,15,26]. Here, initial metal concentrations were highest in the sand fraction (in contrast to the salts); Cr and Mo showed concentrations up to 286 µg/L and 269 µg/L (at L/S = 0.1 L/kg), respectively, followed by Cu (93 µg/L; Figure 7). Generally, metal concentrations decreased again by 90% at L/S 2 in the input material and sand fraction, including As, Ni and Se.

The release pattern of PAHs also varied among the different grain-size fractions of CDW3. Initial concentrations of the 16 PAHs of the input material, silt and sand fractions were >0.2 µg/L (Figure 8). While concentrations of most PAHs were quite constant in the sand and silt fractions, indicating strong sorption; the input material showed a power-law behavior with continuously decreasing concentrations, as already observed for some of the metals and salts (see Figures 6 and 7). The sudden drop in leachate concentrations at L/S 10 of the sand fraction is unclear and possibly reflects an artifact during the sampling procedure.

Generally, PAHs (and metals) may be associated to suspended particles or dissolved organic matter [27–30], but since turbidity and DOC remained well below 100 mg/L (see Figure 6), respectively, this would only affect strongly sorbing compounds with K_d values larger than 10,000 L/kg [13]. DOC was always below 50 mg/L and continuously decreased to less than 10 mg/L (see Figure 6) in all fractions, which is not reflected in the rather stable concentrations of PAHs, e.g., in sand and silt fractions. Similarly, metals such as copper, which are known to form complexes with DOC showed fairly stable concentrations in the silt fraction, while DOC decreased rapidly. High TSS values were observed for silt and the input material, while the sand fraction showed very low and declining TSS values (see Figure 6). TSS in leachate of the silt fraction was quite stable and even showed an increase at L/S 10, while TSS of the input material dropped from 100 mg/L to 10 mg/L, which in principle could have affected leaching of high molecular weight PAHs (Fth, Pyr, BaP). Since all PAHs showed a similar leaching behavior, and concentrations in the sand fraction were higher than in the input material, particle facilitated transport seems not to play a major role (maybe with the exception of BaP, which, however, has the lowest concentrations and does not significantly contribute to the sum of PAHs).

4. Conclusions

Wet processing after crushing of CDW and RB produces approximately 25 % of silt and sand, respectively, whereas the gravel fraction is usually around 50 %. Coarse-grained fractions (gravels) generally fulfilled legal standards for a free reuse in open technical applications (landscaping, etc.), while the sand fractions still showed concentrations which limit their reuse to specific technical applications. Fine-grain fractions (silt) are still contaminated and only allow limited re-use in (semi-) isolated applications, or require landfilling. This is also reflected in concentrations in solids and aqueous leachates up to L/S 2 (Figures 2 and 3).

Results from the short leaching tests showed to be comparable with the cumulative concentrations from the extensive column tests (up to L/S 2 L/kg; Figure 4). Thus, short leaching tests are suitable for compliance testing where concentrations can be compared to threshold values in order to select various material fractions for different recycling applications. Extensive column leaching tests showed, particularly for salts and some metals, a highly dynamic contaminant release with a decline to less than 10% of the initial concentration at *L/S* 2 for the sand fraction and input material. The silt fraction showed quite stable concentrations up to *L/S* 10, probably due to high sorption capacities for metals and PAHs. The leaching behavior of organic and inorganic substances from highly heterogeneous materials (i.e., "input material" of CDW 3) reflects their complex composition, making leaching patterns difficult to predict. As observed in earlier studies, a "typical" leaching behavior of highly soluble substances such as Cl^- and SO_4^{2-}, and metals such as Cu and Mo allows their grouping and can fit with simple transport models. Overall, short column leaching tests provide important information for decision making on the recycling of waste material. Future similar studies may help to optimize processing of mixed solid waste for higher recoveries of material fractions suitable for recycling.

Supplementary Materials: The following supporting information can be downloaded at: https://www.mdpi.com/article/10.3390/ma15030858/s1, Table S1: Chemical analyses of solids and leachates.

Author Contributions: Conceptualization, M.P.-E., B.S. and P.G.; validation, B.S. and P.G.; investigation, M.P.-E., B.S. and P.G.; writing—original draft preparation, M.P.-E.; writing—review and editing, M.P.-E., B.S. and P.G.; visualization, M.P.-E.; supervision, B.S. and P.G. All authors have read and agreed to the published version of the manuscript.

Funding: The first author acknowledges the support by the Mexican National Council for Science and Technology (CONACyT). We acknowledge support by Open Access Publishing Fund of University of Tübingen.

Institutional Review Board Statement: Not applicable.

Informed Consent Statement: Not applicable.

Data Availability Statement: The data presented in this study are available in this article as Supplementary Information.

Acknowledgments: M.P.E. acknowledges Thomas Wendel for his support during the performance of the column tests. We thank Renate Seelig, Sara Cafisso, Larissa Lohmüller, Annegret Walz and Bernice Nisch for their laboratory assistance.

Conflicts of Interest: The authors declare no conflict of interest.

Appendix A. Concentration Measurements

Table A1. Solid concentrations of railway ballast (RB) and three sets of construction and demolition waste (CDW) from different solid fractions. In: input material, U: silt fraction, S: sand fraction and G: gravel fraction. Values exceeding precautionary values in bold.

Samples	Sb	As	Ba	Pb [mg/kg]	B	Cd	Cr	Co	Cu	Mo	Hg [mg/kg]	Ni	Se	Tl	V	Zn [mg/kg]	PHC [a]	PCBs [b]	BaP	16PAHs [mg/kg]
RB-In	3.3	12.9	135	31	68.5	0.64	44.8	15.2	81.4	2.07	0.12	62.9	1.71	0.21	39.5	333	<25	0.003	0.92	12.50
RB-U	3.8	16	163	44.1	81.7	0.65	47.5	16	107	2.07	0.21	63.7	1.6	0.29	40.5	523	63	0.005	1.62	21.83
RB-S	2.9	9.3	139	16.8	49.5	0.35	38.8	13.9	48.5	1.6	0.06	67.8	1.76	0.1	36.4	125	86	0.009	0.26	4.08
CDW1-In	0.23	7.2	39.6	9.91	28	0.23	15.1	4.06	8.1	0.18	<0.05	15.9	2.09	0.11	21.5	34.9	<25	<0.002	0.65	7.91
CDW1-U	0.50	14.9	113	27.7	66.2	0.37	32.5	11.8	21.3	0.34	0.075	43.6	2.07	0.29	43.8	85.7	98	0.016	0.93	10.68
CDW1-S	0.20	7.9	36.9	13.8	22.6	0.12	13.6	3.82	6.6	0.35	0.05	16.2	<0.2	<0.1	17.2	32.6	139	0.004	0.27	3.717
CDW1-G	0.19	5.4	19.5	3.86	19.5	0.25	9.03	2.24	3.6	0.21	<0.05	9.5	1.84	<0.1	11.7	20.6	<25	<0.002	0.0	0.005
CDW2-In	0.44	5.4	90.5	15.8	32.6	0.16	15.2	4.79	17.3	0.33	0.07	16.5	0.31	0.15	19.8	45.4	121	0.011	6.03	81.89
CDW2-U	1.3	12.6	211	49.2	80.1	0.42	10.6	10.3	44.1	1.2	0.21	39	0.69	0.39	36.6	131	191	0.041	3.96	43.31
CDW2-S	0.61	4.6	66.1	15.6	22.8	0.25	13.8	4.31	32.7	0.50	<0.05	15.2	0.23	0.1	15.3	55.8	85	0.016	2.73	38.94
CDW2-G	0.65	4.3	53.4	13	18.1	0.17	9.2	2.4	7.4	0.21	<0.05	9.6	0.2	0.1	11.7	17	104	<0.002	0.08	0.94
CDW3-In	0.65	6.7	62	26.6	38.1	0.43	17.4	6.22	17.2	0.17	0.21	22	0.37	0.14	20.8	68	54	<0.002	0.33	4.18
CDW3-U	1.2	14.2	148	45.6	71	0.51	28.9	11.2	35.3	0.65	0.31	41	0.81	0.37	33.9	120	123	0.012	0.90	17.33
CDW3-S	0.42	9.2	74	26.1	27.2	0.36	13.8	5.13	13.9	0.32	0.05	17.4	0.45	0.11	18.7	35.9	93	<0.002	0.36	6.56
CDW3-G	0.25	6.0	38.5	7.37	24.3	0.20	10.3	3.6	8.4	0.19	<0.05	14.1	0.38	0.1	15.4	27.5	79	<0.002	0.51	6.62
Threshold [c]	-	10	-	40	-	0.4	30	-	20	-	0.2	15	-	0.5	-	60	300 [d]	0.05	0.3	3

[a] Petroleum hydrocarbons of chain C10-C40. [b] Sum of PCBs (PCB28, PCB52, PCB101, PCB138, PCB153, PCB180). Precautionary values are given for PCB6 and PCB118 [19]. [c] Threshold concentration based on precautionary values in soils (BM-0 Sand) [19]. [d] Precautionary value of petroleum hydrocarbons (PHCs of C10-C40) based on limit value of "material class" BM-0* [19].

Table A2. Measured concentrations of herbicides in short leaching test of railway ballast (RB) on different solid fractions. In: input material, U: silt fraction, S: sand fraction and G: gravel fraction.

Samples	Atrazine [μg/L]	Simazine [μg/L]	Bromacil [μg/L]	Desethyl-atrazine [μg/L]	Diuron [μg/L]	Glyphosate [μg/L]	Ampa [μg/L]
RB-In	-	-	-	-	-	-	0.29
RB-U	0.05	0.04	-	0.045	0.046	-	-
RB-S	0.093	0.19	0.026	0.074	0.076	0.31	1.5
RB-G	0.027	0.042	-	-	0.032	0.15	0.43
RB-WW	0.062	0.1	0.029	0.079	0.023	-	0.29
GFS value [a]	0.1	0.1	0.1	0.1	0.05	0.1	0.1
Threshold [b]	0.2	0.2	0.2	-	0.1	0.2	2.5

[a] Insignificant threshold values into groundwater [20]. [b] Threshold concentration for recycling railway ballast material (GS-0) [19].

Appendix B. Long-Term Leaching Behavior Described by the Advection-Dispersion Equation

Column leaching tests represent the percolation of water through different types of solid materials. Initially, the column is saturated so that equilibrium conditions can be achieved rather rapidly (<5 h) [13]. The drop in concentrations is given by a change in non-equilibrium conditions leading to an extended tailing of low concentrations. The advection-dispersion model allows for the description of the movement of the front of clean water through the column as:

$$\frac{\partial C}{\partial t} = \frac{D}{R}\frac{\partial^2 C}{\partial x^2} - \frac{v}{R}\frac{\partial C}{\partial x} \tag{A1}$$

where C is the solute concentration, D is the longitudinal dispersion coefficient [m^2 s^{-1}], v is the average flow velocity [m s^{-1}], t is time, x is the length of the column [m] and R denotes the retardation factor [-], defined as:

$$R = 1 + K_d\frac{\rho}{n} \tag{A2}$$

where K_d denotes the distribution coefficient [L/kg], defined as the ratio of concentrations in the solids to the aqueous concentrations (C_s/C_w), ρ is the dry bulk density [kg/L] and n is the porosity [-]. The advection–dispersion model assumes local equilibrium conditions, but previous studies demonstrated a reasonably well fit with the early leaching behavior [3,4]. This model is solved using the analytical solution for the movement of the front of clean water through the column [31] and expressed based on the dynamic liquid to solid ratio (L/S) as:

$$\frac{C}{C_0} = 1 - 0.5 \left[\mathrm{erfc}\left(\frac{K_d - LS}{2\sqrt{\frac{\alpha}{x}\left(\frac{n}{\rho} + K_d\right)LS}} \right) + \exp\left(\frac{x\left(1 - \frac{1}{R}\right)}{\alpha} \right) \mathrm{erfc}\left(\frac{K_d - Ls}{2\sqrt{\frac{\alpha}{x}\left(\frac{n}{\rho} + K_d\right)LS}} \right)\left(1 - \frac{C_{min}}{C_0}\right) \right] \quad (A3)$$

where C_{min} is the minimum concentrations usually detected at L/S 10 L/kg, and LS is the amount of water percolated through the column after t time relative to the dry weights of the solids in the column ($= v\, n\, t/x\, \rho$). The last term in brackets has been here added to fit the late data of the leaching tests, which show substantial tailing. Eq. B3 accounts for the initial displacement of low sorbing (high soluble) compounds during first flooding of the column. The model is fitted to measured data using MATLAB (v R2021b) and the function *lsqcurvefit*. From the fit, retardation factors (R) and distribution coefficients (K_d) were calculated. In addition, the longitudinal dispersivity was fitted as a function of x (i.e., α/x). Maximum values of K_d and α were set to 100 L/kg and 1 m, respectively.

References

1. BAU. Mineralische Bauabfälle Monitoring 2018. In *Bericht zum Aufkommen und zum Verbleib Mineralischer Bauabfälle im Jahr 2018*; Kreislaufwirtschaft BAU: Berlin, Germany, 2021.
2. Allegrini, E.; Butera, S.; Kosson, D.S.; Van Zomeren, A.; Van der Sloot, H.A.; Astrup, T.F. Life cycle assessment and residue leaching: The importance of parameter, scenario and leaching data selection. *Waste Manag.* **2015**, *38*, 474–485. [CrossRef]
3. Grathwohl, P.; Susset, B. Comparison of percolation to batch and sequential leaching tests: Theory and data. *Waste Manag.* **2009**, *29*, 2681–2688. [CrossRef] [PubMed]
4. Finkel, M.; Grathwohl, P. Impact of pre-equilibration and diffusion limited release kinetics on effluent concentration in column leaching tests: Insights from numerical simulations. *Waste Manag.* **2017**, *63*, 58–73. [CrossRef] [PubMed]
5. Lin, X.; Vollpracht, A.; Markus, P.; Linnemann, V. Optimization of a German short term percolation test to determine the leaching of granular materials. *Waste Manag.* **2020**, *105*, 433–444. [CrossRef] [PubMed]
6. Kalbe, U.; Bandow, N.; Bredow, A.; Mathies, H.; Piechotta, C. Column leaching tests on soils containing less investigated organic pollutants. *J. Geochem. Explor.* **2014**, *147*, 291–297. [CrossRef]
7. Quina, M.J.; Bordado, J.C.M.; Quinta-Ferreira, R.M. Percolation and batch leaching tests to assess release of inorganic pollutants from municipal solid waste incinerator residues. *Waste Manag.* **2011**, *31*, 236–245. [CrossRef]
8. Naka, A.; Yasutaka, T.; Sakanakura, H.; Kalbe, U.; Watanabe, Y.; Inoba, S.; Takeo, M.; Inui, T.; Katsumi, T.; Fujikawa, T.; et al. Column percolation test for contaminated soils: Key factors for standardization. *J. Hazard. Mater.* **2016**, *320*, 326–340. [CrossRef]
9. López Meza, S.; Kalbe, U.; Berger, W.; Simon, F.-G. Effect of contact time on the release of contaminants from granular waste materials during column leaching experiments. *Waste Manag.* **2010**, *30*, 565–571. [CrossRef]
10. DIN 19528. *Elution von Feststoffen–Perkolationsverfahren zur Gemeinsamen Untersuchung des Elutionsverhaltens von Anorganischen und Organischen Stoffen*; Ausgabe 01/2009; Beuth: Berlin, Germany, 2009.
11. Susset, B.; Grathwohl, P. Leaching standards for mineral recycling materials—A harmonized regulatory concept for the upcoming German Recycling Decree. *Waste Manag.* **2011**, *31*, 201–214. [CrossRef]
12. Liu, B.; Finkel, M.; Grathwohl, P. Mass transfer principles in column percolation tests: Initial conditions and tailing in heterogeneous materials. *Materials* **2021**, *14*, 4708. [CrossRef]
13. Grathwohl, P. On equilibration of pore water in column leaching tests. *Waste Manag.* **2014**, *34*, 908–918. [CrossRef]
14. Dabo, D.; Badreddine, R.; De Windt, L.; Drouadaine, I. Ten-year chemical evolution of leachate and municipal solid waste incineration bottom ash used in a test road site. *J. Hazard. Mater.* **2009**, *172*, 904–913. [CrossRef] [PubMed]
15. Quina, M.J.; Bordado, J.C.M.; Quinta-Ferreira, R.M. The influence of pH on the leaching behaviour of inorganic components from municipal solid waste APC residues. *Waste Manag.* **2009**, *29*, 2483–2493. [CrossRef] [PubMed]
16. Van Zomeren, A.; Comans, R.N.J. Contribution of natural organic matter to copper leaching from municipal solid waste incinerator bottom ash. *Environ. Sci. Technol.* **2004**, *38*, 3927–3932. [CrossRef] [PubMed]
17. Ferraz, M.C.M.A.; Lourenço, J.C.N. The influence of organic matter content of contaminated soils on the leaching rate of heavy metals. *Environ. Prog.* **2000**, *19*, 53–58. [CrossRef]

18. Garrabrants, A.C.; Kosson, D.S.; Brown, K.G.; Fagnant, D.P.; Helms, G.; Thorneloe, S.A. Methodology for scenario-based assessments and demonstration of treatment effectiveness using the Leaching Environmental Assessment Framework (LEAF). *J. Hazard. Mater.* **2021**, *406*, 124635. [CrossRef]

19. BGBl. *Verordnung zur Einführung Einer Ersatzbaustoffverordnung, zur Neufassung der Bundes-Bodenschutz- und Altlastenverordnung und zur Änderung der Deponieverordnung und der Gewerbeabfallverordnung*; Bundesrat: Köln, Germany, 2021; p. 155.

20. LAWA. Derivation of insignificant threshold values for groundwater. In *Länderarbeitsgemeinschaft Wasser*; Ministerium für Umwelt, Klima und Energiewirtschaft: Stuttgart, Germany, 2017.

21. Kalbe, U.; Simon, F.-G. Potential use of incineration bottom ash in construction: Evaluation of the environmental impact. *Waste Biomass Valor* **2020**, *11*, 7055–7065. [CrossRef]

22. Zand, A.D.; Grathwohl, P.; Nabibidhendi, G.; Mehrdadi, N. Determination of leaching behaviour of polycyclic aromatic hydrocarbons from contaminated soil by column leaching test. *Waste Manag. Res. J. Sustain. Circ. Econ.* **2010**, *28*, 913–920. [CrossRef]

23. Quenea, K.; Lamy, I.; Winterton, P.; Bermond, A.; Dumat, C. Interactions between metals and soil organic matter in various particle size fractions of soil contaminated with waste water. *Geoderma* **2009**, *149*, 217–223. [CrossRef]

24. Diotti, A.; Perèz Galvin, A.; Piccinali, A.; Plizzari, G.; Sorlini, S. Chemical and Leaching Behavior of Construction and Demolition Wastes and Recycled Aggregates. *Sustainability* **2020**, *12*, 10326. [CrossRef]

25. Butera, S.; Hyks, J.; Christensen, T.H.; Astrup, T.F. Construction and demolition waste: Comparison of standard up-flow column and down-flow lysimeter leaching tests. *Waste Manag.* **2015**, *43*, 386–397. [CrossRef] [PubMed]

26. Dijkstra, J.J.; van der Sloot, H.A.; Comans, R.N.J. The leaching of major and trace elements from MSWI bottom ash as a function of pH and time. *Appl. Geochem.* **2006**, *21*, 335–351. [CrossRef]

27. Schwientek, M.; Rügner, H.; Beckingham, B.; Kuch, B.; Grathwohl, P. Integrated monitoring of particle associated transport of PAHs in contrasting catchments. *Environ. Pollut.* **2013**, *172*, 155–162. [CrossRef] [PubMed]

28. Petruzzelli, L.; Celi, L.; Cignetti, A.; Marsan, F.A. Influence of soil organic matter on the leaching of polycyclic aromatic hydrocarbons in soil. *J. Environ. Sci. Health Part B* **2002**, *37*, 187–199. [CrossRef]

29. Totsche, K.U.; Jann, S.; Kögel-Knabner, I. Release of Polycyclic Aromatic hydrocarbons, dissolved organic carbon, and suspended matter from disturbed NAPL-contaminated gravelly soil material. *Vadose Zone J.* **2006**, *5*, 469–479. [CrossRef]

30. Rügner, H.; Schwientek, M.; Milačič, R.; Zuliani, T.; Vidmar, J.; Paunović, M.; Laschou, S.; Kalogianni, E.; Skoulikidis, N.T.; Diamantini, E.; et al. Particle bound pollutants in rivers: Results from suspended sediment sampling in Globaqua River Basins. *Sci. Total Environ.* **2019**, *647*, 645–652. [CrossRef]

31. Ogata, A.; Banks, R.B. A Solution of the Differential Equation of Longitudinal Dispersion in Porous Media. In *Fluid Movement in Earth Materials*; U.S. Geological Survey: Washington, DC, USA, 1961.

materials

MDPI

Article

Environmental Impact of Geosynthetics in Coastal Protection

Philipp Scholz [1], Ieva Putna-Nimane [2], Ieva Barda [2], Ineta Liepina-Leimane [2], Evita Strode [2], Alexandr Kileso [3,4], Elena Esiukova [3], Boris Chubarenko [3], Ingrida Purina [2] and Franz-Georg Simon [1,*]

[1] BAM Bundesanstalt für Materialforschung und-prüfung, 12200 Berlin, Germany; philipp.scholz@bam.de
[2] Latvian Institute of Aquatic Ecology, 1007 Riga, Latvia; ieva.putna@lhei.lv (I.P.-N.); ieva.barda@lhei.lv (I.B.); ineta.liepina@lhei.lv (I.L.-L.); evita.strode@lhei.lv (E.S.); ingrida.purina@lhei.lv (I.P.)
[3] Shirshov Institute of Oceanology, Russian Academy of Sciences, 117997 Moscow, Russia; aleksandr.kileso@gmail.com (A.K.); elena_esiukova@mail.ru (E.E.); chuboris@mail.ru (B.C.)
[4] Immanuel Kant Baltic Federal University, 236041 Kaliningrad, Russia
*   Correspondence: franz-georg.simon@bam.de

Abstract: Geosynthetic materials are applied in measures for coastal protection. Weathering or any damage of constructions, as shown by a field study in Kaliningrad Oblast (Russia), could lead to the littering of the beach or the sea (marine littering) and the discharge of possibly harmful additives into the marine environment. The ageing behavior of a widely used geotextile made of polypropylene was studied by artificial accelerated ageing in water-filled autoclaves at temperatures of 30 to 80 °C and pressures of 10 to 50 bar. Tensile strength tests were used to evaluate the progress of ageing, concluding that temperature rather than pressure was the main factor influencing the ageing of geotextiles. Using a modified Arrhenius equation, it was possible to calculate the half-life for the loss of 50% of the strain, which corresponds to approximately 330 years. Dynamic surface leaching and ecotoxicological tests were performed to determine the possible release of contaminants. No harmful effects on the test organisms were observed.

Keywords: geosynthetics; geotextiles; dynamic surface leaching test; artificial ageing; marine littering

Citation: Scholz, P.; Putna-Nimane, I.; Barda, I.; Liepina-Leimane, I.; Strode, E.; Kileso, A.; Esiukova, E.; Chubarenko, B.; Purina, I.; Simon, F.-G. Environmental Impact of Geosynthetics in Coastal Protection. *Materials* **2021**, *14*, 634. https://doi.org/10.3390/ma14030634

Academic Editor: Qing-feng Liu
Received: 4 January 2021
Accepted: 25 January 2021
Published: 29 January 2021

Publisher's Note: MDPI stays neutral with regard to jurisdictional claims in published maps and institutional affiliations.

1. Introduction

Geosynthetics are widely used in coastal protection. Their application areas are soil reinforcement, the stabilization of ballast layers, filtration, the waterproofing of dams and canals, and scour protection (e.g., for piles of offshore wind energy plants). The application of geosynthetics in coastal protection has huge economic benefits, such as savings via substitutions of or reductions in selected soil materials, ease of installation, increased speed of construction, life cycle cost savings through improved performance (by increased longevity or reduction in maintenance), and improved sustainability in terms of conserving natural environments as compared to alternative designs [1,2]. It is commonly accepted that geosynthetics which are adequately stabilized with antioxidants (e.g., sterically hindered amines) will last in underwater constructions with limited oxygen supply and temperatures at constantly low levels for at least 100 years.

However, after the end of service lifetime, geosynthetics could be a source of plastic debris in aquatic systems if the construction which the geosynthetic is a part of is not dismantled. Further, additives which are needed as plasticizers or antioxidants could be emitted, with detrimental influence on the environment [3]. The loss of additives is intimately related to the aging of the geosynthetic products. These are the reasons that public authorities are concerned about the approvability of engineering projects using geosynthetics in aquatic systems.

The long-term stability of geotextiles is usually investigated with relation to mechanical stability, which must fulfill certain requirements after aging. Various methodologies are available (e.g., elevated temperatures or increase in oxygen pressure) to accelerate aging in the laboratory [4]. Mechanical properties, such as tensile strength, investigation of chemical oxidation reactions by infrared spectroscopy, and the residual content of stabilizers are typical parameters tested on aged samples [5]. The investigation of the possible environmental impact of the application of geosynthetics in aquatic systems is therefore hardly possible with virgin polymer material. Consequently, polymers must be artificially aged, which is best accomplished with environmental simulation chambers enabling accelerated ageing. In the case of geosynthetics in hydraulic engineering besides oxidation, mechanical stress (e.g., by tidal and wave action, abrasion by sand) and microbiological interactions (the formation of biofilms, etc.) [6] play significant roles and must be considered.

There are only a few investigations on the degradation behavior of geotextiles in marine environments [7,8]. According to these, exposure to UV light has a higher impact on the material properties in comparison to seawater immersion and tidal action. The importance of the stabilization of the polymers was strengthened. It can be expected that the degradation processes of geotextiles are similar to the processes of other plastics reaching the marine environment because they are made from the same types of polymers. Plastic waste exposed to environmental conditions begins to degrade slowly under the impact of temperature and UV radiation [9], generating a large number of macro-, micro, and nano-particles. These particles are freely transported by water flows and have adverse effects on the environment [10,11]. One of the key factors which determines the fate of microplastics in the environment is the density of polymers. The specific density of microplastic can vary significantly depending on the polymer type, technological processes of its production, additives, weathering, and biofouling [12,13]. With time, most floating plastics become negatively buoyant due to both biofouling and the adherence of denser particles and sink to the sea floor [13,14]. Thus, the seabed becomes the ultimate repository for microplastic particles and fibers [15,16]. The evaluation of the contamination level is complicated, not only because of the difficulty of the sampling of sea bottom sediment, but also due to the difficulty of the extraction of small plastic particles from marine deposits.

The project Environmental Impact of Geosynthetics in Aquatic Systems (EI-GEO) [17] aims at the investigation of whether geosynthetics in hydraulic engineering applications could be a source of microplastic or other contaminants in the aquatic environment. Whereas the behavior of geosynthetics in landfill engineering has been well studied and documented for decades [18], little is known regarding applications such as coastal protection or scour protection for off-shore wind energy plants. However, due to the rapid expansion of offshore wind energy, rising water levels, and more extreme weather conditions as a result of climate change, more and more hydraulic engineering projects will be realized in the future.

Construction with geosynthetics boasts various advantages, but it has to be ensured that there is no negative environmental impact from the application of geosynthetics in hydraulic engineering. It is expected that any effect will be visible only in the long term because the virgin raw material used for the production of geosynthetics has almost no release of particles or substances relevant to the environment [19].

Partly from improper material selection and partly from non-professional handling, debris from geosynthetic material can be found on the shore today. Therefore, a field study with sampling and monitoring was performed and the magnitude of this pollution was evaluated (objective 1). Further, an accelerated ageing method was performed to derive the requirements for geosynthetics in hydraulic engineering. The testing of mechanical properties was performed with virgin and artificially aged geosynthetics (objective 2). Finally, leachates of artificially aged geosynthetics were used in ecotoxicological tests, which are essential tools to evaluate the environmental impacts of the pollutants released by geosynthetics during ageing (objective 3).

2. Materials and Methods

The applications of geosynthetics in hydraulic and coastal engineering such as revetments, dyke constructions, or geotextile containers for scour prevention are described in detail elsewhere [1]. For the present study, a multifunctional geotextile for separation, filtration, and protection made of white polypropylene was selected as a test material for the investigations. The mass per unit area was 600 g m^{-2}, the thickness was 5 mm, and the water permeability was 3×10^{-2} m s^{-1}. The material, produced in Germany, is commercially available and widely used for geomembrane protection or for the production of sand container bags.

2.1. Accelerated Ageing Using Autoclave Test

Autoclave tests following DIN EN ISO 13438:2005 (method C) [19] were performed under a pure oxygen atmosphere with pressures between 10 and 50 bar, at temperatures between 30 and 80 °C, and with durations in the range of 14 to 143 days An overview on the performed ageing experiment is given in Table 1. It is important to notice that the test specimens were completely immersed in tap water and the exposure of autoclaves was carried out based on the time-dependent degradation of the mechanical properties of the polypropylene geotextiles. Five PP specimens (250 $\times$ 50 mm^2) were placed in the autoclaves in tap water. The use of artificial seawater was not possible due to the risk of chlorine-induced corrosion at high oxygen pressures. In order to reach thermal equilibrium, the autoclaves were left for 48 h in electronically controlled heating systems before the start of the tests. Hence, single specimens were removed in succession after different ageing periods. Then, the tensile strength was determined accordingly. Two measurements were carried out for each duration of aging. All the tensile test measurements were performed with a Zwick tensile testing machine (Zwick-Roell, Ulm, Germany) (ZPM Model 1464 with testXpert II software (Version 3.31, Zwick, Ulm, Germany)) with a 5 kN force sensor. The tensile tests were performed in an air-conditioned environment at 23 °C and a relative humidity of 50%. For the tensile test measurements, a clamping length of 50 mm and a test speed of 50 mm/min were chosen. Each sample was attached to a sandpaper to avoid sliding during the tensile test.

Table 1. Duration of accelerated ageing in autoclaves in days at 5 different temperatures and pressures.

p (bar)	Temperature (K)				
	303	313	333	343	353
10	-	-	-	-	14, 44, 61
20	-	-	-	-	27, 54, 82, 140
30	-	-	70, 102, 144	-	28, 38, 48, 77
40	-	70, 101, 143	-	-	-
50	70, 101, 143	-	-	70, 101, 143	-

Figure 1 shows a sketch of the autoclave test equipment along with all the instruments and monitoring devices used. The temperature and the pressure were observed and recorded every 15 min using an electronic data recorder (Eurotherm 6100) (Eurotherm, Limburg, Germany). The temperature of the autoclave was controlled by an external heating jacket with a separate PT100 temperature sensor connected to a PID temperature controller (Eurotherm 2216E) (Eurotherm). The heating power line was equipped with an electrical contact controlled by the internal temperature monitoring to prevent overheating of the system. The safe and reliable operation of the autoclaves requires the control and monitoring of the relevant parameters, especially for long-term experiments. All the relevant instruments and transducers were calibrated in order to obtain reliable and reproducible results.

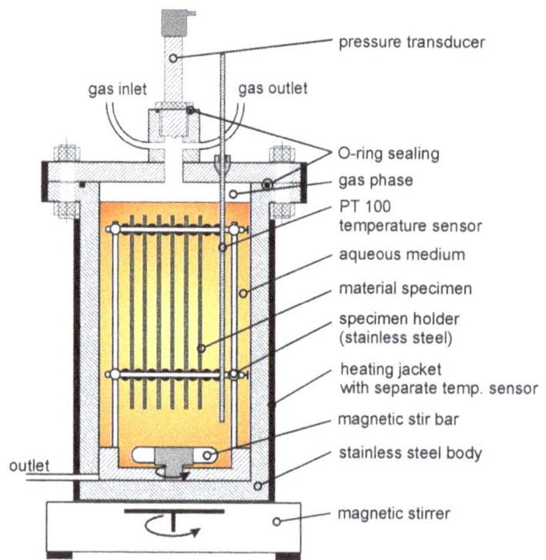

Figure 1. Schematic view of autoclave test equipment (**left**), closing the cover plate of the autoclaves (test rig with two autoclaves).

2.2. Dynamic Surface Leaching Test

Dynamic surface leaching tests (DSLT) were performed on the geosynthetic materials according to the CEN/TS 16637-2 leaching method [20]. The DSLT corresponds to a tank test for the assessment of the surface-dependent release of dangerous substances and is suitable for monolithic construction products. The test specimens were eluted using demineralized water at a defined water/surface ratio (L/A) and a water exchange at several fixed time intervals (6 h, 1 d, 36 d). The L/A ratio was set to 80 L/m^2 in CEN TS 16637-2, but can be reduced to 25 L/m^2 for plate-like products. Tests were performed at 23 ± 2 °C, room humidity $50 \pm 5\%$, in the darkness. Two plates were eluted per coating system to obtain enough eluate volume for all the ecotoxicological tests. Each plate was individually placed in a tank and the eluates of the same fraction were combined and well mixed before aliquoting them for ecotoxicological analysis.

2.3. Ecotoxicological Testing

Internationally agreed and accepted ecotoxicity test methods have been performed to demonstrate the impact of chemicals and other pollutants on the environment and determine the potential damage to organisms and the function of ecosystems [21–23]. Ecotoxicity tests consisted of two acute and one chronic test with organisms from different levels of aquatic food chains. The ecotoxicity test conditions, growth media, dilutions, and replication are summarized in Table 2. The test eligibility criteria for the *Daphnia magna* test is ≤10% immobile organisms in the control treatment and an ≥80% survival for the *Hyalella Azteca* test. For the *Desmodesmus subspicatus* test, control batch absorption measurements should indicate the exponential growth of algal cells, the variation coefficient (CV) of the growth rate in the control replicates should not exceed 5%, and the pH in the control should not increase during the test by more than 1.5 relative to the pH of the growth medium.

Table 2. Ecotoxicity test conditions summary.

Standard	ISO 6341:2012 [21]	ISO 16303:2013 [22]	ISO 8692:2012 [23]
Test organisms	*Daphnia magna*	*Hyalella azteca*	*Desmodesmus subspicatus*
Test duration	48 h	14 days	72 h
Temperature	$20 \pm 1\,^{\circ}\text{C}$	$23 \pm 1\,^{\circ}\text{C}$	$23 \pm 2\,^{\circ}\text{C}$
Growth media	ADaM *	ADaM	BG-11
Test chamber size	6 vial plates	400 mL low form beakers	300 µL
Test volume	15 mL	250 mL	265 µL
Age of test organisms	Less than 24 h old	11 days old at test initiation (1 to 2 day range in age)	Algae culture in exponencial growth phase
Organisms per test chamber	7	10	5 µL (10^4 cells)
Replicates per treatment	4	4	6
Test concentrations	(100%; 50%; 25%; 12.5%; 6.3%	100%; 75%; 50%; 25%; 12.5%; 6.3%; 3.1%	5.9%, 11.8%, 23.6%, 47.2%, 94.3%
Feeding regime	No	YCT food, fed 0.5 mL daily/chamber	Concentrated BG11 (10 µL)/vial in beginning of test
Endpoints	Mortality	Survival (optional, growth by dry weight or length)	Growth inhibition
Reference toxicant	$K_2Cr_2O_7$ 24 h LC 50 0.81 mg/L	$CdCl_2$ (Cd 96 h LC50 = 0.007 mg/L), $CuSO_4$ (Cu 96 h LC 50 = 0.24–0.33 mg/L)	ISO mentioned intercalibration $K_2Cr_2O_7$ 72 h EC 50 = 0.84 mg/L

* ADaM: Aachener Daphnia Medium.

2.4. Continuous Visual Scanning (Field Study)

Since the fragments of plastics and geosynthetic materials were unevenly distributed on the beach, the use of a selective area technique for their search—such as, for example, for anthropogenic debris [24] and microplastics [25]—will not yield results. To analyze the pollution of the beaches at the Southeastern Baltic within the Kaliningrad Oblast (Russia), a continuous visual scanning technique [26] was applied which assumes a continuous passage of a group of several people along the entire coastline, covering the entire width of the beach from the shoreline to the foredune (or cliff).

The width of the beaches of the Kaliningrad Oblast ranges from almost 0 to 188 m and the average value is 30 m, so the group usually included three people. The beach was divided into three control strips, each member of the group controls his strip and even tries to capture the edge of the neighboring zone for a complete scan of the entire beach. During the day, the group could walk 7–10 km, and such monitoring surveys were carried out in 2018.

Each detected plastic or geosynthetic fragment with a size larger than 3–5 cm was attributed to the different type of origin (see Results section), dimension scale (length and area), number of the coastal subsection where this sample was found, and position on the beach (in % of the beach width). Next, photographs were taken and, if necessary, the sample was saved for further laboratory analysis.

3. Results

3.1. Field Study on Kaliningrad Oblast Shore (Russia)

During the surveys of the beaches of the Kaliningrad Oblast (Figure 2) in 2018, a large amount of remnants of geosynthetic materials that are used in coastal protection structures [27] were found. In addition, there was extensive contamination from other building support materials—e.g., geotextile FIBC (Flexible Intermediate Bulk Container) bags ("big bags") and the remains of fishing nets, ropes, and car tires.

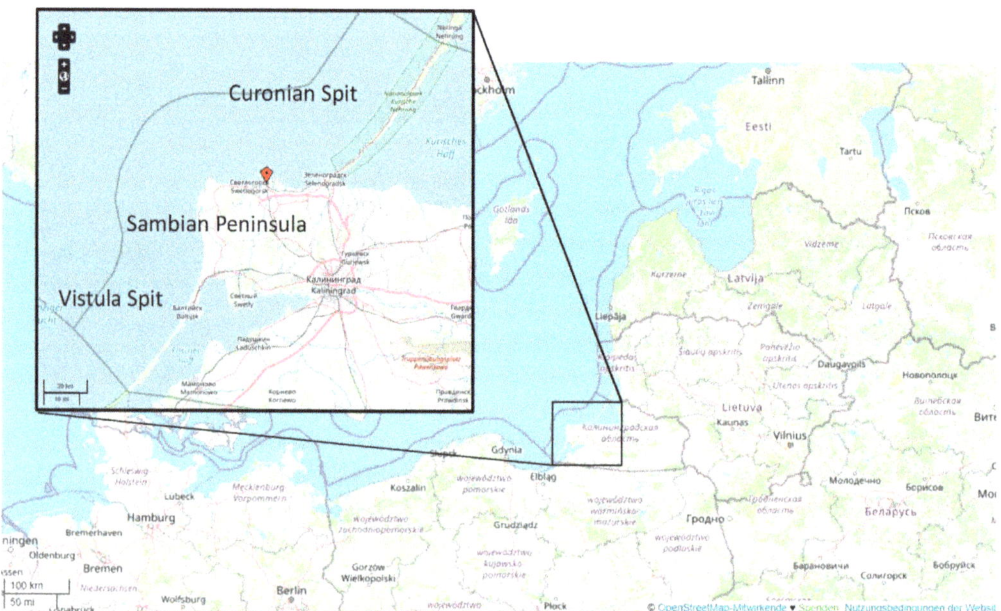

Figure 2. Shoreline of the Kaliningrad Oblast (Russia) in the Baltic Sea including the Sambian Peninsula (quadrangle). Source: OpenStreetMap.

In 2018, 3485 samples were collected from the beaches which, by origin, belonged to several types of materials: geotextiles, geocells, geogrids, plastic coating from gabions, and geotextile big bags. The integral amount of remnants of geotextile objects was more than 190 m^2 and the integral length of the geotextile braids from gabions coating was about 100 m [28].

The occurrence of geosynthetic remnants varies greatly along the entire shore of the Kaliningrad Oblast. The northern shore of the Sambia Peninsula accounts for 66% of the remains found, 31% for the beaches of the Curonian Spit National Park, and only 3% was found on the beaches of the western shore of the Sambia Peninsula and the Vistula Spit. Among the remains of geosynthetic materials found, the largest number was braid from gabions (44%) and geocontainers (43%), pieces of geotextile accounted for only 12%, and the remaining 1% was made up of remnants of geocells and geogrids.

The performed primary statistical analysis on the occurrence of the number of pieces per 1 kilometer for various morphodynamic segments of the coast of the Kaliningrad Oblast (Vistula Spit, western and northern shores of the Sambia Peninsula, Curonian Spit) showed that the main pollution occurs on the northern shore (see Table 3). Considering the average size of one piece of geotextile (0.9 m^2), gabion coating (7.4 cm), big bag (0.3 m^2), and geocell (0.06 m^2), it is obvious that the remnants of geotextile and "big bags" were the mostly visible litter on the beach.

Figure 3. Photographs of samples collected during the field study: (**a**) + (**b**): aged plastic coating of wires in gabions' (**c**) debris from geocell; (**d**): debris from big bag.

Table 3. Occurrence of residues of geosynthetic materials and other large debris in pieces per 1 running kilometer of the coastline in various morphodynamic segments of the Baltic shore of the Kaliningrad Oblast by surveys in 2018.

Morphodynamic Segments of the Shore	Geotextile	Gabion Coating	"Big Bags"	Geocell
Vistula Spit	0.01	0.13	0.25	0
Western shore of the Sambian Peninsula	0	0.18	0.15	0
Northern shore of the Sambian Peninsula	2.90	9.38	5.98	0.13
Curonian Spit	0.24	1.97	4.26	0.09

Note: Numbers are given in pieces/km, while pieces have very different linear sizes (see Figure 3 for examples).

This fact that the northern shore of the Sambian Peninsula is mostly littered correlates well with the location of engineering structures using geosynthetic materials, most of which are located on the northern shore of the Sambian Peninsula [27]. In addition, the main accumulation of residues of geosynthetic materials is observed in the areas adjacent to these engineering structures. On the Curonian Spit (north from the Sambian Peninsula), a large amount of geosynthetic remnants was also found, which were probably brought here by alongshore currents [29]. The occurrence of residues on the Vistula Spit (south from the Sambian Peninsula) and on the western coast of the Sambia Peninsula is low due to the current structure in the eastern part of the Gulf of Gdansk [30].

Gabion coating was found quite often (see Table 3). This came from the plastic coating of the wire used for the gabion's support structure. Obviously, this coating is not weatherproof. A support structure made of stainless steel or Zn-plated wires would not need a plastic coating but is, however, more expensive. Geotextile remnants came from partly destroyed coastal protection structures which stay without proper maintenance during long time. Geocells were found rarely, they were from several locations, where storm events destroyed lawn on the slopes of foredune wall prepared using geocells. Debris from big bags was found often as well. However, these woven geotextiles are rather used for transport of building materials or short-term applications than for coastal protection systems. Occurrence can therefore be attributed to improper waste management.

3.2. Tensile Tests after Accelerated Ageing Using Autoclave Test

The elongation and force of break of the test specimens were measured on a tensile testing machine. The retained elongation R_ε at break is measured as a function of time (and temperature and oxygen pressure) and is expected to be influenced by the ductile–brittle change which is a service lifetime criterion for the geotextile. The retained elongation R_ε is defined as follows:

$$R_\varepsilon = 100\% \, \varepsilon_e / \varepsilon_c, \tag{1}$$

with ε_e then initial elongation at break and ε_c the elongation of the exposed specimen.

The results are displayed in Figure 4. It is clearly visible that increasing temperature leads to a more pronounced decay of the mechanical properties. The loss of retained elongation proceeds with the duration of the exposure, which is visualized in Figure 4 by different gray scales of the respective symbols (bright to dark). The influence of pressure is lower. Experiments performed at 40 and 50 bar show higher values for retained elongation because the temperature was 30 °C and 40 °C, respectively.

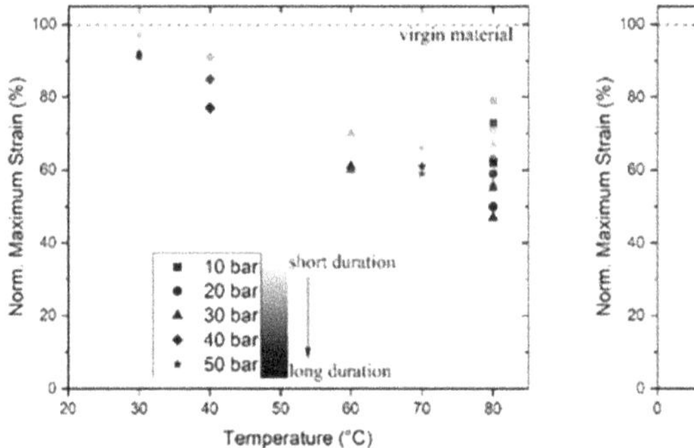

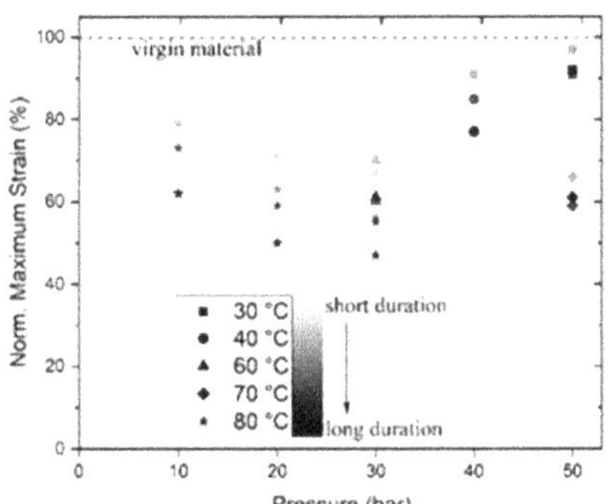

Figure 4. Retained elongation R_ε measured after exposure in autoclaves as a function of temperature (**left**) and pressure (**right**). The duration of exposure is visualized by the gray scale of the symbols. Note that at 80 °C experiments at three different pressures (10, 20, and 30 bar, different symbols) were performed.

The aging of polymers is caused by oxidation. The thermo-oxidation of PP can be defined as an in-chain radical mechanism. The latter generates hydroperoxides more rapidly than they decompose, which strengthen its strong auto-accelerating character. A detailed description of the oxidative aging of polymers is given by Verdu [31]. The accelerated ageing in the autoclaves is a function of temperature and pressure with a (pseudo-)first-order rate constant k (s^{-1}). The temperature and pressure dependence of the oxidation reaction can be approximated by an modified Arrhenius equation (consideration of pressure dependence) [32,33]:

$$\ln \frac{\varepsilon_e}{\varepsilon_c} \sim \ln \frac{c_0}{c} = A \, \exp\left(\frac{-E_a + C\,p}{R\,T}\right) = k\,(T,\,p)\,t, \tag{2}$$

with frequency factor A (s^{-1}), activation energy E_a (J mol^{-1}), pressure factor C (J mol^{-1} bar^{-1}), universal gas constant R, and temperature T (K).

The term $\ln c_0/c$ is usually related to the fate of a substance in a chemical reaction. Here, it is approximated by the loss of mechanical properties and describes the progress of the oxidation and thus degradation of the material without knowing exact concentration of oxidized and non-oxidized polymer material. The experimental data displayed in Figure 4 were fitted with the Solver module in Microsoft Excel (solver method GRG non-linear) (Office 365 for Enterprise). Starting values for activation energy E_a (80,000 J mol^{-1}) and frequency factor A (6×10^8 s^{-1}) were taken from the literature [34]. As a result, k (T, p) was fitted to 0.5 s^{-1} at T = 298 K and p_{O2} = 0.21 bar. The half-life τ at 298 K and 0.21 bar oxygen pressure, i.e., the time were 50% of the mechanical properties are lost under ambient conditions, can be calculated from $\ln2/k$.

$$\tau = \ln2/k = 330 \text{ years} \tag{3}$$

This result is in the same order of magnitude as the results from Hausmann et al. for woven polypropylene geotextiles [34] (483–795 years). Fitted pressure factor C was 146 J mol^{-1} bar^{-1}, so the activation energy E_a in the exponential tern in Equation (2) is reduced by 7300 J mol^{-1} (<10%) at 50 bar oxygen pressure in the autoclave experiment. As stated above, temperature has the strongest influence on the accelerated ageing in the autoclaves, even at highest possible pressure of 50 bar. However, it must be mentioned at this point that the samples are immersed in tap water so that the samples are exposed to the dissolved oxygen in water which is proportional to the partial pressure of oxygen above the liquid (Henry's law). Henry's law solubility constant is substance specific and a function of temperature. An equation to calculate the concentration of dissolved oxygen c_{aq} in water between 273 and 616 K and pressures up to 60 bar was presented by Tromans [35] and reviewed by Sander [36]. For 50 bar and 353 K, the c_{aq} is 3.97×10^{-2} mol kg^{-1}.

3.3. Ecotoxicity Tests

To evaluate the geosynthetic leachate ecotoxicity, a combination of bioassays was applied—both acute and chronic tests and organisms representing two trophic levels were used. Such an approach has advantages over individual component analysis and testing because it can disclose mixture effects.

The algae growth inhibition test was conducted at five volume/volume percent concentrations—5.9%, 11.8%, 23.6%, 47.2%, and 94.3%. Inhibition is evaluated by the reduction in specific growth rate relative to the cultures of the control. Samples Fraction 1 + 2 and Fraction 7 after 72 h exposures did not indicate algae growth inhibition even at the highest test concentration (Figure 5).

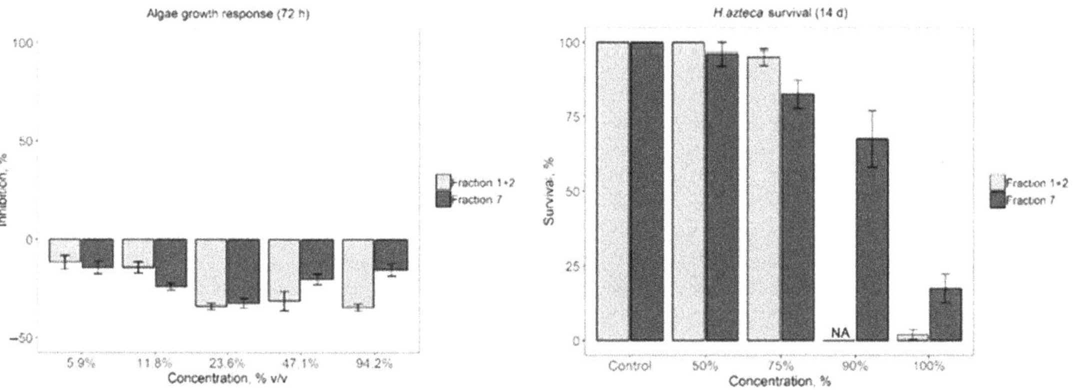

Figure 5. Algae growth response after 72 h (optical density measurements at 680 nm, **left**), *Hyalella. azteca* survival after 14 days (**right**). (NA: not analyzed, right).

The results of an acute *Daphnia magna* test showed the toxicity of Fraction 1 + 2 only at 100% concentration, causing 7.1% daphnia mortality after 24 h and 54% of cladocera mortality after 48 h exposure (Figure 6). However, there was no toxic effect observed when ADaM media microelements were added to the highest concentration. Fraction 7 did not cause any effects on *D. magna* survival during the test.

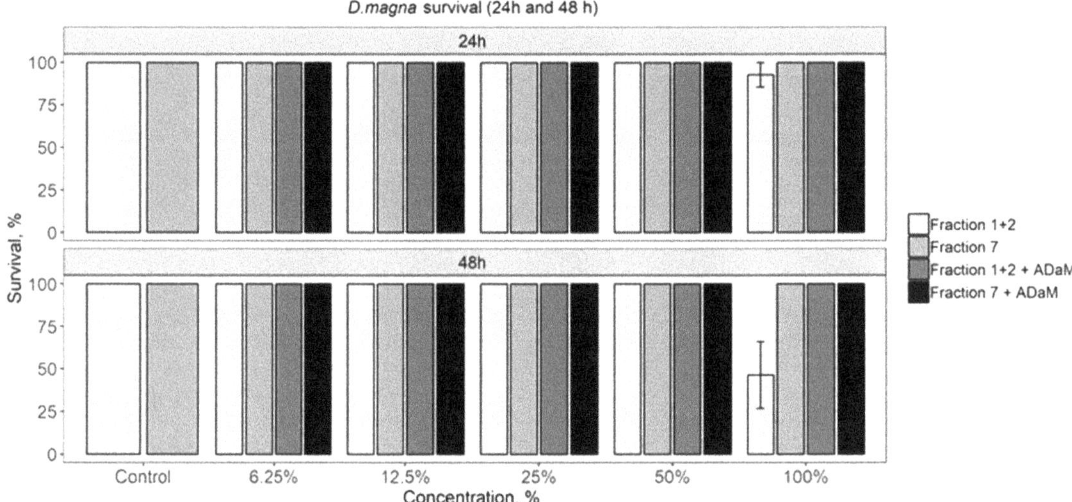

Figure 6. Survival of *Daphnia magna* after 24 h and 48 h.

Although the acute ecotoxicity test results of amphipod *Hyalella azteca* showed the higher toxicity of Fraction 1 + 2 than Fraction 7, no significant differences in toxicity between both samples after 14 days exposure were detected (Figure 5). In the 100% concentrate samples, an 88% mortality of amphipods was detected in Fraction 1 + 2 after 48 h exposure, while toxicity of Sample 7 increased only after one-week exposure. LC_{50} for Fraction 1 + 2 was 83%, while Fraction 7—LC_{50} was at 89%.

Measurements of pH showed an increase by 0.5 units after the 14-day test period, while the oxygen concentration stayed uniform more than 8.00 mg/L all test period. Ammonium concentration during the test did not reach higher than 20 mg/L (ISO 16303:2013 standard mentioned 96 h LC50 ammonium could be 20 mg/L to >200 mg/L [21]).

4. Discussion

The loss of additives, such as plasticizers and antioxidants, during the ageing of geotextiles potentially can add to the concentrations of hazardous substances in the water. This is discussed in a study from South Korea, where more than 200 different chemicals were identified in plastic marine debris and respective new products [37]. Another consideration is that base structure forming polymers gradually degrades to microplastic particles, and as such can be ingested by heterotrophs or interfere with algal photosynthesis [3]. However, ecotoxicological test results in this research did not show significant toxicity of geotextile leachates to water organisms. In case of microalgae, the test samples showed even nutritive properties, as an increase in microalgae concentration was observed during the 72 h of the test. Currently, there is limited research in the field of geosynthetic ecotoxicity, but a study evaluating the environmental safety of construction products also found that geosynthetic PET multifilament yarns and polyamide monofilament with PP fleece coating, have low toxicity [38]. Results indicate that the algae species *Desmodesmus subspicatus* that were also used in our study are slightly less sensitive than the algae *Raphidocelis subcapitata* and daphnia [39].

A concentrated sample of Fraction 1 + 2 (100%) caused mortality of *Daphnia magna*. However, if test sample was spiked with minerals from ADaM growth media, no mortality was observed. No mortality was observed in other test sample dilutions, neither in Fraction 1 + 2, nor Fraction 7. The results indicate that deionized water used in DSL tests might bias the ecotoxicity tests by adding hypoosmotic stress to low toxicity of test media. Concentrated samples (100%) of Fraction 7 did not caused mortality of organisms. These results suggest that the toxicity of additives is decreasing with time and dilution, also indicating that osmotic stress alone does not cause mortality [40].

A lethal concentration (LC_{50}) was calculated only for amphipods *Hyallela Azteca*. However, the LC_{50} at 83% and 89% concentrations can be considered as very low toxicity [41]. As geotextiles in hydraulic engineering are exposed to intensive water exchange, no toxic effects in the environment will be observed. However, even though within the tests with *Daphnia magna*, *Hyalella aztecal* and *Desmodesmus subspicatus* negative effects were not detected, the risk that long-term harsh climate conditions pose an impact on the release and migration of particles as well as hazardous substances cannot be excluded completely (referring to objective 3).

Service lifetime of geotextiles with state-of-the-art stabilization is far above 100 years, which was shown in the present study with accelerated ageing at elevated temperatures and oxygen pressures. The improper installation of the geotextiles and the lack of service and maintenance after extreme weather events could cause the failure of the engineered structures and, as a result, the pollution of the environment by remnants of geosynthetic materials [42]. The successful application of geotextiles in coastal protection depends on the selection of a suitable material and proper installation and maintenance (referring objective 2).

The field study performed at the shore of Kaliningrad Oblast (Russia) demonstrated that debris from plastic and geotextile materials is found in the environment [27,42]. The remnants of the geosynthetic materials are found not only at the beaches of the Kaliningrad Oblast, but at the neighboring beaches of Lithuania [43]. Some of the found objects could be attributed to unsuitable material selection (gabion coating) or improper waste management. Considering that any damage, even partial, of the coastal protective constructions using geosynthetic material could lead to the littering of the beach or the sea, specific attention is needed for the maintenance of such constructions (referring objective 1).

Author Contributions: Conceptualization, F.-G.S. and B.C.; Writing Original Draft, F.-G.S.; Data Acquisition, P.S., A.K., I.P.-N., E.E., I.B., I.L.-L. and E.S.; Visualization, Data Interpretation, Writing—Review & Editing P.S., B.C., I.P.-N. and I.P.; methodology, F.-G.S., B.C. and I.P.; Funding Acquisition, F.-G.S., B.C. and I.P. All authors have read and agreed to the published version of the manuscript.

Funding: This research was funded within the ERANET-RUS plus joint project EI-GEO, ID 212 (RFBR 18-55-76002 ERA_a, BMBF 01DJ18005, FS RTD/2018/21 VIAA Latvia) and supported by ERDF 1.1.1.2. post-doctoral project No. 1.1.1.2/VIAA/3/19/465 and theme 0149-2019-0013 of the Shirshov Institute of Oceanology (instrumental support of the field study).

Institutional Review Board Statement: Not applicable.

Informed Consent Statement: Not applicable.

Data Availability Statement: Data sharing is not applicable to this article.

Conflicts of Interest: The authors declare no conflict of interest.

References

1. Müller, W.W.; Saathoff, F. Geosynthetics in geoenvironmental engineering. *Sci. Technol. Adv. Mater.* **2015**, *16*, 034605. [CrossRef] [PubMed]
2. Wu, H.; Yao, C.; Li, C.; Miao, M.; Zhong, Y.; Lu, Y.; Liu, T. Review of Application and Innovation of Geotextiles in Geotechnical Engineering. *Materials* **2020**, *13*, 1774. [CrossRef] [PubMed]
3. Wiewel, B.V.; Lamoree, M.H. Geotextile composition, application and ecotoxicology—A review. *J. Hazard. Mater.* **2016**, *317*, 640–655. [CrossRef] [PubMed]

4. Bandow, N.; Aitken, M.D.; Geburtig, A.; Kalbe, U.; Piechotta, C.; Schoknecht, U.; Simon, F.-G.; Stephan, I. Using Environmental Simulations to Test the Release of Hazardous Substances from Polymer-Based Products: Are Realism and Pragmatism Mutually Exclusive Objectives? *Materials* **2020**, *13*, 2709. [CrossRef] [PubMed]

5. Simon, F.G.; Wachtendorf, V.; Geburtig, A.; Trubiroha, P. Materials and the environment, Environmental Impact of Materials. In *Springer Handbook of Metrology and Testing*; Czichos, H., Saito, T., Smith, L., Eds.; Springer: Heidelberg, Germany, 2011; pp. 845–860.

6. Oberbeckmann, S.; Löder, M.G.J.; Labrenz, M. Marine microplastic-associated biofilms—A review. *Environ. Chem.* **2015**, *12*, 551–562. [CrossRef]

7. Carneiro, J.R.; Morais, M.; Lopes, M.D.L. Degradation of polypropylene geotextiles with different chemical stabilisations in marine environments. *Constr. Build. Mater.* **2018**, *165*, 877–886. [CrossRef]

8. Hsieh, C.; Chiu, Y.-F.; Wang, J.-B. Weathering properties of geotextiles in ocean environments. *Geosynth. Int.* **2006**, *13*, 210–217. [CrossRef]

9. Tosin, M.; Weber, M.; Esiotto, M.; Lott, C.; Degli-Innocenti, F. Laboratory Test Methods to Determine the Degradation of Plastics in Marine Environmental Conditions. *Front. Microbiol.* **2012**, *3*, 225. [CrossRef]

10. Andrady, A.L. Microplastics in the marine environment. *Mar. Pollut. Bull.* **2011**, *62*, 1596–1605. [CrossRef]

11. Kershaw, R.P. *Biodegradable Plastics and Marine Litter. Misconceptions, Concerns and Impacts on Marine Environments*; United Nations Environment Programme (UNEP): Nairobi, Kenya, 2015; ISBN 978-92-807-3494-2.

12. Chubarenko, I.P.; Bagaev, A.; Zobkova, M.; Esiukova, E.E. On some physical and dynamical properties of microplastic particles in marine environment. *Mar. Pollut. Bull.* **2016**, *108*, 105–112. [CrossRef]

13. Morét-Ferguson, S.; Law, K.L.; Proskurowski, G.; Murphy, E.K.; Peacock, E.E.; Reddy, C.M. The size, mass, and composition of plastic debris in the western North Atlantic Ocean. *Mar. Pollut. Bull.* **2010**, *60*, 1873–1878. [CrossRef] [PubMed]

14. Lobelle, D.; Cunliffe, M. Early microbial biofilm formation on marine plastic debris. *Mar. Pollut. Bull.* **2011**, *62*, 197–200. [CrossRef] [PubMed]

15. Barnes, D.K.A.; Galgani, F.; Thompson, R.C.; Barlaz, M. Accumulation and fragmentation of plastic debris in global environments. *Philos. Trans. R. Soc. B Biol. Sci.* **2009**, *364*, 1985–1998. [CrossRef]

16. Woodall, L.C.; Sanchez-Vidal, A.; Canals, M.; Paterson, G.L.J.; Coppock, R.; Sleight, V.; Calafat, A.; Rogers, A.D.; Narayanaswamy, B.E.; Thompson, R.C. The deep sea is a major sink for microplastic debris. *R. Soc. Open Sci.* **2014**, *1*, 140317. [CrossRef]

17. Simon, F.G.; Barqawi, H.; Chubarenko, B.; Esiukova, E.; Putna-Nimane, I.; Barda, I.; Strode, E.; Purina, I. EI-GEO Environmental Impact of Geosynthetics in aquatic systems. In *3rd Baltic Earth Conference, Earth System Changes and Baltic Sea Coasts, International Baltic Earth Secretariat Publications*; Köppen, S., Reckermann, M., Eds.; International Baltic Earth Secretariat Publications: Geesthacht, Germany, 2020; pp. 151–152.

18. Müller, W.W.; Simon, F.; Wöhlecke, A. 30 Jahre BAM-Zulassung (in der Deponietechnik): Material- und prüftechnische Entwicklungen. *Bautechnik* **2019**, *96*, 912–918. [CrossRef]

19. Holmes, L.A.; Turner, A.; Thompson, R.C. Interactions between trace metals and plastic production pellets under estuarine conditions. *Mar. Chem.* **2014**, *167*, 25–32. [CrossRef]

20. DIN CEN/TS 16637-2. *Construction Products—Assessment of Release of Dangerous Substances—Part 2: Horizontal Dynamic Surface Leaching Test*; German Institute for Standardization, Beuth-Verlag: Berlin, Germany, 2014.

21. ISO 6341. *Water quality–Determination of the Inhibition of the Mobility of Daphnia Magna Straus (Cladocera, Crustacea)–Acute Toxicity Test*; International Standardization Organization: Geneva, Switzerland, 2012.

22. ISO 16303. *Water quality–Determination of Toxicity of Fresh Water Sediments Using Hyalella Azteca*; International Standardization Organization: Geneva, Switzreland, 2013.

23. ISO 8692. *Water quality–Fresh Water Algal Growth Inhibition Test with Unicellular Green Algae*; International Standardization Organization: Geneva, Switzerland, 2012.

24. Haseler, M.; Balciunas, A.; Hauk, R.; Sabaliauskaite, V.; Chubarenko, I.; Ershova, A.; Schernewski, G. Marine Litter Pollution in Baltic Sea Beaches–Application of the Sand Rake Method. *Front. Environ. Sci.* **2020**, *8*, 599978. [CrossRef]

25. Chubarenko, I.P.; Esiukova, E.; Khatmullina, L.; Lobchuk, O.; Grave, A.; Kileso, A.; Haseler, M. From macro to micro, from patchy to uniform: Analyzing plastic contamination along and across a sandy tide-less coast. *Mar. Pollut. Bull.* **2020**, *156*, 111198. [CrossRef]

26. Esiukova, E.E.; Chubarenko, B.V.; Chubarenko, I.P.; Kileso, A.V.; Zhelezova, E.V.; Grave, A.V.; Cukanova, E.S.; Sobaeva, D.A.; Tanurkov, A.G.; Yushmanova, A.V.; et al. Method of selection and accounting of fragments of geosynthetic materials and its testing at the beaches of the South-Eastern Baltic. In *XXVII International Shore Conference Arctic Coast: The Path to Sustainability*; MAGU Publisher: Murmansk, Russia, 2018.

27. Esiukova, E.E.; Chubarenko, B.V.; Simon, F.-G. Debris of geosynthetic materials on the shore of the South-Eastern Baltic (Kaliningrad Oblast, the Russian Federation). In Proceedings of the 2018 IEEE/OES Baltic International Symposium (BALTIC), Klaipeda, Lithuania, 12–15 June 2018; pp. 1–6. [CrossRef]

28. Esiukova, E.E.; Kileso, A.V.; Chubarenko, B.V.; Pinchuk, V.S. Geosynthetic debris on the beaches of Kaliningrad Oblast –result of systematic assessment of 2018. In Proceedings of the Baltic Sea Science Congress (BSSC), Stockholm, Sweden, 23 August 2019.

29. Sokolov, A.; Chubarenko, B.V.; Umgiesser, G. Hydrodynamic conditions near the northern shore of Sambian Peninsula (the Baltic Sea) as a basis of geotextile debris transport analysis. In *Proceedings of the 2018 IEEE/OES Baltic International Symposium (BALTIC)*; Williams, A.J., Ed.; IEEE Xplore: New York, NY, USA, 2018; pp. 1–6. [CrossRef]

30. Babakov, A.N.; Chubarenko, B.V. The Structure of the Net Alongshore Sediment Transport in the Eastern Gulf of Gdansk. *Water Resour.* **2019**, *46*, 515–529. [CrossRef]
31. Verdu, J. *Oxidative Ageing of Polymers*; ISTE Ltd.: London, UK, 2012.
32. Robertson, D. The oxidative resistance of polymeric geosynthetic barriers (GBR-P) used for road and railway tunnels. *Polym. Test.* **2013**, *32*, 1594–1602. [CrossRef]
33. Schröder, H.F.; Munz, M.; Böhning, M. A New Method for Testing and Evaluating the Long-Time Resistance to Oxidation of Polyolefinic Products. *Polym. Polym. Compos.* **2008**, *16*, 71–79. [CrossRef]
34. Hausmann, S.; Zanzinger, H.; Armani, A. Oxidative lifetime prediction of a polypropylene woven geotextile by applying high temperature and moderately increased oxygen pressure. *Geotext. Geomembr.* **2020**, *48*, 479–490. [CrossRef]
35. Tromans, D. Temperature and pressure dependent solubility of oxygen in water: A thermodynamic analysis. *Hydrometallurgy* **1998**, *48*, 327–342. [CrossRef]
36. Sander, R. Compilation of Henry's law constants (version 4.0) for water as solvent. *Atmos. Chem. Phys. Discuss.* **2015**, *15*, 4399–4981. [CrossRef]
37. Rani, M.; Shim, W.J.; Han, G.M.; Jang, M.; Al-Odaini, N.A.; Song, Y.K.; Hong, S.H. Qualitative Analysis of Additives in Plastic Marine Debris and Its New Products. *Arch. Environ. Contam. Toxicol.* **2015**, *69*, 352–366. [CrossRef]
38. Gartiser, S.; Heisterkamp, I.; Schoknecht, U.; Bandow, N.; Burkhardt, N.M.; Ratte, M.; Ilvonen, O. Recommendation for a test battery for the ecotoxicological evaluation of the environmental safety of construction products. *Chemosphere* **2017**, *171*, 580–587. [CrossRef]
39. Hsuan, Y.G.; Olson, M.S.; Spatari, S.; Cairncross, R.; Kilham, S. The roles of geomembranes in algae production at landifills. *Geosynthetics* **2012**, *30*, 34–41.
40. Sachs, F.; Sivaselvan, M.V. Cell volume control in three dimensions: Water movement without solute movement. *J. Gen. Physiol.* **2015**, *145*, 373–380. [CrossRef]
41. US EPA. *Method Guidance and Recommendations for Whole Effluent Toxicity (WET) Testing (40 CFR Part 136), Guidance Document, EPA 821-B-00-004*; United States Environmental Protection Agency, Office of Water: Washington, DC, USA, 2000.
42. Esiukova, E.E.; Chubarenko, B.V. Geosynthetic debris–new pollutant as a result of coastal protection. In Proceedings of the XX International Environmental Forum "Baltic Sea Day", St. Petersburg, Russia, 21–23 March 2019; pp. 80–82.
43. Kileso, A.V.; Esiukova, E.E.; Pinchuk, V.S. Traces of the transboundary pollution of the shore of the Southeastern Baltic by the debris of geosynthetic materials. In Proceedings of the IV International Scientific and Practical Conference (Dedicated to the 1000th Anniversary of the City of Brest), Current Problems of Earth Sciences Research of Cross-Border Regions, Brest, Belarus, 5–6 December 2019; pp. 215–216.

Article

Natural Building Materials for Interior Fitting and Refurbishment—What about Indoor Emissions?

Matthias Richter [1,*], Wolfgang Horn [1], Elevtheria Juritsch [1], Andrea Klinge [2,*], Leon Radeljic [2] and Oliver Jann [1]

1   Materials and Air Pollutants Division, Bundesanstalt für Materialforschung und -prüfung (BAM), Unter den Eichen 44-46, 12203 Berlin, Germany; wolfgang.horn@bam.de (W.H.); ria.juritsch@bam.de (E.J.); oliver.jann@bam.de (O.J.)
2   ZRS Architekten Ingenieure, Schlesische Straße 26, 10997 Berlin, Germany; radeljic@zrs.berlin
*   Correspondence: matthias.richter@bam.de (M.R.); klinge@zrs.berlin (A.K.)

Abstract: Indoor air quality can be adversely affected by emissions from building materials, consequently having a negative impact on human health and well-being. In this study, more than 30 natural building materials (earth dry boards and plasters, bio-based insulation materials, and boards made of wood, flax, reed, straw, etc.) used for interior works were investigated as to their emissions of (semi-)volatile organic compounds ((S)VOC), formaldehyde, and radon. The study focused on the emissions from complete wall build-ups as they can be used for internal partition walls and the internal insulation of external walls. Test chambers were designed, allowing the compounds to release only from the surface of the material facing indoors under testing parameters that were chosen to simulate model room conditions. The emission test results were evaluated using the AgBB evaluation scheme, a procedure for the health-related evaluation of construction products and currently applied for the approval of specific groups of building materials in Germany. Seventeen out of 19 sample build-ups tested in this study would have passed this scheme since they generally proved to be low-emitting and although the combined emissions of multiple materials were tested, 50% of the measurements could be terminated before half of the total testing time.

Keywords: bio-based insulation; earthen building materials; volatile organic compounds; semi-volatile organic compounds; formaldehyde; radon

Citation: Richter, M.; Horn, W.; Juritsch, E.; Klinge, A.; Radeljic, L.; Jann, O. Natural Building Materials for Interior Fitting and Refurbishment—What about Indoor Emissions?. *Materials* **2021**, *14*, 234. https://doi.org/10.3390/ma14010234

Received: 24 November 2020
Accepted: 28 December 2020
Published: 5 January 2021

Publisher's Note: MDPI stays neutral with regard to jurisdictional claims in published maps and institutional affiliations.

1. Introduction

Indoor environment has a significant influence on human health and our perception of well-being. Knowledge of indoor air quality, its significance for our health, and the factors that cause poor air quality are crucial to enable relevant stakeholders—including building owners, developers, users, and occupants—to maintain clean indoor air. Emissions from construction products constitute a significant source of indoor pollution and result, under certain environmental and occupational conditions, in sensory irritation and health complaints [1–4]. This represents a symptom of the fact that in western countries, more than 80% of the time is spent indoors [5,6]. A wide range of volatile organic compounds (VOCs), semi-volatile organic compounds (SVOCs), and carbonyl compounds (including formaldehyde) can be released from construction materials. Emissions concentrations become further elevated in new or refurbished buildings [7], where the rate of air exchange with fresh ambient air may be limited due to improved energy saving aspects [8], which is one of the main reasons for poor indoor air quality. The perpetuation of an air exchange rate of 0.5 h^{-1} is commonly recommended from a hygienic point of view [9]. In Europe, all new buildings should be nearly zero-energy buildings by the end of 2020 [10], which means that air exchange rates, and consequently indoor air quality levels, are set to decrease further. With a view to countering this trend, it is reasonable to suggest that building materials used indoors should be as low-emitting as possible. Furthermore, they should have the capacity to buffer moisture and anthropogenic indoor pollutants (e.g., from cooking, heating, smoking, etc.), as well as being recyclable and sustainable in terms of embodied energy.

Natural building materials, such as wood, cellulose, and earth, meet these criteria. They can easily be dismantled and reused or composted and are therefore the first choice to reduce environmental footprint and life-cycle costs. Moreover, they have excellent hygrothermal and acoustic properties, and show higher adsorptive capacities related to VOCs than other building materials [11,12].

Since the use of modern bio-based insulation materials is not widespread, only a few investigations on material emissions have been published [13]. In the study presented here, the emissions of formaldehyde, VOCs, SVOCs, and radon from different combinations of more than 30 natural materials (earthen dry boards and plasters, bio-based insulation materials, and boards made of wood, flax, reed, straw, etc.) underwent emission testing in specially designed test chambers. In standardised test procedures, it is normal to investigate materials individually. The results from individualised emission rate testing helps derive potential VOC indoor concentrations but does not take into account the combination effects arising from different source's emissions. The test series were arranged such that either single materials (first and foremost earth plasters) or complete wall assemblies in different combinations were tested. An evaluation of the results according to the AgBB scheme [9] was carried out to decide if the materials or material combinations are suitable for indoor use.

2. Materials and Methods

2.1. Materials

In Tables 1 and 2, the tested materials as well as the different material combinations tested for emissions are listed. The combinations are selected as to structural-physical considerations. Based on a literature study (scientific journals, conference proceedings, research reports), building materials suitable for flat separation walls, internal partition walls, and internal insulation of external walls have been identified and selected. Special emphasis was placed on those earth plasters that demonstrate an increased moisture buffer capacity as well as on wood fibre boards, not only because they are good moisture adsorbers, but also since they can prevent against overheating in summer and contribute to improved room acoustics. In addition, innovative construction materials, e.g., strawboards or sandwich boards made out of earth plaster and wood fibre boards that bear the potential to speed up the construction process, have been selected: in many cases, earthen materials (a9–a11) with different additives (b1–b4) to enhance their functionality are used for wall finishing. It is well known that earthen materials contribute to a healthy room climate [14,15]. Furthermore, it is a sustainable building material that can be used as a structural [16] or non-structural material (e.g., in plasters) [17]. Once at the end of its service life, earthen materials can be recovered and reused without loss of performance [18]. The capacity for moisture buffering of earthen and wooden materials is four to five times higher than that of conventional building materials [19]. In order to increase this effect, the earth mortars were modified with silica aerogel material Quartzene®. Aerogels have not only low thermal conductivities, but due to their high specific surface area, they provide excellent sorption properties. The optimisation of these materials was part of another investigation and therefore sample build-ups finished with earth plaster containing different amounts of aerogel were also tested for emissions.

Soil-based materials like clay contain naturally occurring radionuclides, such as U-238, Th-232, and K-40 [20], and tend to emit the radioactive noble gas radon (Rn-222) [21], which is classified by the World Health Organization (WHO) to be the second most common cause for lung cancer after smoking [22]. In order to accommodate this fact, radon was added to the emitters of interest.

Besides earthen materials, dry boards and insulation materials made of renewable resources including cellulose, flax, reed, straw, and wood were tested. Expected emissions from wood-based materials are terpenes, acids, and aldehydes. Though wood-based materials are often perceived as pleasantly odorous, their emissions can also cause allergies, e.g., irritation of the mucous membrane and skin [23]. Numerous studies have shown that terpenes emitted from wood can combine naturally with ozone in the atmosphere to produce irritants which could be responsible for eye and airway complaints [4,24,25]. Besides hexanal, one of the most often identified wood-based aldehydes, formaldehyde, can also be emitted depending on the type of timber [26]. Formaldehyde is also an ingredient in synthetic resins used in the production of mineral wool [27,28]. Two types of mineral wool (f1, f2) with different production technologies were tested.

In this study, since the final emissions of the combined materials were of interest, for most of the single materials, no emission tests were carried out, except for the samples 1–6 (Table 2). For such data, it is referred to measurements carried out by Hofbauer, W., Krueger, N. [13].

Table 1. Matrix of materials that are combined (combinations see Table 2).

	Type	Material	Thickness (mm)	Identifier
		Dispersion type	0.25	a1
	Paint	Casein	-	a2
		Earth basis	-	a3
	Flour	Marble	-	a4
	Chalk	-	-	a5
Coating	Primer	Casein	-	a6
	Joint filler	Gypsum	0.25	a7
	Deep penetrating primer	-	-	a8
	System compatible filler	Earth basis	3.0	a9
		Earth coarse grained (final coat)	s. Table 2	a10
	Plaster	Earth coarse grained (base coat)	s. Table 2	a11
		Earth fine grained (final coat)	s. Table 2	a12
	-	Straw	-	b1
Additives	-	Cellulose	-	b2
	-	Aerogel	-	b3
	-	Naturally coloured clay (red)	-	b4
Adhesives	-	Earth	4.0	c1
	System compatible	-	1.0	c2
Reinforcement	Fibre	Flax	0.5	d1
	Fibre	Glass	0.5	d2
		Gypsum	12.5	e1
	Dry board	Earth	15.0	e2
Panels			20.0	e3
		Straw	60.0	e4
		Gypsum	18.0	e5
	Fibre board	Wood	20.0	e6
		Wood fibre/flax core	70.0	e7
	Yellow (standard)	Mineral wool	60.0	f1
	Brown (eco technology)		40.0	f2
			40.0	f3
Insulation	Blocks	Wood fibre	60.0	f4
		Wood fibre (conifer)/cellulose core	120.0	f5
		Reed	100.0	f6
	Dry board	Calcium silicate	30.0	f7
Studs		Wood	60.0	g1
		Sheet steel	50.0	g2
Other	Blocks	Earth	115.0	h1
		Autoclaved aerated concrete	100.0	h2

Table 2. Matrix of sample build-ups. Material identifiers (Reference to Table 1) are listed in the order of their installation in the sample, starting with the side facing the interior.

	No.	Name	Materials	Thickness (mm)	Comment
Renders	1	Earthen render (final coat) A	a10/b1	5	benchmark
	2	Earthen render (final coat) B	a10/b1 *(4 parts)*, b3 *(1 part)*	5	new development
	3	Earthen render (base coat) A	a11/b1	10	benchmark
	4	Earthen render (base coat) B	a11/b1 *(4 parts)*, b3 *(1 part)*	10	new development
	5	Earthen render (final coat) C	a12/b2	3	benchmark
	6	Earthen render (final coat) D	a12/b2 *(4 parts)*, b3 *(1 part)*	3	new development
Internal partition walls	7	Dry lining wall—gypsum plaster board	a1, a7, a6, e1, f1, g2	123	benchmark
	8	Dry lining wall—gypsum fibre board	a4, a5, e5, a6, f2, g1	118	benchmark
	9	Dry lining wall—earth dry board	a10/b1	6	benchmark
			c1, d1, e3, f3, g1	124	
	10	Dry lining wall—wood fibre board	a10/b1	6	protection against overheating in summer, low emissions adsorption capacity, acoustic protection
			c1, d1, e6, f3, g1	124	
	11	Dry lining wall—earth block, dry stacked	a3, a12/b2	4	
			c1, d1, e3, h1, g1	258	
	15	Dry lining wall, Earth cellulose board	a9, d2, e2/b2, f3, g1	119	market innovation
	19	Non-load bearing, solid wall A—autoclaved aerated concrete	b4	2	cost efficient construction
			c1, d1, h2	105	
			c1 *(bonding of blocks)*	1	
	12	Non-load bearing, solid wall B—strawboards	a10/b1	6	market innovation
			c1, d1, e4	64	
	13	Non-load bearing, solid wall C—wood fibre board	a10/b1	6	market innovation
			c1, d1, f5, c2	126	
	14	Non-load bearing, solid wall D—wood fibre sandwich board with flax core	a10/b1	6	market innovation
			c1, d1, e7	75	
	16	Non-load bearing, solid wall E—reed blocks	a12/b2	3	cost efficient construction
			c1, d1, f6	105	
Internal insulation of external walls	17	Internal insulation external wall A—mineral insulation	a12/b2	3.0	high standard, refurbishment, mould resistant solution
			c1, d1, f7	35	
	18	Internal insulation external wall B—wood fibre insulation	a10/b1	6	high standard, refurbishment
			c1, d1, f4	65	

2.2. Sample Preparation

All sample materials were installed in sample intakes completely made of stainless steel, as depicted in Figure 1. Each one consists of several parts designed to fit into each other and which can be arranged to different heights, aiming to produce test specimens of varying thickness. The bottom is made of a plate with a furrow at its edge to allow the stacking of various sections (concentric rings) with an inner diameter of 314 mm, which were produced from a commercially available tube. Abutting edges were sealed with self-adhesive aluminium tape to ensure air tightness. All sample materials were built up layer by layer into the intakes, taking care that they exactly fitted the dimensions of the rings. The last layer was applied flush with the top edge of the last ring and was in most cases a plaster. The plasters were applied with the manufacturer-recommended water content using a plastering trowel.

Solid samples were prepared by cutting a circular piece from the centre area of the board. Build-ups made of natural building materials and finished with earth render were pre-conditioned at 23 °C and a relative humidity (RH) of 50% in order to prevent the samples elevating the RH in the emission test chamber above the required test conditions during testing as described in Section 2.3. As a rule of thumb, earthen plasters have reached their state of "intended use" after a drying time of about 14 days. Hence, this period was chosen as the maximum possible conditioning time because there are no product standards available yet specifying requirements for emissions testing.

The described sample preparation procedure is seen as beneficial since no additional sealing of edges is required. The sealing is represented by the wall of the concentric rings and it is ensured that only the surface normally facing indoors is exposed. Thus, the impact of all combined materials on the indoor air quality can be accounted for.

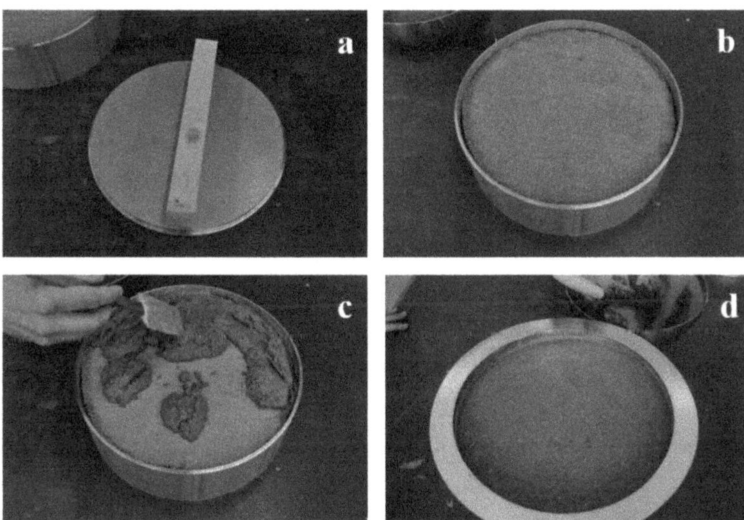

Figure 1. Sample installation into sample intake: (**a**) timber stud, (**b**) mounted earth dry board (wood fibre insulation below), (**c**) earth plaster final coat, (**d**) mounting of flanged section ring for installation into test chamber.

2.3. Emissions Testing

The emission tests were carried out in specially designed test chambers as depicted in Figure 2. Mounted onto the sample intake (4) is the actual test chamber comprising a cylinder with 420 mm in height (3), a stainless steel connection ring (2) with ports for air supply and sampling, and a glass lid (1) equipped with an agitator for the homogenisation of the test chamber air. The overall chamber volume above the surface of the test specimen is 38.5 L. The tests were conducted according to the requirements of EN 16516 [29] at a temperature of (23 ± 1) °C and a relative humidity (RH) of (50 ± 5)%.

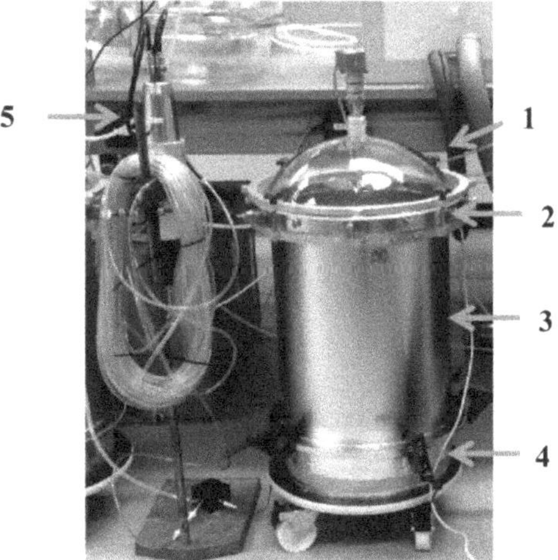

Figure 2. Emission test chamber assembly with Lucas scintillation cell for radon measurement: (**1**) glass lid with agitator, (**2**) connection ring, (**3**) hollow cylinder, (**4**) sample intake, (**5**) Lucas cell.

Radon exhalation was performed in parallel on the basis of the procedure published by Hofmann et al. [30–32].

The test chambers were operated dynamically by applying an air change rate n; n is defined as the ratio of air volume fed into the test chamber to the free volume of the test chamber and should be set to 0.5 h^{-1} in order to simulate normal indoor air conditions. Representativeness with regards to the intended use of the test samples is assured by applying a product-loading factor L of 1.0 m^2/m^3 for wall materials. L is defined as the ratio of the exposed area of the test specimen and the volume of the test facility. The ratio of n and L makes the area specific air flow rate q (0.5 m^3/m^{2}h). Considering an exposed surface area of 0.025 m^2, the air flow through the test chamber was adjusted to 0.62 L/h.

Emission tests for the evaluation of construction products normally last 28 days. After this period, either steady-state emissions have been reached or the decay of emissions has slowed down significantly. If such a situation was achieved before the end of the testing period, the test was terminated, which was in some cases after the 10th sampling day.

2.4. Sampling and Analysis

Air sampling and analysis were carried out according to ISO 16000-3 [33] and ISO 16000-6 [34] at days 3, 7, 10, 14, and 28 after sample installation in the test chamber. Sampling took place at 23 °C and the analysis immediately after sampling.

Cartridges filled with the adsorbent 2,4-Dinitrophenylhydrazine (DNPH) (Supelco, Bellefonte, PA, USA) were used for the determination of carbonyl compounds (aldehydes and ketones), particularly formaldehyde; while for the determination of VOCs and SVOCs, glass tubes filled with the adsorbent Tenax TA$^{®}$ (Supelco, Bellefonte, PA, USA) were used. The sampling volume on DNPH cartridges was 30 L (60 min at a sampling flow rate of 500 mL/min). Afterwards, they were eluted with a mixture of acetonitrile and water (volume ratio 4:1). This solution was then analysed using high pressure liquid chromatography equipped with a diode array detector (HPLC-DAD) on a ZORBAX Eclipse XDB-C18 column (4.6 mm × 150 mm, 5 µm, Agilent, Santa Clara, CA, USA) with methanol and water as mobile phase.

For Tenax sampling, 1 L was taken at 100 mL/min for 10 min, followed by thermal desorption of the tubes and analysis using gas chromatography on a slightly polar column (Rxi$^{®}$-5 ms, 30 m × 0.25 mm × 1.0 µm, Restek, Bellefonte, PA, USA) and with a mass selective detector (GC-MS) (Agilent, Santa Clara, CA, USA). In general, duplicate sampling was carried out. All identifiable VOCs that can be found on the LCI-list (cf. Section 2.5) were quantified by using their individual response factor. Compounds that are not listed or show a mass spectrum that cannot specifically be assigned to a certain compound were quantified by use of the response factor for toluene (toluene equivalent). The blank value of the adsorbent tube was determined by analysis of the unloaded tube prior to sampling and subtracted. SVOCs were quantified by integrating the chromatogram between the elution range of hexadecane and docosane with toluene equivalents, giving the sum parameter ΣSVOC.

Radon (Rn-222) measurement was performed using a calibrated, self-made Lucas scintillation cell. The cell itself is a small chamber of about 250 mL, whose inside walls are coated with silver-activated zinc sulphide, ZnS(Ag). In use, a filtered sample of chamber air continuously enters the cylinder. The Lucas cell is placed on a photomultiplier tube (PMT) inside a light-tight enclosure. Alpha particles emitted by radon and radon decay products strike the ZnS(Ag) phosphor, which emits light pulses that are amplified by the PMT and then counted by an alpha counter (Ortec Digibase) (Ortec, Oak Ridge, TN, USA). For more information on the measurement procedure, refer to Quindos-Poncela, L.S., Fernandez, P.L. [35]. A shortcoming of the procedure is the non-selectivity towards Rn-222 and its isotope Rn-220 (thoron). In order to be able to discriminate both, chamber air is led through a PVC hose (delay line) before entering the measuring cell. The tube is as long as three half-lives of the short-lived Rn-220 ($\lambda_{\text{Rn-220}}$ = 55 s, $\lambda_{\text{Rn-222}}$ = 3.8 d), making 24 m at an inner diameter of 4 mm and a continuous sampling flow of 110 mL/min (Figure 3). This procedure was described by Hofmann, M., Richter, M. [32].

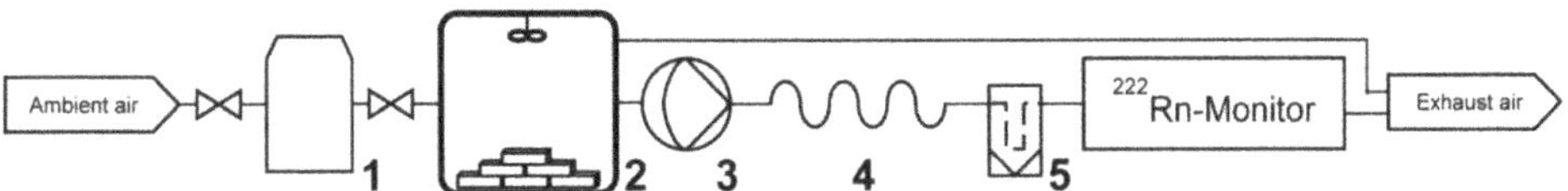

Figure 3. Sampling procedure with thoron delay line: (**1**) air supply system, (**2**) loaded emission test chamber, (**3**) sampling pump, (**4**) delay line, (**5**) radon progeny filter [32].

2.5. Evaluation of Measurement Results

The measurement results were evaluated against the German AgBB scheme [9]. The procedure is based on the analysis of test chamber air sampled on at least the 3rd and 28th day after loading. The following parameters are monitored:

- TVOC (total VOC): sum of the concentration of all individual substances with concentrations equal to or greater than 5 µg/m^3 within the retention range C6–C16.
- ΣSVOC: sum of the concentration of all individual substances with concentrations equal to or greater than 5 µg/m^3 within the retention range > C16–C22.
- Carcinogenic substances belonging to EU categories 1 and 2 or EU categories 1A and 1B.
- Assessable compounds: all VOCs with an LCI value; those compounds are listed in the appendix of the scheme; R ≤ 1.
- Non-assessable compounds: sum of VOC with, which cannot be identified, or do not have an LCI value.

The so-called R-value is based on the results of the assessable compounds on the 28th sampling day, or earlier in the case that the test can be prematurely terminated. It is a sum parameter calculated according to Equation (1) and may not be greater than 1.

$$R_i = \sum_i (c_i / LCI_i), \tag{1}$$

with c_i as the chamber air concentration of compound i and LCI_i as the Lowest Concentration of Interest of compound i as listed in the annex of the AgBB evaluation scheme [9]. This value must not be exceeded by any of the listed analytes.

For the evaluation of the radon exhalation, no criteria are actually defined. European council directive 2013/59/EURATOM [36] sets limits for a maximum indoor radon concentration in new and existing buildings (300 Bq/m^3), but how much the contribution of the building material can be is not yet defined. In Germany, the Federal Office for Radiation Protection (BfS) proposed to limit the total indoor radon concentration to 100 Bq/m^3, whereby building materials should contribute at most 20 Bq/m^3 [37], taking into account that the main source of radon is from soil. This criterion was adopted for the evaluation of the tested materials in this study.

3. Results and Discussion

3.1. (S)VOC Emissions and Formaldehyde

In general, with all tested materials and material combinations, low to very low indoor formaldehyde, VOC, and SVOC concentrations were determined. Table 3 gives an overview of the TVOC and ΣSVOC for all sample build-ups at the last sampling day, which was in most cases the 28th. However, in some cases, the testing was terminated after the 10th day, when the values determined were less than half the requirements for the 28th-day values and no significant increase in the concentration of individual substances was observed in comparison to the measurement on day 3.

Table 3. Results of evaluation of investigated sample build-ups according to the AgBB scheme.

Sample No.	TVOC ($\mu g/m^3$)			ΣSVOC ($\mu g/m^3$)	ΣR_i	Evaluation
	Assessable Compounds	Non-Assessable Compounds	Σ			
1	0	7 [a]	7 [b]	0 [a]	0.0 [c]	passed
2	0	91	91	53	0.0	passed
3	0	49 [a]	49 [b]	0 [a]	0.0 [c]	passed
4	0	66	66	46	0.0	passed
5	0	151 [a]	151 [b]	60 [a]	0.0 [c]	passed
6	0	28	28	11	0.0	passed
7	0	17 [a]	17 [b]	0 [a]	0.0 [c]	passed
8	0	16 [a]	16 [b]	10 [a]	0.0 [c]	passed
9	7	61	68	0	0.0	passed
10	17	76	93	0	0.0	passed
11	62	196	258	32	0.0	failed
12	0	7	7	1	0.0	passed
13	342	452	794	0	0.1	failed
14	92	41	133	9	0.0	passed
15	42	64	106	42	0.0	passed
16	17	79 [a]	96 [b]	10 [a]	0.0 [c]	passed
17	0	35 [a]	35 [b]	0 [a]	0.0 [c]	passed
18	7	89	96	0	0.0	passed
19	0	50	50	15	0.0	passed

[a] Tests terminated after 10th sampling day. At this time, the value may not exceed 50 $\mu g/m^3$; [b] tests terminated after 10th sampling day. At this time, the value may not exceed 500 $\mu g/m^3$; [c] tests terminated after 10th sampling day. At this time, ΣR_i must be ≤ 0.5.

According to the AgBB scheme, the value shall not exceed 1000 $\mu g/m^3$ for the TVOC and 100 $\mu g/m^3$ for the ΣSVOC on the 28th day after sample installation into the emission test chamber. The TVOC is furthermore divided into assessable and non-assessable compounds. The former are taken into account when calculating the R-value, the latter may not exceed 100 $\mu g/m^3$ after 28 days or 50 $\mu g/m^3$ after 10 days. Due to this criterion, samples no. 11 and 13 would not have passed the evaluation criteria and would therefore not be considered suitable for indoor use. However, due to the fact that the AgBB criteria were developed for individual building materials, it is supposed that for the analysis of wall systems, a different set of criteria would be more appropriate in further studies. In Figure 4, six detailed results of samples finished with earthen and non-earthen coatings are exemplarily depicted. All assessable and quantified VOCs are presented in the stacked bar charts, added to the non-assessable compounds, thus resulting in the TVOC. For all other results, please refer to the Supplementary Materials section (Figure S1).

All measurements were carried out in the same way as simulating conditions in a reference room, as described in Section 2.3. Therefore, all values are given in mass concentrations in $\mu g/m^3$. It can be assumed that those concentrations would result from installing the sample materials in an equivalent environment.

With the exception of samples no. 13 and 14, the expected decrease of the concentrations over time can be observed. Sample no. 13 showed a very constant emission profile over the whole 28 days, with a significant increase of non-assessable compounds on day 10. This build-up was the one with the highest emissions over the whole testing period. Sample no. 14 emitted compounds at a much lower level, with hexanal as main VOC, presumably released by the wood fibre part of the sandwich board with a flax core. The course of the emission profile of sample no. 14 was not observed until the 28th sampling day, since the termination criterion was reached at day 10 (cf. Table 3).

With the exception of samples no. 7 and 13, where 2-furaldehyde emissions were found on day 3 in concentrations of 21 and 6 $\mu g/m^3$, respectively, no further carcinogenic compounds were identified in any of the samples. 2-furaldehyde emissions had completely disappeared at day 10 in both cases.

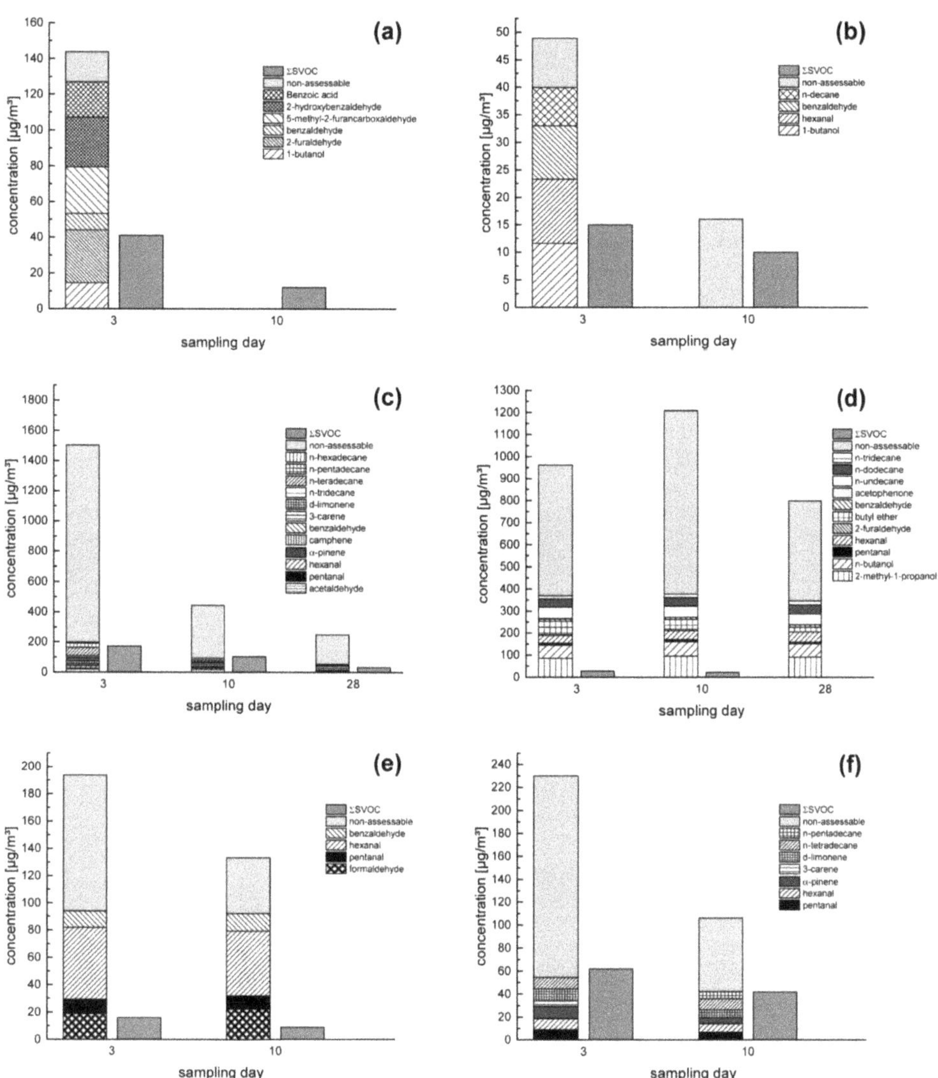

Figure 4. Measured volatile organic compounds (VOC) and sum parameter of semi-volatile organic compound (ΣSVOC) concentrations of six selected samples showing single VOC concentrations: samples no. 7 (**a**), no. 8 (**b**), no.11 (**c**), no. 13 (**d**), no. 14 (**e**), no. 15 (**f**).

Sample no. 8 showed medium initial concentrations, which rapidly decreased until day 10 when the tests were aborted. In this case, the increase of the non-assessable compounds from day 3 to day 10 is noticeable. Sample no. 15 showed only a moderate decrease of VOC concentrations, with a relatively high portion of non-assessable compounds at the beginning. The testing was aborted after day 10, although the sum of the non-assessable compounds was 64 µg/m^3, slightly higher than the termination criterion of 50 µg/m^3. However, the decrease between day 3 and day 10 was relatively fast and the LCI of the relevant compounds was low.

In most of the cases, the high proportion of not significantly identifiable or assessable VOCs relative to the identified ones is remarkable. One reason for this could be that in the inorganic earthen materials, a relatively small amount of organic material is present, either naturally or artificially. This undefined mixture could lead to an accumulation of different organic compounds with individually low concentrations. The relatively small amount of non-assessable VOCs in samples no. 7 and 8 might underpin this assumption as they were not coated with earth plasters.

The values for ΣSVOC were at an overall low level. At the beginning, samples no. 5 and 19 showed comparatively high SVOC releases that decreased rapidly until the last sampling day. Although SVOC concentrations may increase over time [38], there was no indication that this would occur during the testing, so the experiments were terminated before the 28th sampling day.

The sample build-ups containing wood-based insulation performed surprisingly well. In their study, Hofbauer, W., Krueger, N. [13] found that, in two of three investigated wooden samples, considerably high values for acetic acid (346 and 724 $\mu g/m^3$, LCI-value: 1250 $\mu g/m^3$) were present. In the third wooden material, they found a 2-furaldehyde concentration of 84 $\mu g/m^3$ (LCI-value: 20 $\mu g/m^3$). Other typical wood emissions, such as terpenes and aldehydes, particularly formaldehyde, were not reported by the authors or were found in negligible amounts. From all sample build-ups investigated in the current study that contained wood fibre insulations, i.e., samples no. 9, 10, 13, 14, 15, and 18, acetic acid and terpenes were found in negligible amounts on the 28th sampling day; aldehydes, e.g., hexanal, were found in concentrations between 7 and 48 $\mu g/m^3$, which, compared to its LCI-value of 900 $\mu g/m^3$, is very low as well. This difference could either be explained by low-emission properties in general or by a possible buffering effect from the earthen materials mounted in front of the insulation. Since clay has a significant sorptive capacity for water (e.g., McGregor, F., Heath, A. [14]), it seems that this is also true for VOCs, particularly the polar ones.

Formaldehyde was also measured in negligible concentrations ranging between 4 $\mu g/m^3$ (samples no. 15, 16, and 19), 13 $\mu g/m^3$ (sample no. 12), and 22 $\mu g/m^3$ (sample no. 14). From sample no. 7 (glass wool insulation), no formaldehyde was emitted. The LCI value is set to 100 $\mu g/m^3$.

When comparing the results of all measurements with the AgBB criteria (Table 3), all build-ups investigated in this study, with the exception of samples no. 11 and 13, would have passed the test. The reason in both cases is the same. The portion of non-assessable VOCs may not be higher than 10% of the TVOC, which was exceeded in both cases, most significantly by sample no. 13. The reason for these high amounts cannot be satisfactorily clarified, since measurements on individual materials were not planned in the study for time reasons. However, due to the fact that no harmful substances had been found in both cases on the 28th sampling day, there is no evidence that those build-ups would have any undesirable impact on occupants. This especially applies for sample no. 11, where indication for a further decrease of the emissions was given. Conversely, in sample no. 13, this trend was not observed. Particularly in buildings with air exchange rates lower than the 0.5 h^{-1} used in this study, the overall concentration level could be too high in the end.

3.2. Radon Exhalation

In Figure 5, the results of the radon measurements are presented. The concentrations ranging between 2 and 13 Bq/m^3 were considerably lower than the recommended maximum contribution of building materials to the indoor radon concentration of 20 Bq/m^3 (dashed line) that corresponds to an annual dose of about 1 mSv. In the diagram in Figure 5, the expanded measurement uncertainty is shown (k = 2). It is ranging between 5.0 and 5.3 Bq/m^3, indicating that the concentrations are very close to, or in some cases below, the limit of quantification, which is about 2 Bq/m^3. The expectable indoor radon concentration is therefore almost negligible. In their study, Richter, M., Jann, O. [39] found the correlation between installed mass of radon-exhaling building materials and radon concentration; therefore, those low concentrations may result from the low amount of earthen material proportional to the complete wall build-up. Consequently, higher indoor radon concentrations are expected when higher amounts of earthen material are applied, which can be observed for sample no. 11, almost completely consisting of earth, however, still well below the recommended threshold.

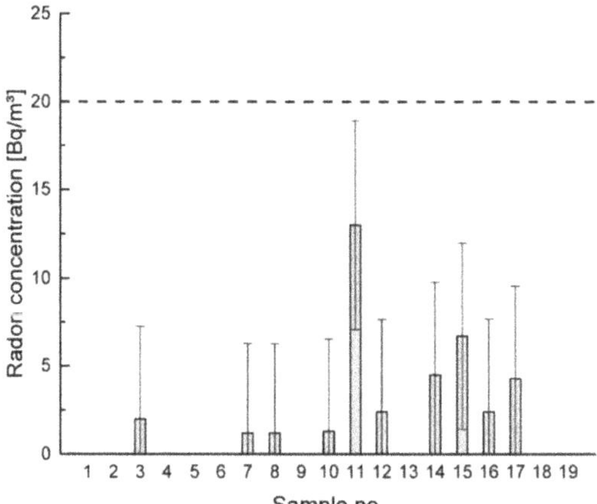

Figure 5. Measured radon concentrations of samples foremostly containing earthen materials.

4. Conclusions

This study pursued two aims, the evaluation of natural building materials in terms of their potential impact on indoor air quality, as well as the development of a test procedure for the reliable measurement of composite materials under standardised conditions. It revealed a good applicability of the new test chamber design in accordance with the established emission test chamber standard ISO 16000-9. Furthermore, all tested natural building materials were found uncritical with respect to their emission properties and could be installed in buildings in almost any combinations. The results of previous studies, which attested to the low emissions generated by insulation materials made from renewable raw materials, were confirmed. However, two sample build-ups did not meet the requirements of the AgBB evaluation scheme (samples no. 11 and 13) due to the emission of non-assessable compounds. However, it is noticed that in the reported work, the combined emissions from the different build-up layers were investigated. In standard tests, only the emissions from single materials are taken into consideration, disregarding the combined effect. For such investigations, an adaptation of the guideline values would be necessary. It was furthermore shown that the radon exhalation from the earthen materials was consistently below the recommended threshold of 20 Bq/m^3. The beneficial influence of these materials on the indoor environment in terms of being an active agent to regulate indoor RH and prospectively to reduce the presence of indoor pollutants helps to produce a healthier environment, prevailing over any potential health risks, as have been investigated and published in the past.

Supplementary Materials: The following are available online at https://www.mdpi.com/1996-194 4/14/1/234/s1, Figure S1: Measured VOC and ΣSVOC concentrations: Samples no.: (a) 1; (b) 2; (c) 3; (d) 4; (e) 5; (f) 6; (g) 9; (h) 10; (i) 12; (j) 16; (k) 17; (l) 18; (m) 19.

Author Contributions: Conceptualization, A.K., M.R., O.J. and W.H.; methodology, A.K., E.J., L.R. and M.R.; validation, E.J. and M.R.; formal analysis, E.J. and M.R.; investigation, E.J. and M.R.; resources, O.J.; data curation, A.K. and M.R.; writing—original draft preparation, M.R.; writing—review and editing, A.K., L.R., M.R., O.J. and W.H.; funding acquisition, O.J. All authors have read and agreed to the published version of the manuscript.

Funding: This research was funded by the European Union's Seventh Framework Programme for research, technological development and demonstration, grant number 608893.

Data Availability Statement: The data presented in this study are available on request from the corresponding author. At the time the project was carried out, there was no obligation to make the data publicly available.

Acknowledgments: The authors would like to thank Michael Hofmann for carrying out the radon measurements.

Conflicts of Interest: The authors declare no conflict of interest.

References

1. Redlich, C.A.; Sparer, J.; Cullen, M.R. Sick-building syndrome. *Lancet* **1997**, *349*, 1013–1016. [CrossRef]
2. Burge, P. Sick building syndrome. *Occup. Environ. Med.* **2004**, *61*, 185–190. [CrossRef]
3. Hodgson, M. Indoor environmental exposures and symptoms. *Environ. Health Perspect.* **2002**, *110*, 663–667. [CrossRef]
4. Wolkoff, P. Indoor air pollutants in office environments: Assessment of comfort, health, and performance. *Int. J. Hyg. Environ. Health* **2013**, *216*, 371–394. [CrossRef]
5. Brasche, S.; Bischof, W. Daily time spent indoors in German homes-Baseline data for the assessment of indoor exposure of German occupants. *Int. J. Hyg. Environ. Health* **2005**, *208*, 247–253. [CrossRef] [PubMed]
6. Hoppe, P.; Martinac, I. Indoor climate and air quality-Review of current and future topics in the field of ISB study group 10. *Int. J. Biometeorol.* **1998**, *42*, 1–7. [CrossRef] [PubMed]
7. Kephalopoulos, K.; Crump, D.; Däumling, C.; Winther-Funch, L.; Horn, W.; Keirsbulck, M.; Maupetit, F.; Säteri, J.; Saarela, K.; Scutaru, A.M.; et al. *ECA Report No. 27-Harmonisation Framework for Indoor Products Labelling Schemes in the EU, 27*; Publications Office of the European Union: Luxembourg, 2012.
8. Müller, B.; Mertes, A. *Indoor Air Quality after Installation of Building Products in Energy-Efficient Buildings*; Umweltbundesamt (UBA): Berlin, Germany, 2017.
9. AgBB. Requirements for the Indoor Air Quality in Buildings: Health-related Evaluation Procedure for Emissions of Volatile Organic Compounds (VVOC, VOC and SVOC) from Building Products. 2018. Available online: https://www.umweltbundesamt. de/sites/default/files/medien/360/dokumente/agbb_evaluation_scheme_2018.pdf (accessed on 20 December 2020).
10. EU. *Directive 2010/31/EU-Energy Performance of Buildings (2010)*; EU: Brussels, Belgium, 2010.
11. Eckermann, W.; Röhlen, U.; Venzmer, H.; Ziegert, C. *About the Influence of Earthen Building Materials on Indoor Air Humidity in German: Zum Einfluss von Lehmbaustoffen auf die Raumluftfeuchte, Europäischer Sanierungskalender (2008)*; Beuth Jahrbuch: Berlin, Germany, 2007.
12. Klinge, A.; Roswag-Klinge, E.; Ziegert, C.; Fontana, P.; Richter, M.; Hoppe, J. Naturally ventilated earth timber constructions. *Expand. Boundaries Syst. Think. Built Environ.* **2016**, 674–681. [CrossRef]
13. Hofbauer, W.; Krueger, N.; Mayer, F.; Scherer, C. *Untersuchungen zur Optimierung und Standardisierung von Dämmstoffen aus Nachwachsenden Rohstoffen-Final Report on Research Project-FKZ: 22013802*; Fraunhofer-Institut für Bauphysik: Stuttgart, Germany, 2007.
14. McGregor, F.; Heath, A.; Shea, A.; Lawrence, M. The moisture buffering capacity of unfired clay masonry. *Build. Environ.* **2014**, *82*, 599–607. [CrossRef]
15. Osanyintola, O.F.; Simonson, C.J. Moisture buffering capacity of hygroscopic building materials: Experimental facilities and energy impact. *Energy Build.* **2006**, *38*, 1270–1282. [CrossRef]
16. Miccoli, L.; Muller, U.; Fontana, P. Mechanical behaviour of earthen materials: A comparison between earth block masonry, rammed earth and cob. *Constr. Build. Mater.* **2014**, *61*, 327–339. [CrossRef]
17. Röhlen, U.; Ziegert, C. *Earth Building Practice*; Bauwerk-Beuth Verlag: Berlin, Germany, 2011; ISBN 978-3-410-21737-4.
18. Müller, U.; Ziegert, C.; Roehlen, U.; Schroeder, H. Standardisation of a Sustainable and Eco-Efficient Building Materia: Earth—A German Perspective. In *Cesb 10: Central Europe Towards Sustainable Building-from Theory to Practice*; Hajek, P., Tywoniak, J., Lupisek, A., Ruzicka, J., Sojkova, K., Eds.; CESB: Prague, Czech Republic, 2010; ISBN 978-80-247-3624-2.
19. Müller, U. Standards for Earthen Building Products—A new perspective for an old material. *ENBRI Newsl. Constr. Technol. Eur.* **2010**, *42*, 6–7.
20. Moharram, B.M.; Suliman, M.N.; Zahran, N.F.; Shennawy, S.E.; El Sayed, A.R. External exposure doses due to gamma emitting natural radionuclides in some Egyptian building materials. *Appl. Radiat. Isot.* **2012**, *70*, 241–248. [CrossRef]
21. Szabo, Z.; Jordan, G.; Szabo, C.; Horvath, A.; Holm, O.; Kocsy, G.; Csige, I.; Szabo, P.; Homoki, Z. Radon and thoron levels, their spatial and seasonal variations in adobe dwellings—A case study at the great Hungarian plain. *Isot. Environ. Health Stud.* **2014**, *50*, 211–225. [CrossRef] [PubMed]
22. WHO. *Handbook on Indoor Radon—A Public Health Perspective*; World Health Organization: Geneva, Switzerland, 2009; ISBN 978-92-4-154767-3.
23. Risholm-Sundman, M.; Lundgren, M.; Vestin, E.; Herder, P. Emissions of acetic acid and other volatile organic compounds from different species of solid wood. *Holz Roh Werkst.* **1998**, *56*, 125–129. [CrossRef]
24. Wolkoff, P.; Clausen, P.A.; Larsen, S.T.; Hammer, M.; Nielsen, G.D. Airway effects of repeated exposures to ozone-initiated limonene oxidation products as model of indoor air mixtures. *Toxicol. Lett.* **2012**, *209*, 166–172. [CrossRef]

25. Wilkins, C.K.; Wolkoff, P.; Clausen, P.A.; Hammer, M.; Nielsen, G.D. Upper airway irritation of terpene/ozone oxidation products (TOPS). Dependence on reaction time, relative humidity and initial ozone concentration. *Toxicol. Lett.* **2003**, *143*, 109–114. [CrossRef]
26. Roffael, E. Volatile organic compounds and formaldehyde in nature, wood and wood based panels. *Holz Roh Werkst.* **2006**, *64*, 144–149. [CrossRef]
27. Gellert, R.; Horn, W. Europäische Dämmstoffnormen der 2. Generation: Prüfmethoden zur Ermittlung flüchtiger organischer Komponenten (VOC). *Bauphysik* **2005**, *27*, 202–207. [CrossRef]
28. Kelly, T.J.; Smith, D.L.; Satola, J. Emission Rates of Formaldehyde from Materials and Consumer Products Found in California Homes. *Environ. Sci. Technol.* **1999**, *33*, 81–88. [CrossRef]
29. DIN. *Construction Products-Assessment of Release of Dangerous Substances-Determination of Emissions into Indoor Air*; EN 16516:2017; DIN: Berlin, Germany, 2018.
30. Hofmann, M.; Richter, M.; Jann, O. Use of commercial radon monitors for low level radon measurements in dynamically operated VOC emission test chambers. *Radiat. Prot. Dosim.* **2017**, *177*, 16–20. [CrossRef]
31. Hofmann, M.; Richter, M.; Jann, O. Robustness validation of a test procedure for the determination of the radon-222 exhalation rate from construction products in VOC emission test chambers. *Appl. Radiat. Isot.* **2020**, *166*, 109372. [CrossRef]
32. Hofmann, M.; Richter, M.; Jann, O. Determination of radon exhalation from building materials in dynamically operated test chambers by use of commercially available measuring devices. In Proceedings of the Healthy Buildings Europe 2015, Eindhoven, The Netherlands, 18–20 May 2015.
33. International Organization for Standardization. *Indoor Air-Part 3: Determination of Formaldehyde and Other Carbonyl Compounds-Active Sampling Method*; ISO 16000-3:2011; International Organization for Standardization: Geneva, Switzerland, 2011.
34. International Organization for Standardization. *Indoor Air-Part 6: Determination of Volatile Organic Compounds in Indoor and Test Chamber Air by Active Sampling on Tenax TA® Sorbent, Thermal Desorption and Gas Chromatography Using MS/FID*; ISO 16000-6:2011; International Organization for Standardization: Geneva, Switzerland, 2011.
35. Quindos-Poncela, L.S.; Fernandez, P.L.; Sainz, C.; Arteche, J.; Arozamena, J.G.; George, A.C. An improved scintillation cell for radon measurements. *Nucl. Instrum. Methods Phys. Res. Sect. A Accel. Spectrometers Detect. Assoc. Equip.* **2003**, *512*, 606–609. [CrossRef]
36. EU. *Directive 2013/59/EURATOM, Euratom Basic Safety Standards Directive (2013)*; EU: Brussels, Belgium, 2013.
37. BfS. *Jahresbericht 2009*; Bundesamt für Strahlenschutz: Salzgitter, Germany, 2009.
38. Wilke, O.; Jann, O.; Brödner, D. VOC- and SVOC-emissions from adhesives, floor coverings and complete floor structures. *Indoor Air* **2004**, *14*, 98–107. [CrossRef] [PubMed]
39. Richter, M.; Jann, O.; Kemski, J.; Schneider, U.; Krocker, C.; Hoffmann, B. Determination of radon exhalation from construction materials using VOC emission test chambers. *Indoor Air* **2013**, *23*, 397–405. [CrossRef] [PubMed]

Article

Formaldehyde Emissions from Wooden Toys: Comparison of Different Measurement Methods and Assessment of Exposure

Morgane Even [1,2,3,*], Olaf Wilke [2], Sabine Kalus [2], Petra Schultes [4], Christoph Hutzler [1] and Andreas Luch [1,3]

1 German Federal Institute for Risk Assessment (BfR), Department of Chemical and Product Safety, Max-Dohrn-Strasse 8-10, 10589 Berlin, Germany; christoph.hutzler@bfr.bund.de (C.H.); andreas.luch@bfr.bund.de (A.L.)

2 Bundesanstalt für Materialforschung und -prüfung (BAM), Division 4.2–Materials and Air Pollutants, Unter den Eichen 44-46, 12203 Berlin, Germany; olaf.wilke@bam.de (O.W.); sabine.kalus@bam.de (S.K.)

3 Freie Universität Berlin, Department of Biology, Chemistry, Pharmacy, Institute of Pharmacy, Königin-Luise-Strasse 2-4, 14195 Berlin, Germany

4 Chemical and Veterinary Analytical Institute Münsterland-Emscher-Lippe (CVUA-MEL), Joseph-König-Str. 40, 48147 Münster, Germany; petra.schultes@cvua-mel.de

* Correspondence: morgane.even@bam.de

Abstract: Formaldehyde is considered as carcinogenic and is emitted from particleboards and plywood used in toy manufacturing. Currently, the flask method is frequently used in Europe for market surveillance purposes to assess formaldehyde release from toys, but its concordance to levels measured in emission test chambers is poor. Surveillance laboratories are unable to afford laborious and expensive emission chamber testing to comply with a new amendment of the European Toy Directive; they need an alternative method that can provide reliable results. Therefore, the application of miniaturised emission test chambers was tested. Comparisons between a 1 m^3 emission test chamber and 44 mL microchambers with two particleboards over 28 days and between a 24 L desiccator chamber and the microchambers with three puzzle samples over 10 days resulted in a correlation coefficient r^2 of 0.834 for formaldehyde at steady state. The correlation between the results obtained in microchambers vs. flask showed a high variability over 10 samples (r^2: 0.145), thereby demonstrating the error-proneness of the flask method in comparison to methods carried out under ambient parameters. An exposure assessment was also performed for three toy puzzles: indoor formaldehyde concentrations caused by puzzles were not negligible (up to 8 $\mu g/m^3$), especially when more conservative exposure scenarios were considered.

Keywords: formaldehyde; wooden toys; emission test chamber; flask method; EN 717-3; microchamber

Citation: Even, M.; Wilke, O.; Kalus, S.; Schultes, P.; Hutzler, C.; Luch, A. Formaldehyde Emissions from Wooden Toys: Comparison of Different Measurement Methods and Assessment of Exposure. *Materials* **2021**, *14*, 262. https://doi.org/10.3390/ma14020262

Received: 19 November 2020
Accepted: 31 December 2020
Published: 7 January 2021

Publisher's Note: MDPI stays neutral with regard to jurisdictional claims in published maps and institutional affiliations.

1. Introduction

Formaldehyde, the simplest aldehyde (HCHO), is colourless and detectable in the gas phase at ambient temperature. It is mainly used in the production of industrial resins, adhesives, and coatings. Since it was demonstrated to induce tumours in the nasopharynx of rodents [1], it has been classified as a carcinogen of category 1B since 2016 [2]. Formaldehyde scored highly as one of the top chemicals for both exposure and toxicity in Washington, USA [3], and in the European Union [4].

The German committee on indoor guideline values determined a guideline value of 100 $\mu g/m^3$ based on toxicological data [5], which is in line with the WHO guideline [6]. An initial German survey in the years 1985–1986 revealed indoor formaldehyde concentrations of up to 309 $\mu g/m^3$, with a mean concentration of 59 $\mu g/m^3$ from 329 measurements [7]. In the following years, great efforts were made to reduce the formaldehyde sources and lower indoor air concentrations were measured, with a maximum of 68.9 $\mu g/m^3$ during 2003–2006, for example [8]. A recent statistical review analysis from 2019 indicates that average concentrations of formaldehyde are within the range of 20–30 $\mu g/m^3$ for European households under typical residential conditions [9].

Wood-based materials made of urea-formaldehyde resins may emit high formaldehyde concentrations [10,11]. They are mainly used as building materials or in the manufacturing of furniture, which caused 70% of formaldehyde indoor air concentrations in newly built timber-frame houses [12]. Urea-formaldehyde adhesives have poor water resistance: the presence of water causes hydrolysis and, consequently, the release of formaldehyde [13]. The European standard EN 717-1 suggests determining the release of formaldehyde from wood-based panels through the emission test chamber method [14]. The test chamber method is regarded as the method of choice for emission measurements as it mimics a real indoor environment (air exchange, temperature and humidity). Since 2017, the new standard method EN 16516 is in place in Europe: it describes emission testing with lower air change rate, higher relative humidity and higher chamber loading factor than EN 717-1 [15]. Since January 2020, the German national chemicals prohibition ordinance sets stricter requirements as EN 16516 must now be used instead of EN 717-1 to comply with the concentration limit of 0.1 ppm (corresponding to 124 μg/m^3) for formaldehyde [16]. For the same chamber loading, EN 16516 leads to measured concentrations being a factor of 1.6 higher compared to EN 717-1 [10]. With a higher chamber loading of 1.8 m^2/m^3 instead of 1.0 m^2/m^3, a factor of 2 could be expected. According to EN 717-1, the air samples from test chamber measurements are analysed by photometry after reaction with acetylacetone or with liquid chromatography (HPLC) after derivatisation with 2,4-dinitrophenylhydrazine (DNPH), following ISO 16000-3 [17].

Toys made of wood-based panels may also emit formaldehyde. However, their origin and quality are not typically controlled in the same way as particleboards because they are usually directly imported from distant countries. The European Toy Safety Directive 2009/48/EC [18] specifies a general maximum level of 0.1% (1000 mg/kg$_{toy}$) for carcinogenic compounds such as formaldehyde; however, this represents only a content limit and does not account for its emission behaviour. As formaldehyde is usually present in a chemically bound form and only emitted after hydrolysis, a content analysis for formaldehyde does not give any indications on the inhalation exposure assessment.

The so-called flask method is widely used by official control laboratories (OCLs) which are responsible for the toy market surveillance in the EU member states to measure formaldehyde emission of products [19]. It was developed by Roffael in the 1970s [20] and adapted into the European standard EN 717-3 [21]. The tested material is placed into the headspace of a 500 mL bottle filled with 50 mL water. After the incubation period of 3 h at 40 °C, the amount of formaldehyde dissolved in the water is determined by photometry. The method is still in use for wooden toys because of a lack of alternative methods, although it has been proven that the correlation to emission chamber testing is poor [22]. Moreover, different limits are used in the practice: EN 71-9 stipulates a maximum level of 80 mg/kg$_{toy}$ if EN 717-3 is used (3 h experiment) [23], whereas the former German Federal Health Agency (BGA) recommended a limit of 110 mg/kg$_{toy}$ for a 24 h flask experiment [24]. Using different materials, a study demonstrated that the values obtained by the flask method remained linear over time for at least 30 h [22], meaning that the two different limits are not comparable. The same study also suggested using an emission chamber test for more realistic results. There were several discussions at the subcommittees of analytics and toys related to the BfR's committee for consumer products where German OCLs asked for advice and developments of reliable measurement methods for formaldehyde in wooden toys with respect to children's safety [25].

In November 2019, a new European directive was adopted, amending 2009/48/EC for the purpose of specific limit values for chemicals used in certain toys [26]: here, in compliance with the German Chemicals Prohibition Ordinance [27], an emission limit of 0.1 ppm was stipulated for formaldehyde from resin-bonded material, starting from May 2021. In addition, the working group recommended emission testing by following EN 717-1 (i.e., a standardised method for wood-based panels) [14]. However, the OCLs will not be able to afford emission chamber testing for every toy and are therefore in need of an alternative method which provides reliable results. Smaller test chambers are cheaper, adapted to the typical size of toys and enable a higher sample capacity; their comparability to the standard chambers should be assessed considering the results obtained by the flask method.

Several studies have compared methods for determining formaldehyde emissions in the past. Firstly, the Field and Laboratory Emission Cell (FLEC) was compared to a standard 1 m^3 emission chamber and provided good correlation [28]. Unfortunately, this method cannot be used for toys, which do in most cases do not have flat surfaces. In another study, most standard methods were compared and showed sample-dependent results [29]. This may have been influenced by the fact that test conditions also vary between different standards. Three environmental chambers of different sizes were also compared for formaldehyde emissions from carpets [30]. In this case, the test conditions (temperature, humidity, air change rate and loading factor) were kept constant but considerable differences in formaldehyde emissions could still be observed. These previous studies did not consider the use of microchambers (μ-CTE) which allow cheaper measurements of small products in replicates and already showed good correlation for VOC emissions from a polymeric material [31]. The μ-CTE is a device with six 44 mL (or four 114 mL) miniaturised emission test chambers where the temperature, humidity and air change rate are controlled: the air can be sampled at the chamber outlet [32]. To our knowledge, microchambers have so far never been compared to large and regular emission chambers in terms of formaldehyde emission testing. The so-called "Dynamic microchambers" (DMC) were used on particleboards by Hemmilä et al. (2018) [11] and compared with a 1 m^3 emission test chamber and the perforator method (ISO 12460-5 [33]). However, DMC have a much higher volume (44 L) than the microchambers used in this study and are therefore linked with higher operating costs. Another micro-scaled chamber (1 L) that allowed process automation was tested for formaldehyde emission. However, no correlation with standard emission chambers could be demonstrated [34].

A standard cost-effective routine method usable for formaldehyde emission testing of toys and other consumer products in OCLs still needs to be established. Thus, we tested the comparability of formaldehyde emissions from wooden products in emission test chambers of different sizes and with the flask method: we demonstrated that microchambers can be used as a good alternative to the existing methods. Finally, we estimated the corresponding inhalative exposure against formaldehyde from wooden toys and showed that it was not negligible.

2. Materials and Methods

2.1. Samples

An overview of the samples used is given in Table 1, the exact dimensions are provided in Table S1. Two particleboards were initially studied. They were bought from a local do-it-yourself store and had already shown relatively high formaldehyde emissions during previous tests two years earlier [10]. Eight different wooden toys were also investigated. They were bought in local stores and had shown (except for Sample #9) flask method values (40 °C, 24 h) beyond the limit of 110 mg/kg$_{toy}$ recommended by the former German Federal Health Agency (BGA) [24] during market surveillance (see Table S2 for the exact values). Their country of origin was always China if it could be identified, meaning that the initial wood-based materials had not necessarily been controlled according to European standards [14,15]. Until usage, the samples were kept at room temperature in their original packaging or covered with aluminium foil. Pictures of the samples are provided in Figure S1.

Table 1. Overview and dimensions of the samples studied; ¥: the puzzle pieces were cut to fit in the microchambers, open edges were partly (#1 and #2) or entirely (#5, #6, #8 and #10) covered.

No.	Description	Origin	Sample Surface Area (cm^2)		Desiccator		Number of Pieces Per Set
			Microchambers	1 m^3 Chamber	Plate	Pieces	
#1	Particleboard	E.U.	16.0 ¥	9300			
#2	Particleboard	E.U.					
#3	Block set	China	34.4				
#4	Hammer and nail set	unknown	28.9				
#5	Puzzle birds	China	19.7 ¥		1456	651	12
#6	Puzzle fish	China	12.6 ¥				
#7	Puzzle shapes	unknown	24.0		1426	711	12
#8	Play set meal	China	23.9 ¥		1475	828	5
#9	Puzzle numbers	unknown	29.0				
#10	Plug set garden	unknown	16.9 ¥				

2.2. Emission Test Chambers

Three different types of emission test chambers (1 m^3, 24 L, and 44 mL) were used for emission testing, along with a clean air supply system. The 1 m^3 chamber was the standard VOC emission test chamber model from Heraeus-Vötsch Industrietechnik (Balingen-Frommern, Germany) with an inner chamber made of electro-polished stainless steel and a ventilator to ensure homogeneous air distribution. The 24 L chambers were desiccators made of glass and equipped with a ventilator from the BAM (Bundesanstalt für Materialforschung und -prüfung, Berlin, Germany). They were used instead of the 1 m^3 test chambers as standard chambers for the wooden puzzles because some samples were too small to obtain meaningful concentrations in the bigger chambers. The 44 mL chambers were part of a micro-chamber/thermal extractor device (μCTE®) produced by Markes (Llantrisant, UK).

The edges of the particleboard pieces (two plates of 0.5 m × 0.5 m and 0.43 m × 0.5 m in the 1 m^3 chamber and 1 piece of 2 cm × 4 cm in the microchambers) were covered with an emission-free aluminium-coated tape according to EN 717-1 [14]. The ratio between the open edge and the total surface was adjusted to 1.5 m/m^2. Some toy samples had to be cut with a saw to fit into the microchambers (#5, #6, #8 and #10). In this case, the freshly cut edges were covered completely with tape; indeed, the non-geometrical form of the toy makes it difficult to cover a defined ratio of the edges.

The two particleboards were placed upright in the 1 m^3 chamber. The puzzle and toy pieces were placed on metal carriers in the desiccators and on small plastic carriers in the microchamber if air would not otherwise circulate under the sample. Pictures of chamber loading are presented in Figures S2 and S3. Replicates were used for the microchambers: two or three chambers were always loaded with similar pieces of the same sample.

The systems were set to a temperature of 23 ± 1 °C and 50 ± 5% relative humidity. The microchambers were operated at a flow of 23.1–29.3 mL/min, while the desiccators were operated with 1.80 and 1.88 L/min. Similar to our previous work [31], the air change rate in the 1 m^3 chamber was adapted to the chamber loading to obtain a similar area-specific airflow rate (ratio of air change rate to loading) as applied for the microchamber, resulting in a flow of 14.5 L/min. Evidently, this represents a crucial parameter for such studies [35] and should be kept as constant as possible. Despite the maximum possible loading of the desiccator (all the puzzle pieces with the exception of the one placed in the microchambers), the area-specific air flow for chamber comparison was lower in the microchamber but still in the same order of magnitude. The area-specific values for air flow used during chamber comparisons are summarised in Table 2. To compensate the discrepancies, the results of method comparisons are presented as surface area specific.

Table 2. Sample area-specific air flow (m^3/m^2·h) values for chamber comparisons (n.u., not used; −, range due to flow fluctuation).

No.	Microchamber	Desiccator	1 m^3 Chamber
#1	0.97−1.06	n.u.	0.94
#2	0.95−1.03	n.u.	0.94
#5	0.72−0.79	1.73	n.u.
#7	0.61−0.63	1.59	n.u.
#8	0.68−0.74	1.36	n.u.

2.3. Air Sampling and Analysis of Air Samples

Air sampling was performed using DNPH cartridges (Supelco, St. Louis, MO, USA). The DNPH method [17] was preferred to the photometry method [14] for sample analysis because it was already widely used and validated in our laboratory. Active sampling was carried out for the 1 m^3 chamber and desiccators following ISO 16000-3 [17] using an air check 3000 sample pump (SKC Ltd., Dorset, UK) at 1 L/min for 30 min. Two samples were collected simultaneously for each time point in the 1 m^3 chamber: a self-designed sampling pump was used for the second sample. For the microchambers, the sampling lasted 20 h at the outlet to allow a sampling volume of around 30 l. Several samples were taken before the actual measurements started to control for blank values of the chambers and the DNPH cartridges. Air samples were regularly collected over 28 or 10 days after loading of the chambers.

The cartridges were refrigerated before and after sampling and eluted with 2 mL acetonitrile within two weeks after sampling. The solutions were analysed using HPLC (HP1100 from Hewlett-Packard, Waldbronn, Germany) in accordance with ISO 16000-3 [17]. An UltraSep ES ALD column (125 mm × 2.0 mm) and a pre-column (10 mm × 2 mm) from SepServ (Berlin, Germany) were used. The gradient of acetonitrile to water + 6% tetrahydrofuran varied between 30% and 83% (30% hold for 5 min, to 32% in 5 min and hold for 20 min, to 83% in 25 min). The mobile phase flow was 0.5–0.6 mL/min and the Diode Array Detector was used at 365 nm. Formaldehyde was quantified via external calibration with a commercial solution of its derivative from Sigma-Aldrich (Darmstadt, Germany) with a maximum concentration of 50 ng/µL. Samples were diluted if they did not fit into the calibration range. Data was processed using the OpenLab Data Analysis A.01.02 software from Agilent (Waldbronn, Germany). The results are provided as area-specific emission rates (SER$_A$), weight-specific emission rates (SER$_W$) or indoor air concentrations (C$_{indoor}$):

$$SER_A = \frac{C_{CH} * V_{CH} * n_{CH}}{A} \tag{1}$$

where SER$_A$ is the area-specific emission rate (mg/h·m^2); C$_{CH}$ is the chamber concentration (mg/m^3); V$_{CH}$ is the chamber volume (m^3); n$_{CH}$ is the chamber air change rate (/h); and A is the sample surface area (m^2).

$$SER_W = \frac{C_{CH} * V_{CH} * n_{CH}}{m} \tag{2}$$

where SER$_W$ is the weight-specific emission rate (mg/h·g); and m is the sample weight (g).

$$C_{indoor} = \frac{SER_A * A}{V_{room} * n_{room}} = C_{CH} * \frac{V_{CH} * n_{CH}}{V_{room} * n_{room}} \tag{3}$$

where C$_{indoor}$ is the indoor air concentration (mg/m^3); V$_{room}$ is the room volume (30 m^3 [15]); and n$_{room}$ is the room air change rate (0.5/h [15]).

Surface areas of the samples were determined by approximating their shape to geometrical forms (e.g., ellipse and triangle, see Table S1) if they were not already geometrical. For Sample #1, #2, #5, #7 and #8, all surface areas were determined. For the other samples, only the surface areas of the pieces placed in the microchambers were determined; the last approximation (3.4) with the whole sample surface area was done with the mean surface area of Sample #5, #7 and #8.

When two chambers were compared, an offset was calculated:

$$\text{Offset} = \frac{(\text{Highest SER}_A - \text{Lowest SER}_A)}{\text{Lowest SER}_A} \; (\%) \tag{4}$$

The use of the offset allows a direct comparison of the differences between emission test chambers for different samples.

The linearity of the correlation between SER_A at steady state in different emission test chambers was investigated. The coefficient of determination (R^2) and the p-values were considered for statistical analysis of the linear regressions. P-values were computed with the mean of each data point and were considered statistically significant when < 0.05 and highly statistically significant when < 0.001.

2.4. Flask Method

The flask method was carried out the same way as it is done by toy market surveillance [24]: in accordance with EN 717-3 [21] at 40 °C but for 24 h. The results are given in mg formaldehyde released per kg toy ($\text{mg/kg}_{\text{toy}}$). The linearity of the correlation between the flask method values and the emission rates after 10 or 11 days in the microchamber was investigated.

For Samples #3–#7, the test was conducted again after microchamber testing to study the influence of emission testing on the flask method values.

Except for the samples (particleboards and wooden toys), which are possibly only purchasable for a restricted time frame, and the desiccators which were self-made, all the materials and equipment used in this study are available commercially.

3. Results and Discussion

3.1. Chamber Comparison Using Particleboards

The 1 m^3 chamber and the microchamber were first loaded with pieces from the same particleboards and air samples were collected regularly over 28 days. Area-specific emission rates are depicted in Figure 1.

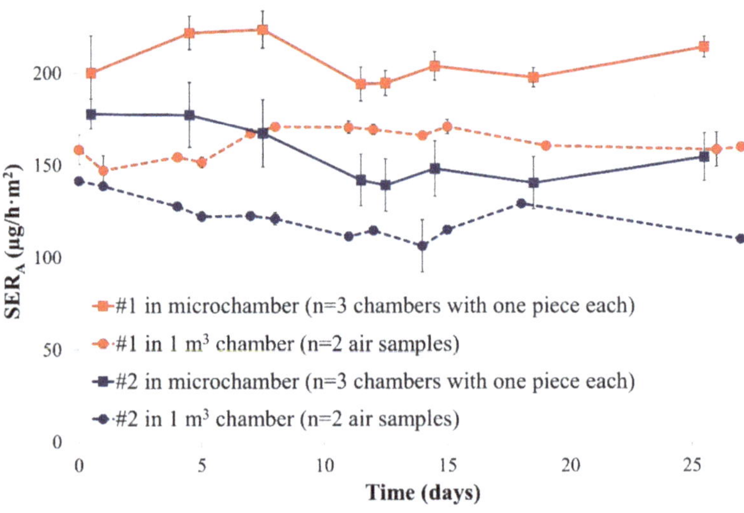

Figure 1. Emission profiles of formaldehyde from the two particleboards in two different emission chambers over four weeks (SER$_A$, area-specific emission rate). Error bars represent SD (standard deviation).

Firstly, it was observed that both chamber types led to similar emission profiles for formaldehyde: area-specific emission rates were relatively constant over 28 days, probably due to a year-long storage under chamber climate similar conditions. Emission rates were always used for test chamber comparisons because it is directly related to the indoor air concentration (see Equation (1)) but normalised to the area-specific air flow rate. Secondly, a relatively stable offset was observed between both chamber types: the emission levels measured in the microchamber were in mean about +27% and +28% (offset calculated according to Equation (4)) compared to those of the 1 m^3 chamber. A possible reason for the observed discrepancies could be the covering of the open edges with a ratio to the total surface of 1.5 m/m^2 as stipulated by EN 717-1 [14]: it represented 2.4 mm of open edges for the 2 cm × 4 cm pieces placed in the microchambers, which is difficult to accurately achieve using tape. Differences in air velocities at the sample's surface could also explain this deviation between both chambers. However, it is not possible to measure the air velocity in the microchambers.

These data still indicate that a good correlation between both chamber sizes was observed under the selected conditions. Thus, small chamber sizes might be a promising alternative for cost-effective emission measurement of formaldehyde from particleboards.

It has been demonstrated in previous work that the flow circulation in the microchamber is heterogeneous [32]. The height of the sample could disturb the air flow and thus would have an influence on the emission. This is of particular importance in the case of specimens that represent one-third of the chamber volume. For this reason, formaldehyde emission was analysed for different positions of both Samples #1 and #2 in the microchamber. The results are provided in Figure 2.

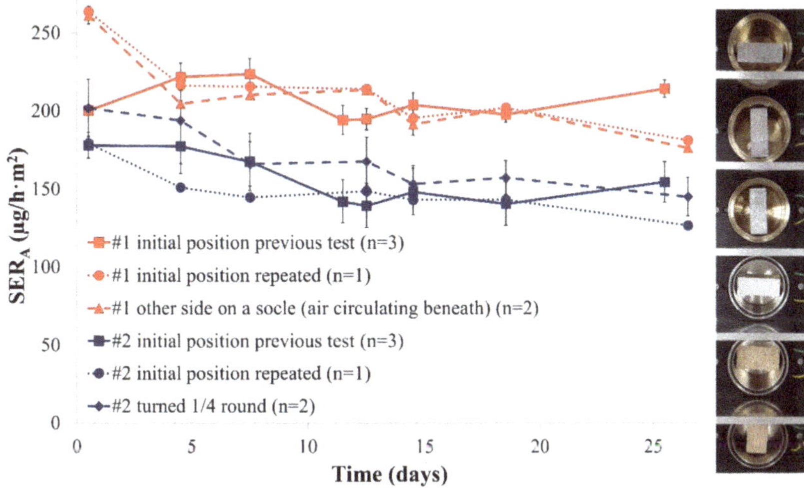

Figure 2. (**Left**) Emission profiles of formaldehyde released from two particleboards for different sample positions in the microchamber (SER$_A$, area-specific emission rate); and (**Right**) picture of the different testing positions. Error bars represent SD.

This experiment revealed that the position of the sample in the microchamber is only of low importance: irrespective of the exact position, the area-specific emission rate was similar (same position repeated or new position tested). This is an important result as it means that the exact position of the sample in the emission chamber would not be a crucial parameter in market control experiments. Furthermore, Figures 1 and 2 both show that an emission rate equilibrium is reached very quickly. For this reason and to allow fast and efficient investigations the following experiments were limited to 10 days.

3.2. Chamber Comparison for Toy Samples

A similar experiment to the one presented in Section 3.1 was conducted using wooden puzzles. Most puzzle pieces fit into the microchamber or easily fit when cut, and the cutting edge was covered by aluminium tape. Puzzle or play set pieces were chosen for this comparison; one piece was loaded into the microchamber while all other available pieces (8–18) of the same sample were loaded into a desiccator chamber, resulting in area-specific air flow rates that were slightly higher in the desiccator (see Table 2) compared to the microchamber. Three samples were studied over 10 days. The formaldehyde emission results are shown in Figure 3.

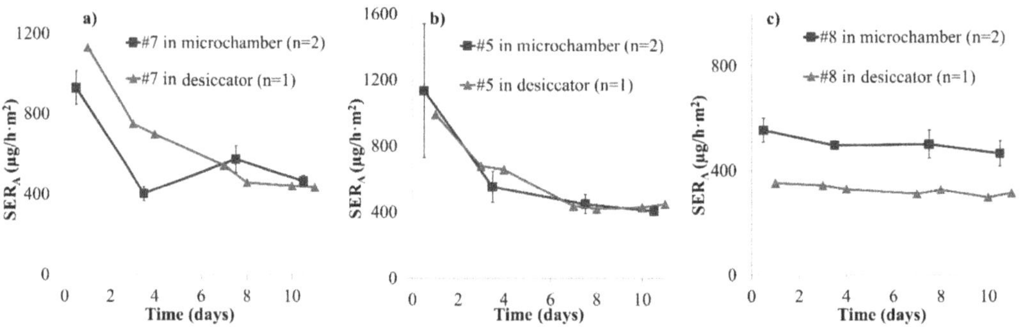

Figure 3. Emission profiles of formaldehyde from puzzle pieces in two different emission chambers over 10 days (SER$_A$, area-specific emission rate). (**a**) Sample #7; (**b**) Sample #5; (**c**) Sample #8. Error bars represent SD.

Formaldehyde concentrations were found to be fairly constant after seven days. Over the course of 10 days, the area-specific emission rates were similar in both chambers for all three samples. For Samples #5 and #7, there was no significant offset between both chamber results in contrast to Figure 1. For Sample #8, the average offset between microchamber and desiccator results was +53%, slightly higher than for Samples #1 and #2. For Sample #8, the area-specific air flow rate was only 1.9 times higher in the desiccator than in the microchamber while the ratio between both chambers was 2.3 and 2.6 for Samples #5 and #7, respectively (see Table 1). This finding may contribute to the fact that Sample #8 behaved similarly to Samples #1 and #2 (for which the area-specific air flow was constant between chambers). Moreover, the shape of the pieces from Sample #8 (play set) were thicker and approached the shape of Samples #1 and #2 more than the puzzle pieces from Samples #5 and #7. This may lead to differences in air velocities at sample surface. Additionally, for Samples #5 and #7, a decrease of the formaldehyde emission rate is observed over the first few days. Such a decrease was not observed for Sample #8 or for the particleboards in Figure 1. These differences can be due to more consistent conditions during storage or to the fact that Samples #5 and #7 were coloured with stickers while Sample #8 was painted (see Figure S1). Stickers could emit high formaldehyde concentrations during the first hours. The decrease was also observed for Sample #6 (see Figure S4). Hemmilä et al. (2018) also observed that the formaldehyde profile before steady state depended on the sample: no linear correlation was found between DMC results at Days 1 and 7 for different samples [11].

The results obtained from the microchamber and the desiccator experiments are similar, especially after 10 days. Again, the microchamber seems to provide reliable and comparable formaldehyde emission results. When using the microchamber, temperature and humidity are regulated according to EN 717-1 [14], but the chamber loading factor and the air change rate are higher to achieve similar area-specific air flows (see Table 2). In consequence, this leads to the conclusion that the standard method cannot be applied word by word. Furthermore, it will be difficult in practice to comply with the standard

EN 717-1 [14] as it requires that sample edges should be partly covered by a special ratio and that a certain loading factor should be used. This is much more complicated for toy samples than for large and rectangular wood-based panels. A practical suggestion would be to completely cover cut edges with tape before sample loading.

The linear correlation between the emission rates measured at steady state in the emission test chambers and the microchambers is shown in Figure 4. For Samples #1, #2 and #8, the whole measurement period was considered as steady state. For Samples #5 and #7, only the measurements at Days 7 and 10 were considered.

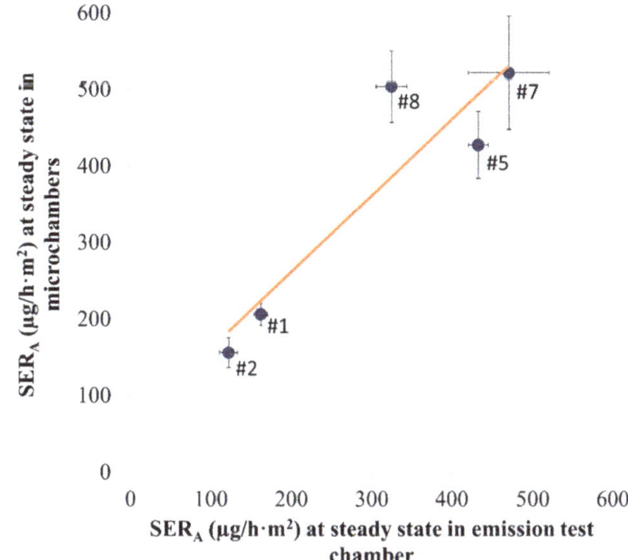

Figure 4. Study of the linearity between the area-specific emission rate (SER_A in $\mu g/h \cdot m^2$) in the emission test chamber and in the microchambers for Samples #1, #2, #5, #7 and #8. R^2: 0.834; p-value: 0.0304 (n = 4–24, depending on the number of air samples during steady state). y = 0.9944x + 62.113. Error bars represent SD.

A good correlation was observed between the microchamber and the test chamber (1 m³ or desiccator) results, with a R^2 of 0.834 and a significant p-value of 0.0304 (<0.05). A slope of 0.9944 indicates a good matching between chambers of different sizes.

3.3. Feasibility of the Flask Method

For all investigated samples, microchamber and flask experiments were carried out using two pieces of the same sample that were as similar as possible. The correlation between the flask method value and the emission rates at Day 10 (or Day 11 for Samples #1 and #2) is presented in Figure 5.

In Figure 5a, both the R^2 (0.145) and the p-value (0.2775 > 0.05) indicate a poor correlation between flask method values and area-specific emission rates (SER_A) of the samples. If Samples #3, #4 and #8 are removed, a R^2 of 0.956 is obtained, indicating a good linear correlation with a highly significant p-value (0.00014 < 0.001). As observed by Hemmilä et al. (2018) with the perforator method, the correlation between the emission chamber and the flask method result is consistent only for a selection of samples [11]. The results seem to depend on the geometry of the toy. Samples #3 and #8 were the thickest samples (1.4 and 1.8 cm) and yielded the lowest ratios of the flask method value to SER_A, while Sample #4, as the thinnest sample (0.3 cm), was found to result in the highest ratio. Hemmilä et al. (2018) also observed a differentiation between samples with varying thickness [11].

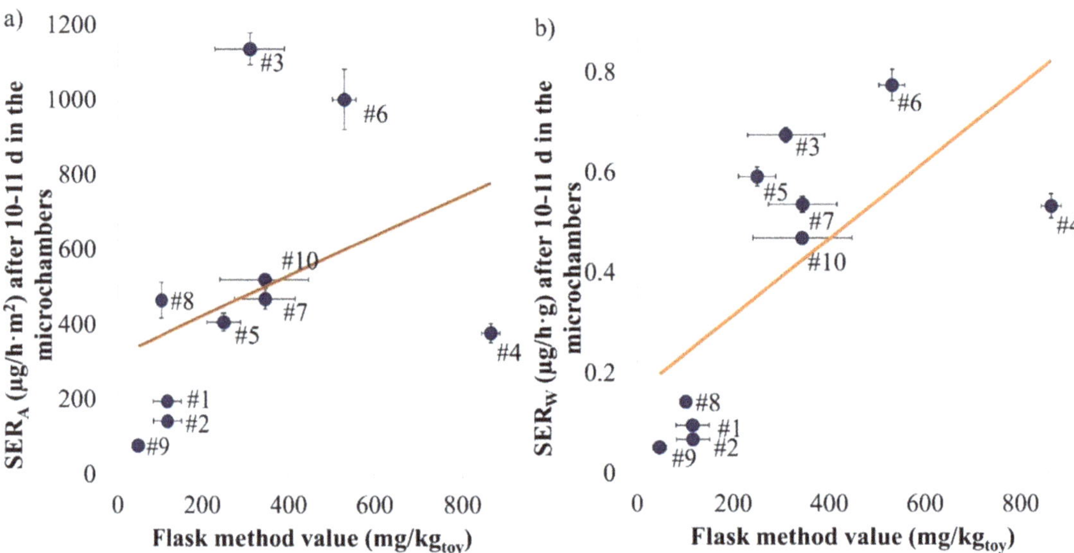

Figure 5. Study of the linear correlation between the flask method values and the area-specific ((**a**) R^2: 0.145; *p*-value: 0.27775) or weight-specific ((**b**) R^2: 0.470; *p*-value: 0.02876) emission rates after 10 or 11 days in the microchamber for each sample investigated. Error bars represent SD (n = 2 for microchamber samples, for flask method see Table S2).

SER$_A$ is a common unit for material emission measurements but as the flask method value is based on the weight of the toy, a correlation with the weight-specific emission rate (SER$_W$) was also considered and is presented in Figure 5b. In this case, a better correlation is obtained between the flask method values and the weight-specific emission rate, with a statistically significant correlation (*p*-value: 0.02876 < 0.05) but a poor R^2 (0.470). If only Sample #4 is removed, a R^2 of 0.845 and a highly statistically significant *p*-value of 0.0005 (<0.001) are obtained. The correlation between the microchamber and the test chamber (1 m^3 or desiccator) results is better (R^2: 0.834, see Figure 4). An interesting observation was that Sample #3 seemed to be made of massive wood and still emitted as much formaldehyde (1.4 mg/h·m^2 at Day 10) as the other samples. Such results have also been observed in the past [36] and may be attributed to the lacquer. The area-specific emission rate curves from the toy samples over 10 days in the microchamber are provided in Figure S4.

Overall, the flask method is not a good way to predict the emission measurements performed under realistic environmental conditions for varying toys. This mirrors the evidence obtained by other studies [22,29]. The flask method also has the significant disadvantage that the sides cannot be covered permanently (the humidity is too high for the tape) if the sample needs to be cut.

Additionally, the influence of the time point of the flask method test on the results was studied. The results presented above consider the flask experiment conducted with samples similar to those used in the microchamber. The flask method was carried out again for some samples that had undergone the microchamber experiment (Samples #3–#7) and the observed values are shown in Table S2 together with those of the similar samples. Generally, no significant difference was observed between both values (margins of error overlap). Significant differences were only observed for Sample #4: this would enhance the correlation with the microchamber results, which will however stay poor (R^2: 0.517).

Samples #5 and #6 resulted in similar values in both scenarios: it seemed to be of minor significance whether they had an open edge (following the chamber experiment).

3.4. Exposure Assessment from Whole Toy Samples

As shown in Section 3.2, puzzle or play set pieces were studied in a desiccator for comparison with the formaldehyde emission in the microchamber in which they could often fit without further sample preparation. The puzzle plates can also emit formaldehyde, but they are not investigated by OCLs because they also do not fit in the flask. It may be possible that the plate is the part with the greatest emissions. However, this is not necessarily considered for market surveillance or exposure assessment purposes. An evaluation of the contribution of both sample parts is presented: puzzle plates for each of the three samples (Samples #5, #7 and #8) were also studied in a desiccator for 10 days. The results of the piece-specific emission rates, normalised to one plate or to the number of associated pieces, are presented in Figure 6.

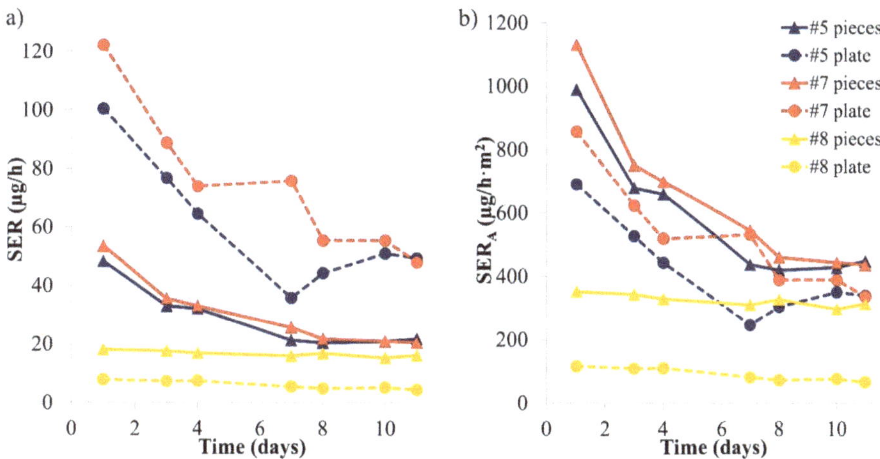

Figure 6. Emission rates of formaldehyde from the plate and the pieces of three puzzles: SER (sample-specific emission rate) (**a**); and SER$_A$ (area-specific emission rate (**b**)). The results were obtained with the desiccator method ($n = 1$).

For Samples #5 and #7, the plate was emitting higher amounts of formaldehyde compared to the puzzle pieces. However, for play set Sample #8, the pieces were emitting more formaldehyde compared to the plate. These differences are likely due to the different geometries of the samples: the play set plate (682 cm^2) is smaller compared to the puzzle plates (1426 and 1456 cm^2), whereas the sum of the piece surface areas was similar between samples (474–518 cm^2). The results in area-specific emission rates are similar for the pieces and the plate for Samples #5 and #7. For Sample #8, pieces emitted much more formaldehyde per surface unit than the plate: this is probably partly due to the fact that they were thicker (typically 1.8 cm) than the plate (0.5 cm). This shows that every part of a toy should be investigated when exposure needs to be assessed.

Furthermore, an exposure assessment of formaldehyde was carried out using the results of the desiccator experiments. The influence of a whole puzzle set (plate and corresponding number of pieces, as a consumer would buy it; see number of pieces in Table 1) on the formaldehyde room concentration has been considered. Evaluating indoor air concentrations allows a direct comparison with the indoor air guideline and therefore a reliable risk assessment [37]. They were calculated as shown in Equation (3), and the results are displayed in Figure 7.

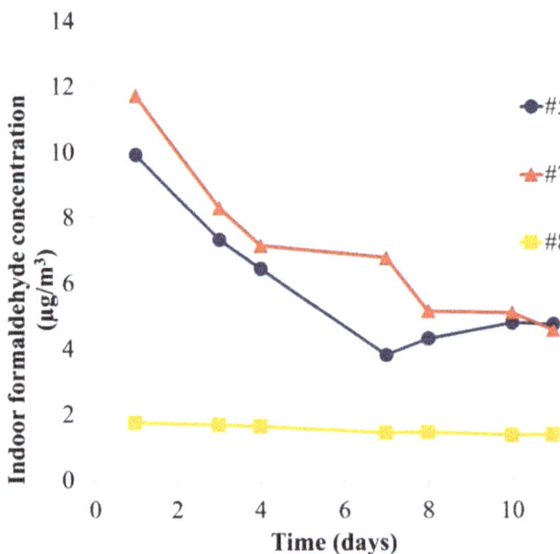

Figure 7. Calculated formaldehyde concentration for a 30 m^3 room resulting from each puzzle set sample (plate and pieces) ($n = 1$).

In the indoor air scenario (Figure 7), the influence of one sample on the formaldehyde concentration is considered in a 30 m^3 room with an air change rate of 0.5/h [15]. The assessment reveals that whole puzzle samples could induce indoor air concentrations of formaldehyde of up to 12 µg/m^3 on the first day and 5 µg/m^3 after 10 days. In comparison, the indoor air guideline value [5] is 100 µg/m^3. However, the children may play in very close proximity to the product in a poorly ventilated space with a concentration gradient. The concentration in the child's breathing zone will then be higher than the average room concentration. Moreover, there may also be other formaldehyde sources in the indoor air environment, meaning that the contribution of such products cannot be considered negligible. The BfR stated in 2007 that emissions from toys should only contribute to 10% of the overall indoor formaldehyde guideline concentration [38]. As an example, Samples #5, #7 and #8 represent, respectively, 4.8%, 4.6% and 1.7% of the 100 µg/m^3 guideline at steady state. An exceedance is easily possible if a room containing several toys and other formaldehyde sources such as furniture or building products is considered. It should also be considered that an increased temperature and/or humidity can enhance formaldehyde emissions drastically [39] and therefore cause even higher exposures. Lower air change rates indoors can also lead to higher VOC concentrations. A reduction of the formaldehyde emission limit is currently under discussion in Europe. Lower limits are already in effect for certain types of samples in the USA [40] and in Japan [41].

A similar exposure assessment for formaldehyde could be carried out directly using the area-specific emission rates from microchamber studies as they correlated with those obtained from bigger emission test chambers. The results from this approximation are shown in Table 3. This would lead to slightly higher concentrations for Samples #3 and #6 than for the previous considerations.

Table 3. Room concentration at Day 10 or 11 approximated from microchamber measurements.

	#3	#4	#5	#6	#7	#8	#9	#10
Approximated C_{room} from microchamber ($\mu g/m^3$) ($n = 2$)	8.27 ±0.31	2.75 ±0.18	3.29 ±0.18	7.29 ±0.59	3.71 ±0.20	3.71 ±0.35	0.56 ±0.01	3.78 ±0.12
C_{room} from desiccator ($\mu g/m^3$)			4.78		4.84	1.71		
Percentage of the WHO guideline	8.3%	2.7%	3.3%	7.3%	3.7%	3.5%	0.6%	3.8%

The difference between the approximated formaldehyde concentrations of the desiccator and the microchamber for Sample #8 is due to the plate, which emitted less formaldehyde than the pieces and was not considered in the microchamber approximation. This underlines the fact that representative pieces (e.g., a cut piece of the plate with covered edges) should be analysed if using the microchamber. With this consideration, the microchamber seems to be an appropriate method for market surveillance. The approximation carried out with only pieces of Sample #5 and #7 are close to the values obtained with the desiccator.

4. Summary and Conclusions

Formaldehyde emission is a key aspect when ensuring wooden toy safety. The emission test chamber method described in EN 717-1 [14] is not practicably and economically feasible for measurement purposes of toys. There is an urgent need for a standardised measurement method which demonstrates a good correlation to the existing emission test chamber methods whilst being more cost-effective. A possible method was investigated in our present study: the capacity of miniaturised emission test chambers (44 mL) to facilitate the surveillance of formaldehyde emissions from wooden toys was evaluated. It was shown that the emission results obtained (area-specific emission rates) were comparable to those from bigger chambers for both particleboards and wooden toys. In contrast, the currently used flask method showed a bad correlation with emission test chamber results. Its further use for market control of wooden toys is highly questionable. The main drawback in suggesting large-volume emission test chambers for toy market surveillance are higher costs and low sample capacity. Therefore, microchambers represent an affordable alternative for reliable market surveillance by the OCLs. They show a statistically significant correlation with bigger chambers (overtime and at steady state with $p < 0.05$ and R^2: 0.834), but further investigations with a larger number of samples and a validation using a homogeneous material are suggested to support these findings before standardisation.

As the sample may be heterogeneous, it is necessary to analyse representative pieces of the toy. Moreover, the standard EN 717-1 [14] is not directly suited to microchamber testing of toys. The air change rate will be higher than 1/h and the toy edges cannot be covered with a specific ratio. In addition, it is impossible to use a defined chamber loading factor for wooden toys due to the variable shapes. In the most recent standard EN 16516, different loading factors are stipulated for different sample types [15]. The new amendment of the toy safety directive [26] could indicate either an area-specific emission limit or a concentration limit per toy piece instead of following EN 717-1 with a defined chamber loading. Additionally, other analytical techniques, such as photometry [14], could be considered for air sample analysis to further reduce measurement costs.

An exposure assessment led to notable indoor air concentration values, indicating that formaldehyde emissions from a single wooden play set could represent up to 8% of the WHO formaldehyde guideline for indoor air. These products should therefore be considered as important emission sources, especially if numerous wooden toys are kept in smaller rooms or if children play with such toys and keep them in close proximity to their breathing zone.

Supplementary Materials: The following are available online at https://www.mdpi.com/1996-1 944/14/2/262/s1, Figure S1. Representative pictures of the investigated toy samples, Figure S2: Example image of chamber loading for Sample #1 in: the 1 m^3 chamber (a); and the microchamber (b), Figure S3: Example image of chamber loading for Sample #7: pieces in the microchamber (a); pieces in the desiccator (b); and plate in the desiccator (c), Figure S4: Emission profiles of formaldehyde from wooden toy pieces in the microchamber over 10 days, Table S1: Dimensions of the samples for chamber studies, Table S2: Flask method values of the samples similar to those studied in the microchamber and of samples after microchamber testing (mg/kg$_{toy}$).

Author Contributions: Conceptualisation, M.E., O.W., P.S., and C.H.; methodology, M.E., O.W., S.K., and C.H.; validation, M.E. and S.K.; formal analysis, M.E.; investigation, M.E. and S.K.; resources, P.S.; data curation, M.E.; writing—original draft preparation, M.E.; writing—review and editing, O.W., S.K., P.S., C.H., and A.L.; visualisation, M.E.; supervision, M.E., O.W., P.S., C.H., and A.L.; project administration, M.E. and C.H.; and funding acquisition, O.W., P.S., C.H., and A.L. All authors have read and agreed to the published version of the manuscript.

Funding: We acknowledge the BfR intramural financial support of our studies (project grants 1329-559 and 1322–631).

Institutional Review Board Statement: Not applicable.

Informed Consent Statement: Not applicable.

Data Availability Statement: The data presented in this study are available on request from the corresponding author. The data are not publicly available due to the absence of such requirements at the time the project was carried out.

Acknowledgments: The authors additionally thank Ankur Midha for valuable comments and suggestions on the manuscript and Stefan Miethig and Susann Schulz for excellent technical assistance.

Conflicts of Interest: The authors declare no conflict of interest.

References and Note

1. Schulte, A.; Bernauer, U.; Madle, S.; Mielke, H.; Herbst, U.; Richter-Reichhelm, H.-B.; Appel, K.-E.; Gundert-Remy, U. Assessment of the Carcinogenicity of Formaldehyde [CAS No. 50-00-0]. 2006. Available online: https://bfr.bund.de/cm/350/assessment_of_the_carcinogenicity_of_formaldehyde.pdf (accessed on 21 December 2020).
2. Commission Regulation (EU) No 605/2014 of 5 June 2014 amending, for the purposes of introducing hazard and precautionary statements in the Croatian language and its adaptation to technical and scientific progress, Regulation (EC) No 1272/2008 of the European Parliament and of the Council on classification, labelling and packaging of substances and mixtures. *Off. J. Eur. Union* **2014**, *167*, 36–49.
3. Smith, M.N.; Grice, J.; Cullen, A.C.; Faustman, E.M. A toxicological framework for the prioritization of children's safe product act data. *Int. J. Environ. Res. Public Health* **2016**, *13*, 431. [CrossRef]
4. Kotzias, D.; Kephalopoulos, S.; Schlitt, C.; Carrer, P.; Maroni, M.; Jantunen, M.C.; Cochet, C.H.; Kirchner, S.; Lindvall, T.; McLaughlin, J.; et al. *The INDEX Project: Critical Appraisal of the Setting and Implementation of Indoor Exposure Limits in the EU*; European Commission: Luxembourg, 2005.
5. German Committee on Indoor Guide Values. Available online: https://www.umweltbundesamt.de/en/topics/health/commissions-working-groups/german-committee-on-indoor-guide-values#textpart-1 (accessed on 21 December 2020).
6. Kaden, D.B.; Mandin, C.; Nielsen, G.D.; Wolkoff, P. *WHO Guidelines for Indoor Air Quality: Selected Pollutants—Part. 3 Formaldehyde*; World Health Organization Regional Office for Europe: Copenhagen, Denmark, 2010.
7. Krause, C.; Chutsch, M.; Henke, M.; Huber, M.; Kliem, C.; Leiske, M.; Mailahn, W.; Schulz, C.; Schwarz, E.; Seifert, B.; et al. *Umwelt-Survey 1985/1986 Band IIIc: Wohn-Innenraum: Raumluft*; (German); Umweltbundesamt: Berlin, Germany, 1991.
8. Schulz, C.; Detlef, U.; Pick-Fuß, H.; Seiwert, M.; Conrad, A.; Brenske, K.-R.; Hünken, A.; Lehmann, A.; Kolossa-Gehring, M. *Kinder-Umwelt-Survey (KUS) 2003/06 Innenraumluft—Flüchtige Organische Verbindungen in der Innenraumluft in Haushalten mit Kindern in Deutschland*; (German); Umweltbundesamt: Berlin, Germany, 2010.
9. Salthammer, T. Data on formaldehyde sources, formaldehyde concentrations and air exchange rates in European housings. *Data Brief* **2019**, *22*, 400–435. [CrossRef] [PubMed]
10. Wilke, O.; Jann, O. Comparison of Formaldehyde Concentrations in Emission Test Chambers Using EN 717-1 and EN 16516. In Proceedings of the 15th Conference of the International Society of Indoor Air Quality and Climate Conference, Philadelphia, PA, USA, 22–27 July 2018. paper 627.
11. Hemmilä, V.; Zabka, M.; Adamopoulos, S. Evaluation of dynamic microchamber as a quick factory Formaldehyde emission control. Method for industrial particleboards. *Adv. Mater. Sci. Eng.* **2018**, 1–9. [CrossRef]

12. Plaisance, H.; Mocho, P.; Sauvat, N.; Vignau-Laulhere, J.; Raulin, K.; Desauziers, V. Using the chemical mass balance model to estimate VOC source contributions in newly built timber frame houses: A case study. *Environ. Sci. Pollut. Res. Int.* **2017**, *24*, 24156–24166. [CrossRef] [PubMed]

13. Salthammer, T.; Mentese, S.; Marutzky, R. Formaldehyde in the indoor environment. *Chem. Rev.* **2010**, *110*, 2536–2572. [CrossRef] [PubMed]

14. *DIN EN 717-1:2005-01, Wood-Based Panels—Determination of Formaldehyde Release—Part 1: Formaldehyde Emission by the Chamber Method*, German version EN 717-1:2004; Beuth Verlag: Berlin, Germany, 2005.

15. *DIN EN 16516:2020, Construction Products—Assessment of Release of Dangerous Substances—Determination of Emissions into Indoor Air*, German version EN 16516:2017+A1:2020; Beuth Verlag: Berlin, Germany, 2020.

16. *Bundesministerium für Umwelt, Naturschutz und nukleare Sicherheit, Bekanntmachung analytischer Verfahren für Probenahmen und Untersuchungen für die in Anlage 1 der Chemikalien-Verbotsverordnung genannten Stoffe und Stoffgruppen;* (German); German Federal Ministry for the Environment: Bonn, Germany, 2018.

17. *DIN EN ISO 16000-3:2013-01, Indoor Air—Part 3: Determination of Formaldehyde and Other Carbonyl Compounds in Indoor Air and Test Chamber air—Active Sampling Method (ISO 16000-3:2011)*, German version ISO 16000-3:2011; Beuth Verlag: Berlin, Germany, 2013.

18. Directive 2009/48/EC of the European Parliament and of the Council of 18 June 2009 on the safety of toys. *Off. J. Eur. Union* **2009**, *170*, 1–37.

19. European Commission. Safety Gate: Rapid Alert System for Dangerous Non-Food Products—Most Recent Alerts. Available online: https://ec.europa.eu/consumers/consumers_safety/safety_products/rapex/ (accessed on 21 December 2020).

20. Roffael, E.; Rauch, W.; von Bismarck, C. Formaldehydabgabe und Festigkeitsausbildung bei der Verleimung von Eichenspänen mit Harnstofformaldehydharzen. *Holz als Roh- und Werkstoff* **1975**, *33*, 271–275. (In German) [CrossRef]

21. *DIN EN 717-3:1996-05, Wood-Based Panels—Determination of Formaldehyde Release—Part 3: Formaldehyde Release by the Flask Method*, German version EN 717-3:1996; Beuth Verlag: Berlin, Germany, 1996.

22. Maciej, B.H.U.; Schelle, C. Beeinflussung der Formaldehydemission aus unterschiedlichen Holzwerkstoffen in Abhängigkeit physikalischer Einflussgrößen, Korrelation generierter Messergebnisse auf Grundlage verfügbarer normativer Prüfverfahren und Vorstellung einer Prüfkammermethode mit alternativem Bewertungsansatz zur Charakterisierung der Formaldehydbelastung an Holzspielzeug. *Deutsche Lebensmittel Rundschau* **2011**, *107*, 169–176. (In German)

23. *DIN, EN 71-9:2007-09 Safety of Toys—Part 9: Organic Chemical Compounds—Requirements*, German version EN 71-9:2005+A1:2007; Beuth Verlag: Berlin, Germany, 2007.

24. German Federal Authorities BGA and BMJFFG, and Regional Authorities Responsible for the Safety of Consumer Goods from the Market. Internal Communications Dated 25/02/1989 and 15/03/1989. Not Publicly Available.

25. Protokoll der 9. Sitzung der BfR-Kommission für Bedarfsgegenstände. In German. 2012. Available online: https://www.bfr.bund.de/cm/343/9-sitzung-der-bfr-kommission-fuer-bedarfsgegenstaende.pdf (accessed on 21 December 2020).

26. Commission Directive (EU) 2019/1929 of 19 November 2019 amending Appendix C to Annex II to Directive 2009/48/EC of the European Parliament and of the Council for the purpose of adopting specific limit values for chemicals used in certain toys, as regards formaldehyde. *Off. J. Eur. Union* **2019**, *299*, 51–54.

27. Verordnung über Verbote und Beschränkungen des Inverkehrbringens und über die Abgabe bestimmter Stoffe, Gemische und Erzeugnisse nach dem Chemikaliengesetz. Available online: https://www.gesetze-im-internet.de/chemverbotsv_2017/ChemVerbotsV.pdf (accessed on 21 December 2020).

28. Risholm-Sundman, M. Determination of formaldehyde emission with field and laboratory emission cell (FLEC)–recovery and correlation to the chamber method. *Indoor Air* **1999**, *9*, 268–272. [CrossRef] [PubMed]

29. Risholm-Sundman, M.; Larsen, A.; Vestin, E.; Weibull, A. Formaldehyde emission—Comparison of different standard methods. *Atmos. Environ.* **2007**, *41*, 3193–3202. [CrossRef]

30. Katsoyiannis, A.; Leva, P.; Kotzias, D. VOC and carbonyl emissions from carpets: A comparative study using four types of environmental chambers. *J. Hazard. Mater.* **2008**, *152*, 669–676. [CrossRef] [PubMed]

31. Even, M.; Hutzler, C.; Wilke, O.; Luch, A. Emissions of volatile organic compounds from polymer-based consumer products: Comparison of three emission chamber sizes. *Indoor Air* **2020**, *30*, 40–48. [CrossRef] [PubMed]

32. Schripp, T.; Nachtwey, B.; Toelke, J.; Salthammer, T.; Uhde, E.; Wensing, M.; Bahadir, M. A microscale device for measuring emissions from materials for indoor use. *Anal. Bioanal. Chem.* **2007**, *387*, 1907–1919. [CrossRef] [PubMed]

33. *DIN EN ISO 12460-5:2015, Wood-Based Panels—Determination of Formaldehyde Release—Part 5: Extraction Method (Called the Perforator Method)*, German version EN ISO 12460-5:2015; Beuth Verlag: Berlin, Germany, 2015.

34. Nie, Y.; Lerch, O.; Kleine-Benne, E. *Determination of Formaldehyde and VOCs in Wood-Based Products using an Automated Micro-Scale Chamber*; GERSTEL Application Note; Gerstel: Mülheim, Germany, 2017.

35. Gunnarsen, L.; Nielsen, P.A.; Wolkoff, P. Design and characterization of the CLIMPAQ, Chamber for Laboratory Investigations of Materials, Pollution and Air Quality. *Indoor Air* **1994**, *4*, 56–62. [CrossRef]

36. Roffael, E. Formaldehydbestimmung nach der WKI-Flaschen-Methode und hiervon abgeleiteten Verfahren. *Holz als Roh- und Werkstoff.* **1988**, *46*, 369–376. [CrossRef]

37. Even, M.; Girard, M.; Rich, A.; Hutzler, C.; Luch, A. Emissions of VOCs from polymer-based consumer products: From emission data of real samples to the assessment of inhalation exposure. *Front. Public Health* **2019**, *7*, 202. [CrossRef] [PubMed]

38. BfR Opinion. BfR schlägt die Überprüfung des Grenzwertes der DIN-Norm für die Formaldehydausgasung aus Holzspielzeug vor (in German). 2007. Available online: https://www.bfr.bund.de/cm/343/bfr_schlaegt_die_ueberpruefung_des_grenzwertes_der_din_norm_fuer_die_formaldehydausgasung_aus_holzspielzeug_vor.pdf (accessed on 21 December 2020).
39. Parthasarathy, S.; Maddalena, R.L.; Russell, M.L.; Apte, M.G. Effect of temperature and humidity on formaldehyde emissions in temporary housing units. *J. Air Waste Manag. Assoc.* **2011**, *61*, 689–695. [CrossRef] [PubMed]
40. *Code of Federal Regulations (CFR) 40 CFR § 770 on Formaldehyde Standards for Composite Wood Products, Subpart C—Composite Wood Products, §770.10 Formaldehyde Emission Standards*; US EPA: Washington, DC, USA, 2017.
41. JIS A 1460. *Building Boards. Determination of Formaldehyde Emission—Desiccator Method*; Japanese Industrial Standard Committee: Tokyo, Japan, 2001.

MDPI
St. Alban-Anlage 66
4052 Basel
Switzerland
Tel. +41 61 683 77 34
Fax +41 61 302 89 18
www.mdpi.com

Materials Editorial Office
E-mail: materials@mdpi.com
www.mdpi.com/journal/materials